Third Edition

Digital

Production

Donald W. Connelly

Western Carolina University

Long Grove, Illinois

For information about this book, contact:
Waveland Press, Inc.
4180 IL Route 83, Suite 101
Long Grove, IL 60047-9580
(847) 634-0081
info@waveland.com
www.waveland.com

10-digit ISBN 1-4786-3418-9
13-digit ISBN 978-1-4786-3418-8

Printed in the United States of America

7 6 5 4 3

To my family,

Debie, Amanda and Mike, Sarah Grace, Nolan, and Benjamin

Contents

About the Author

Now professor and head of the Department of Communication at Western Carolina University, Don Connelly began his broadcasting career as an undergraduate student with a morning radio show, winning production awards from the Missouri Broadcasters Association while still in school. His first job out of school was teaching and simultaneously managing a public radio station. The station attracted national attention for its large audience and its consistent ability to generate more community funding per capita than any public station in the country.

Photo courtesy of Western Carolina University

Connelly later ventured to leave behind a tenured teaching position and enter commercial broadcasting. This led to 18 years of commercial experience managing and developing radio stations in Florida. Prior to his return to teaching and joining the faculty at Western Carolina, Connelly was with iHeartMedia, Inc. in Orlando, Florida, as affiliate relations director for the Florida News Network.

Connelly is a Society of Broadcast Engineers Certified Broadcast Technologist and holds the highest level of professional certification from the Radio Advertising Bureau, Certified Radio Marketing Expert. He is also a Certified Digital Marketing Consultant.

Throughout his career, Connelly has always maintained a keen interest in production. He is a three-time recipient of the Broadcast Education Association's Best of Competition award and a four-time recipient of the Best of Festival King Foundation award. Connelly's work in radio production has also earned him a prestigious Gabriel award from the Catholic Academy of Communication Arts Professionals. Above all he says his greatest satisfaction and reward as a faculty member is seeing his students excel as they start their professional careers in some of the top markets in the country.

Preface

Scope and Focus

Radio is over 100 years old. Students often ask about the future of the medium. At one time I would have speculated an answer. I now respond with something that Tim Berners-Lee (creator of the World Wide Web) is said to have remarked when asked about the future of the Internet. "I don't give predictions . . . the next thing you find is somebody has come around to your office, knocked on the door, and said that, by the way, they've done it."

The history of the radio industry is its creative adaptability, demonstrated through radio's success in overcoming every new medium that has threatened to replace it. *Digital Radio Production* offers a broadcaster's approach to a constantly evolving industry. Now, more than ever, the old axiom of "You are either moving forward, or you are moving backward—remaining stationary is not an option" is true.

Written for an industry in perpetual motion, *Digital Radio Production* is designed to prepare students with a well-rounded and comprehensive background in radio production. Although the production person's job is to produce commercial and programming elements, broadcasting demands a broad knowledge of sales, promotion, programming, the web, social media, and other key areas of the station. That is why *Digital Radio Production* takes a holistic approach to radio and the production person. Radio is tightly focused on communicating with listeners—its technology and business environment cannot be ignored though. A production person must focus on transforming entertainment and advertising into content across multiple digital platforms. The core element of radio production is communicating with the listener regardless of the device the message is received on.

While holding the humanity of radio communication center stage, in its technical approach *Digital Radio Production* is first and foremost digital. Today's broadcasting students have never used a typewriter, have never seen a reel-to-reel tape recorder, and have likely never purchased music on a CD. This text focuses on new technologies and trends with chapters that introduce students to core concepts vital to a successful career in radio. Some of the special features of this text include:

- An explanation of the role of the production person in the broadcast industry.
- Easy-to-understand, cutting-edge presentations about digital audio, recording, storage, manipulation, audio processing, and special effects.
- Critical elements of commercial writing and production techniques that recognize the role the sales department plays in production.
- Chapters about station promotion, imaging, and the station on location.
- Digital audio transmission methods for everything from transferring a commercial to another station to high-definition radio.
- A no-nonsense approach to how sales, programming, and production go hand in hand.
- The key role radio plays in the development of rich content for the web and the constant expansion of radio's interactive role and reach in the community through social media.
- A realistic approach of how to get a first job in radio, including the use of social media.

While each chapter covers a different group of core skills and background information, all are interrelated. Each chapter features suggested activities to involve the student outside of class, industry websites for more information, and a chapter-by-chapter glossary of industry terms.

In addition, *Digital Radio Production* has two special features that make it truly unique and place it in a league of its own in terms of making technology accessible to students. A collection of nearly 100 audio examples of virtually every aspect of radio production, from microphone techniques to commercial production samples, is available to students on the web. In addition, a music section features some 40 cuts of production music and a custom studio-tracking session written for this text with suggested activities.

Structure

In the first chapter, The Production Person, the stage is set for the rest of the text, explaining the business and programming aspects of radio, identifying the key players and departments, and how the production person is an integral part of each area of the station.

Radio is a technology and hardware/software-based industry; chapters 2 through 5 focus on the science and hardware of digital radio production. Students are introduced to Basic Science: Analog and Digital Audio in chapter 2, including analog sound and how humans hear as the foundation for analog-to-digital conversion and the practical uses of digital audio. Chapter 3, Microphones and Their Role in Radio Production, is an in-depth examination of microphones, digital microphones, mounts, cables, windscreens, and preamplifiers in addition to special purpose and wireless microphones. Chapter 4, Control, Mixing, and Monitoring, focuses on control boards, mixing, and monitoring across the on-air, production, and portable console platforms. There is also an in-depth discussion of the importance of metering in digital production. Basic Concepts in Digital Recording (chapter 5) introduces students to recording, mixing, and mastering using digital audio workstations and a variety of digital media. Chapter 5 also features instruction in the basic setup and operation of Adobe Audition Creative Cloud and Pro Tools digital recording software.

The foundation of knowledge developed in the first five chapters is a base to advance into the artistic and creative side of production in chapters 6 through 10. Audio Processing (chapter 6) is an introduction to the creative use of dynamic, dimensional, and frequency processing including voice, effects, and broadcast audio processors. In chapter 6, care is taken to emphasize not only what can be created using audio processing, but the perceptual effect it can have on listeners.

The Art of 60-Second Storytelling (chapter 7) introduces commercial production as a part of the sales process and approaches commercial creation from the point of moving a project from a sales order to an on-air and web-based advertising product. This chapter is not about writing so much as it is about how to create well-crafted solutions to a client's marketing problems. Producing Commercials, Promos, and News (chapter 8) begins with the talent-selection process and advances through production preplanning, production, and delivery to the client. Chapter 8 also has in-depth discussions on the use of music and sound effects. The chapter also includes a section on the special challenges of news production. Chapter 9, Communicating with the Listener: Announcing, begins with the philosophy of how to communicate one-on-one with listeners and advances to show planning, delivery, voice tracking, news, and sports delivery. The chapter also includes sections on working with producers and production directors as well as a section on developing your voice.

Promotion and Station Imaging, chapter 10, introduces the concepts of branding and positioning for station promotion and imaging across a station's on-air and web platforms. There is in-depth examination of effective production for liners, sweepers, station IDs, jingles, and imaging placement. The section on contest promotion moves from the initial tease to postpromotion.

Chapter 11 begins by explaining the unique relationship between the sales, promotion, and production departments that results in successful live broadcasts and web events. Fieldwork: Taking the Station on Location includes detailed sections on digital remote transmission methods and producing commercial, sports, and news remotes.

The Web is a straightforward chapter that addresses audio production for the web and explores the unique programming and business relationship radio enjoys with the web and social media. While covering streaming audio, the focus of chapter 12 is on audio production and radio's more creative uses of the web in building a station community.

Chapter 13, From Here to There: Radio and Audio Transmission in a Digital World, introduces the student to the many digital transmission methods radio has available for everything from transferring commercials from one station to another to using ISP technology for voice tracking and live remotes. The student is introduced to the basics of HD Radio.

Programming, Production, and Measuring Success, chapter 14, explores the relationship between radio production and radio programming. Included are sections on the radio production person's role in programming, music and format selection, and format delivery methods ranging from live to automated voice tracking. The chapter also discusses Nielsen Audio audience ratings and how they are used to measure a radio station's success.

Getting Your First Job in Radio provides students with the key information they need to secure their first position in radio from someone who spent over 20 years in radio management. Chapter 15 has the information a university's career services office does not usually provide to broadcasting students. This tightly focused chapter's topics range from establishing an internship to completing an audition, using social media in a job search and application, and a special section on electronic resumes and auditions.

Demonstration Audio and Music Files

Digital Radio Production is accompanied by two powerful learning tools: audio and music demonstration files that are available for download at http://www.waveland.com/Connelly/. With these tools students will have the advantage of being able to experience what they are reading about 24/7. The audio files feature nearly 100 audio concepts that are discussed throughout the text. The audio cuts include a number of unedited recordings created for this text, such as a shuttle launch at the Kennedy Space Center, a real Baldwin steam locomotive captured in the Great Smoky Mountains, and a recording made deep in a national forest demonstrating some of the most basic sound principles. The audition examples from chapter 15 are actual student and professional auditions.

The music files that accompany *Digital Radio Production* feature 40 cuts of production music written especially for this text. What is so unique is that each selection has been written as a production exercise. Students are treated to a raw multitrack recording session that they can edit, mix, and master over 100 new musical selections with. Beginners to masters will find something in the music files that excites their creativity.

Digital Radio Production, with its demonstration audio and music files, is a complete production package featuring the information and tools for students to become a successful production person.

Acknowledgements

With this third edition of *Digital Radio Production,* I once again bid a fond farewell to more outdated analog technologies. I am sure some readers will wince reading that ISDN lines are gone and many early digital transmission methods are also outdated. Just like broadcasters, *Digital Radio Production* must move forward—remaining stationary is not an option.

When my wife and I fell in love with the mountains of western North Carolina we made the tough decision to leave our professional careers and do something we wanted to do. There were so many terrific people who helped in our transition from the industry to academics. My sincere thanks are due Dr. Kathleen Wright, Professor Emeritus, at Western Carolina University. At the time she was head of the communication department; she took a chance and gave me the opportunity to excel at the profession of teaching. When she retired I had the honor of helping to establish and fund a scholarship in her name. I will never forget my faculty mentor, Dr. "Newt" Smith of Western's English department (since retired). When I told Newt I had been approached to write the first edition of this text he looked me square in the eye, pointed his finger at me, and in his best North Carolina drawl said, "If you don't do this, I will kick your ass." People who know him will tell you he wasn't kidding.

I have worked with a lot of great people in my broadcasting career and I would sincerely like to thank everyone who assisted me with this project. I would particularly like to thank Jeff Davis, Senior Vice President of Programming at iHeartMedia Greenville, South Carolina, & Asheville, North Carolina, for allowing me to "invade" his extremely successful operation, which includes a Country Music Station of the Year award winner. Also, a big thanks to John Anderson, services director at iHeartMedia in Asheville for his assistance and willingness to always be of assistance. I must also mention the brothers D'Innocenzi. Josh and Aaron were both students of mine and have gone on to successful major-market careers, one as a program director and the other as a national award wining music director.

Dr. Bruce H. Frazier is the Carol Grotnes Belk Distinguished Professor in Commercial and Electronic Music at Western Carolina University. Bruce is a good friend and an incredibly talented professional. The Academy of Television Arts and Science twice recognized him for his contributions to dramatic underscore and sound mixing for television programs. He has also been nominated for several Emmys for his role as music editor on the TV series *Quantum Leap*, and a Golden Reel nomination for his work on *JAG*.

I am still amazed at how he took my basic concept for music to accompany the text and created 40 "musical exercises" for students to work with. My sincerest thanks go to Bruce for his contribution to the text.

I must thank Tim Neese, president of MultiTech Consulting in Asheville, North Carolina. Tim is a Certified Senior Radio Engineer (Society of Broadcast Engineers) and a Microsoft Certified Systems Engineer with more than 15 years of field experience. He always has an answer (the correct one) and is always willing to spend time with my students.

I owe a special note of thanks to one of the best voices and imaging talents in the country, Jeff Laurence and his Autumn Hill Studios. Jeff's voice is heard on over 125 radio stations worldwide and, additionally, he has voiced nearly 3,000 regional, national, and international television commercials. His credits look like a who's who of advertising. Jeff provided valuable insight, help, and guidance in developing the chapter on promotion and station imaging.

I also want to extend my thanks to Mark Levy. Mark is the president and CEO at Revenue Development Resources, Inc. of Dallas. He is known nationally and internationally as one of those people who broadcasters turn to to inspire employees to do incredible things in broadcasting. For many years he has held the title of EOC (Emperor of Change). Mark is a dynamo of energy (his nickname really is Sparky) and a wonderful resource for my students.

I owe a great deal to Diane Evans at Waveland Press. Diane has an incredible eye for detail and moving projects forward. She was always there with guidance and great suggestions and she always asked the right questions. My sincerest thanks to Diane and all of the staff at Waveland for their assistance in making this third edition possible.

Finally, I must express my deepest gratitude to my wife Debie for her incredible support during this project. For the last year most of my spare time has been spent in front of a computer writing and developing the third edition. Her support and encouragement are what kept me going. With the publication of the third edition of *Digital Radio Production* I plan on joining her for dinner and evenings at home.

Donald W. Connelly

1

The Production Person

Highlights

Introduction

Since the first voice broadcast from the Outer Banks of North Carolina in 1902 by Reginald Fessenden, the radio industry has been in a constant state of change. Along the way there have been numerous new media that were heralded as the "death of radio." Yet each time radio's obituary was written, the industry transformed to overcome the challenge. Soon radio will be celebrating the 100th anniversary of the first commercial radio station (KDKA, Pittsburgh).

There are three words that keep the radio industry moving forward: consolidation, convergence, and the word that makes it all possible, digital. Of the three, digital continues to have the most impact. Every day technology reshapes radio and increases its ability to reach people in every aspect of their lives. Interestingly these changes are not all the work of the large broadcast companies, small-market broadcasters were some of the earliest adopters of digital technology in order to reduce their operating costs and make themselves more profitable. For many small-market broadcasters adopting new technology is a matter of survival. Small-market owners can more effectively serve their communities by accessing station operational systems remotely via smart phones or tablets. That access allows them to produce live news and sports programs much more efficiently. Remote access also permits a station to maintain all of its social media platforms. The employees of a radio station are no longer confined to the station's studios and can work virtually from just about anywhere.

There is another interesting aspect to the ongoing change in radio—how the listener receives a station's signal. Radio is a consumer-driven platform and listeners have many choices with which to receive a station, including traditional radios; high-definition radios featuring multiple channels on the same frequency; web streams of the on-air signal on various devices; or smart phones equipped with FM receivers. Forward thinking stations are experimenting with software to transform the station's music library into localized custom music streaming services for listeners interspersed with station content.

Through convergence of technologies and new production techniques, radio stations have become entertainment and advertising platforms that are expected to offer products and services on air, online, on demand, on your phone, and on your tablet. Simply stated, radio is reaching out to listeners no matter where they are on whatever device they are wearing or carrying. To meet this challenge broadcasters are seeking multitalented people with a broad range of skills because production people work in every aspect of station operations.

Parts of this chapter may sound more like a management text to you; however, it is really important to your success that you understand how a radio station operates and the role a production person plays in it.

One reason people are drawn to production is that the job is challenging in its broad scope. Production people are involved in just about every aspect of a station's operation and on-air sound. Through creative commercials, promotional announcements, rich content for social media, and services offered to the announcers and the promotion, news, sports, and social media departments, a production person is an integral part of the station.

Another, less obvious, reason people are drawn to production is the mastery of technology. Typically, new technology creates panic among the masses. And while many run from new technology, production people see the opportunities new technology can bring to radio. Everyone desires to be the master of something, and technology offers a certain mystery and excitement with which to be creative (see figure 1.1). There is always the opportunity to seemingly create something from nothing.

Before entering the radio industry, though, it is imperative that you understand what a radio station is, how it operates, and what role production people play in each station's success.

As a former co-chief operating officer of CBS Radio, and now senior advisor to MultiBrand Media International, David Pearlman once summed up his view of what it takes for a career in broadcasting:

> To best prepare for a career in the new world of radio, one must refine a total knowledge of the business. For example, although a successful PD [program director] must understand the musical nuances of the format, he or she is now asked to manage the P&L [profit and loss] of their department and initiate revenue-generating opportunities. . . . It is imperative to develop effective communications skills and an understanding of marketing and sales, even if your ultimate goal is an on-air shift.

Figure 1.1 An **integrated studio audio system** offers the opportunity to create commercials, promotional announcements, and creative features for the on-air staff. Production people often assist the promotion, news, and sports departments as well.

The Radio Station as a Business: The Profit Perspective

Every radio station has a specific purpose for being on the air. However, as a production person, before that purpose can be accomplished, you must clearly understand what a radio station is. First and foremost, a radio station is a business. It has one primary purpose: to make a profit for its owner. Even noncommercial public radio stations must have a steady source of income with which to pay their bills.

Radio stations are unique businesses that operate at the pleasure of the people of the United States. Stations are given an eight-year license, or franchise, by the Federal Communications Commission (FCC) to use a frequency or dial position. A radio station owner never owns or takes possession of the radio frequency; it remains the property of the people of the United States. There are FCC rules and regulations with which the owner and employees of a radio station must comply. The FCC has the power to issue large forfeitures (fines) and even revoke the license of a radio station for a serious violation of its rules and regulations.

The FCC long ago established that the primary responsibility of a radio station is to serve the public's interest, convenience, and necessity in the community to which the station is licensed; this is often referred to as a station's social responsibility. If the station owner serves the community during the license period, the station license is typically renewed for another eight-year period.

Like any business, a radio station has operational expenses, such as retirement of debt or paying off the mortgage, radio **spectrum use fees**, building and tower rental, music copyright fees, payroll, and so on. If a radio station cannot meet its expenses it will be sold, leased to someone else to operate, or transferred to the entity holding the largest debt against the radio station. In a worst-case scenario, the station will go dark (go off the air), existing in some form of bankruptcy until someone purchases the remaining assets of the radio station. In some situations the station's license might have to be returned to the FCC.

If a radio station does not produce a profit, it is very difficult for it to meet its legal and service obligations to the community. That is why, as a business, a radio station must make a profit. For a radio station to grow and expand its services, it must produce financial gains.

Although many think of radio as an art form, the reality is that radio stations are not referred to as belonging to the radio arts. Radio stations are referred to as belonging to the radio industry. The purpose of any industry is to produce profits for its owners and contribute to the social welfare and economic well being of the community in which it operates.

Radio Broadcasting Revenues

The radio industry is a big business. Annual revenues from on-air advertising, digital platforms (social media), and nontraditional events like concerts are approximately $18 billion a year. As stations quickly expand their digital offerings, a growing area of income for radio stations is online and social media advertising sales. A radio station represents a substantial investment, often costing millions of dollars to purchase. At the beginning of this chapter, it was made clear that the purpose of a radio station is to make a profit so that the station can meet its legal and social obligations to the community. These expectations require significant contributions and effort by the station's staff. There is a saying in management that "you are only as good as your last quarter's balance sheet." For the most part, this is true.

Station managers have two obligations to their ownership: to make a return on the investment (profit) and to show continuing growth or new business. As a part of this business environment, managers who oversee several stations in a single market or region employ strict cost-accounting measures to ensure that the return on investment and projected annual growth are met. Successful small-market managers follow similar types of business plans to ensure steady income and annual growth.

As a part of this cost-accounting process, a manager examines every line item of the station budget and asks a simple question: Does this line item produce profit or significantly contribute to the profitability of the radio station? If the answer is yes, then the item is left in place. If the answer, on the other hand, is no, then some serious questions have to be answered. Sometimes a line item involves a person or a position. Can the line item be eliminated as a cost? If it cannot be eliminated entirely, can the line item be transferred to another department capable of accomplishing the same task more efficiently, thus lowering station costs and increasing profits? It is very difficult for a manager to justify keeping a line item that does not produce a profit or significantly contribute to the profitability of the radio station.

That brings us to the production person. The production person is a line item on someone's spreadsheet. If you are a production person, how do you demonstrate profitability? How do you keep yourself in the budget? After all, the production person is not like the account executive that sells the ads and brings in cash to the radio station. You are also not like the on-air talent who is achieving high audience ratings, and thus increasing the value of the station's **commercial inventory**. So how do you show profitability? One of the keys to your success in radio is an understanding that the production director and the production people work for the sales department. In fact, everyone in a radio station works for the sales department. With this basic understanding of radio as a business and the necessity to make a profit, you can enter the industry a step ahead of many.

The Role of the Radio Production Person

The production person is in effect the Renaissance person of the radio station. Production people must master a number of different technologies and skills, and serve in several roles to several different departments. As you prepare for a career in radio it is very important that you understand the vital role the production person plays in the industry.

The production department is the central hub around which the radio station revolves; its spokes reach to every department and provide critical support services and creative products, such as commercials, promotional announcements, social media content, and assistance in producing news, sports, and public affairs programming (see figure 1.2). The power that turns the wheel is the sales department. Just about every project in a radio station involves an account executive and a

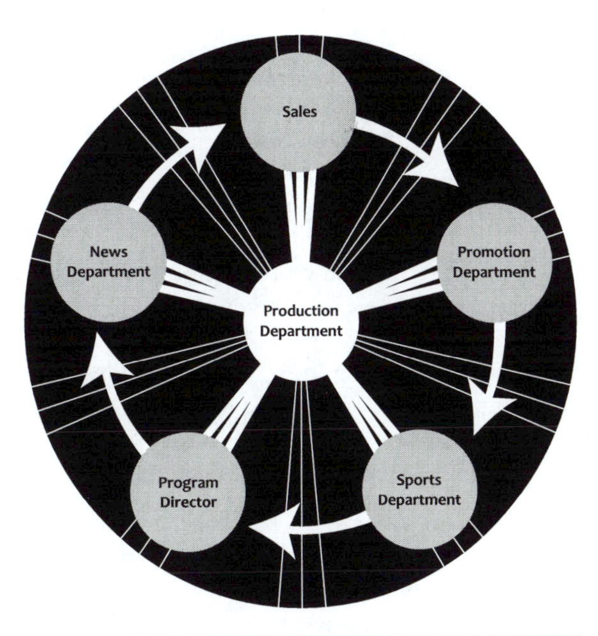

Figure 1.2 The production department is the hub around which the radio station revolves; its spokes reach to every department and provide critical support services and creative products.

production person in some respect. The air staff depends on the support of production people, the account executives in the sales department depend on production people, and the programming and promotion departments depend on production people. In general, production people support the station and its format with a variety of creative services.

Although production people do not get the listeners' attention like the on-air staff, and they don't get the glory of bringing in a huge new sales contract, what they do get—if they have done their job properly and supported the radio team—is respect as talented craftspeople whose skills are vital in making the radio station profitable.

What Makes a Good Production Person?

Great production people know the power of radio as a medium. More importantly, they are enthusiastic about the station they're working for and confident in their abilities to assist the station in achieving market dominance. This level of job satisfaction comes with education, mastery of technology and skills, confidence in your abilities, and a genuine desire to see both you and everyone around you succeed.

Good production people have many qualities, including computer and technical skills, creativity, good writing skills, a knowledge of music, and an ability to express themselves artistically through the medium.

Computer and Technical Skills. A large part of a production person's job involves the use of computers and software in the recording, storage, and manipulation of audio. Additionally, many production people are called upon to edit video clips, take photographs, and post on various social media platforms as a part of their job. A production person is required to be technically skilled with the hardware and software of the industry. The production person must constantly be learning and be looking to new technologies and techniques that help deliver a more creative and effective product in a more cost-efficient manner.

Thanks to technology, there is always going to be a new piece of hardware, software, or a new production style to help you be more creative and efficient. Every day brings you a new and different challenge. For example, there was a time when doing a remote broadcast involved lugging around a 30-pound remote broadcast transmitter, antennas, cables, etc. That gear can now be replaced with a smart phone, a broadcast app, and a microphone; in essence, a shirt pocket remote! It's positive changes like this that the production person can look forward to; they have made your job easier and improve the quality of the product.

Creativity. Do you doodle or have a coloring book? Do you ever color outside the lines? Do you take time to play "what if" and just have some fun every day? The production person is charged with the seemingly impossible task of taking words, using hardware and software, and creating a 30- or 60-second "theater-of-the-mind experience" so exciting that listeners are stimulated to go somewhere they have never been or consider buying something they never knew they needed.

As a production person you are asked, multiple times each day, to take an idea from a brainstorming session the salesperson had with the client and turn that idea into a great radio commercial that gets the listeners' immediate attention and projects an image of products or services onto their mind's eye. Your task is to create something that transforms a listener into a customer walking into a sponsor's business; it's what retailers refer to as traffic.

There was a time when it was popular to say that production people must work "outside the box." The problem though was that the box was still taking up space in the room. Forget the stupid box—take it apart and remove it from the building! Create some open real estate for your mind. You will be required to challenge yourself on a regular basis to come up with concepts or approaches that are something other than bland meat-and-potatoes concepts for commercials or promotions.

Writing Skills. You want to go into radio, so why do you have to learn to write? Although many stations employ copywriters, many do not; in that case, you're it. A good production person must be able to write a commercial or station promotional announcement when necessary. These writing skills need to be matched with knowledge gained from working alongside salespeople, so that the production person can assemble a powerful commercial to increase store traffic and deliver the results the client expects. You need to have a good understanding of what makes a commercial "work" for the client and which elements are necessary for a great commercial and which are not.

There are also going to be those times when the client has given the salesperson a piece of commercial copy and you are assigned the task of editing four-and-a-half pages of handwritten copy into an effective 60-second commercial.

From a basic writing standpoint, how many times does the client's name need to be mentioned in the ad? How do you position the price and product within the ad? Where does the address go? Does the client's phone number need to be in the ad? Does the client's web address need to be in the ad? Do the client's social media platforms need to be in the ad? What does it take to create an effective commercial? By working with the salesperson, you'll develop the knowledge of what kinds of ads produce results for a particular client and what kinds of ads do not. This even includes becoming familiar with the kinds of people the client is trying to attract to his or her business, including such things as their lifestyle, consumer habits, and product preferences.

Musicology. Music is powerful. Combined with just the right words it can be even more powerful. It can motivate people to make buying decisions, to take part in community activities, and to continue listening to your radio station. A production person works with an exclusive library of music (see figure 1.3 on the next page). This library consists of specially written, copyright-cleared production music, just like the production music included with this text. What makes it so special is that the musical selections are 30 or 60 seconds long. The music has been written specifically for use as a part of a commercial, even to the point that some music is industry-specific. In other words, it has the qualities or thematic style associated with a particular type of business.

Figure 1.3 Production music is specially written, copyright-cleared music intended for use in commercials. Each station purchases, subscribes, or "pays per use" for production music that complements its musical format. (Courtesy of TM Studios, Inc.)

A production person needs to be familiar with his or her production music library and have an excellent knowledge of the radio station's musical format, even if it is not his or her favorite style of music. Production music has to complement and enhance not only a client's commercial, but also the station's musical format as well. Music can give a commercial terrific impact when matched to the right copy. Likewise, it can create a serious distraction when a mismatch occurs (see audio demo, cuts 1-1 and 1-2). Selecting the right production music for a commercial or station promotional announcement (promo) is a skill that comes with experience, but it also requires taste and an ear for selecting music that complements the advertiser's business and the station's format.

Artistic Expression. A painter selects brushes, paints, and canvas. A production person selects microphones, computers, software, and music and voice talent, blending them to create commercials and program elements to support the station format, create interest, and motivate listeners. A lot of production people become truly talented and gifted artists—the person who has the ability to mix just the right amount of each of the preceding elements to trigger the wow factor in a listener.

The Radio Station as an Organization

The production person is only one member of a team that is charged with making the radio station a success. The sooner you understand how you fit within the radio station's organization, the quicker you are going to understand your responsibilities with regard to the radio station's success. By understanding your position and how you relate to the rest of the radio station, you can avoid many time-wasting and costly misadventures.

The Business Side

Most successful radio stations, even many small-market stations, typically have an organizational chart clearly outlining the radio station's structure (see figure 1.4). To complement the organizational chart, there should be written job descriptions for each of the positions on the organizational chart. As a new employee, you should ask for a copy of each so you know where you fit in and what your specific duties are.

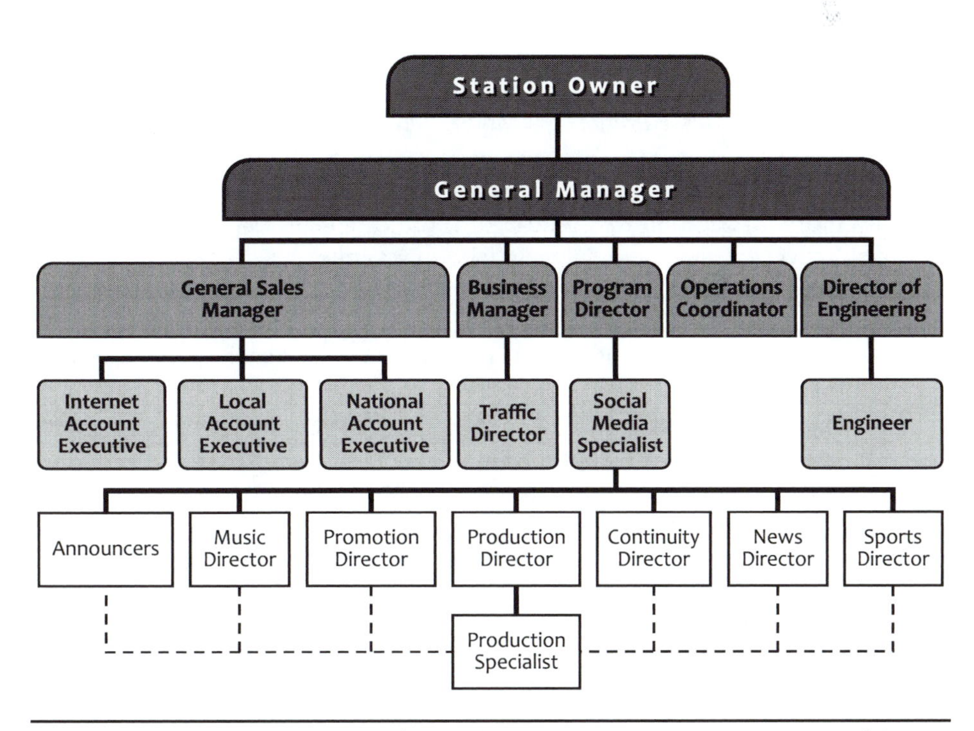

Figure 1.4 An organizational chart for a radio station is a graphic representation of the station staff and who each individual reports to. Depending on the market size, some of the jobs represented here may be combined and assigned to one person. For example, the general sales manager may also be the national account executive.

General Manager. Just below the station owner at the top of the organizational chart is the general manager, or GM, as he or she is often called. The GM is responsible for *everything.* Along with staff, the GM develops a business plan to guide the station to a position of sales and audience dominance in the marketplace. Using the resources of the station department heads, the GM oversees and coordinates all planning, including growth and income projection, programming, marketing, and execution of the station's business plan.

The GM is the parent figure that guides the station and settles any disputes that sometimes arise between departments. In addition, the GM is a headhunter, with the responsibility to search for the new staff necessary to meet the station's needs as it grows. The GM is the person who makes the ultimate decision to hire or terminate employees.

You may also find yourself working in an environment where there are multiple radio stations under one roof. Typically, there is one group or market manager over all the stations. In some cases, the stations in a particular geographic area may be located close together, and the GM for all of the stations is known as a regional manager. The exact wording of the title and responsibilities varies from company to company, but the end result is the same: the GM, or market manager, is near the top of the organizational chart.

General Sales Manager. Working very closely with the GM is the general sales manager, or GSM. The general sales manager is responsible for helping to project and generate station income through the sale of radio commercials, or inventory. In addition to a return, or profit, the general sales manager's business plan reflects growth or an increase in sales revenues.

Advertising dollars come from four primary sources: local advertising sales, regional/national advertising sales, Internet sales, and nontraditional revenue. Nontraditional revenue includes events and projects such as job fairs, boat shows, station magazines, or other radio station sidelines that produce revenue for the station. Depending on market size, the GSM may have a local, a national, or an Internet sales manager who reports to him or her. Further, the GSM may employ the services of a national firm to represent the station to national clients, and is often referred to as a **national rep**.

Business Manager. A business manager does just what the name implies. This is the person held fiscally responsible for the operation of the radio station, controlling every aspect of the money that flows in and out. The business manager advises the general manager and general sales manager on financial matters and keeps them abreast of the station's billing and collections. Business managers also track the station's financial performance to determine profitability and whether the station is meeting sales and growth projections. The business manager is responsible for establishing the station's commercial billing and collection practices. It is the business manager who determines who gets credit and who gets taken off the air because they did not pay their advertising bill. In some stations the business manager is referred to as the controller.

Traffic Director. Working alongside the business manager is the traffic director, or TD. The TD is the person who keeps track of every salable minute of advertising inventory the station has to offer both on air and on the station's web stream. Additionally, the TD may be responsible for the ad content on the station's social media sites. Account executives consult the traffic director before signing a contract with an advertiser to make sure the commercial inventory that they want to sell to the client is available. The TD schedules every commercial, station promo, or event that is to run on the air and on the station's web stream and social media (see figure 1.5). Often the commercials airing on the station's web stream and social media are different from the commercials airing at the same time on the radio station.

The TD advises the business manager as to what percentage of available inventory has been sold so that the business manager can alert the GM and GSM. This information is used to determine advertising rates. If the station is consistently sold out and demand for commercials is high, rates go up. If the station has a large amount of available inventory, rates are adjusted so that the inventory can be sold.

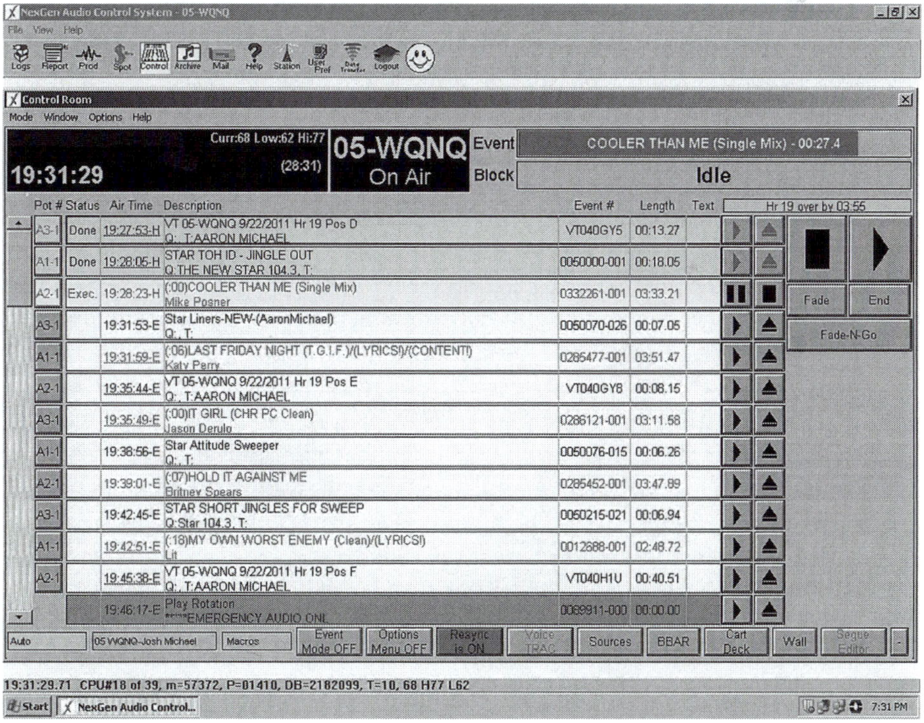

Figure 1.5 The traffic director creates the station's daily program log, which lists every commercial, station promotional announcement, or event that is to run on the air. Like most program logs, this log is electronic and appears on a computer monitor. It includes additional information like the current time and weather in the upper left corner. Notice that every event is timed to the second.

Account Executive. At the street level is the account executive, or AE, also sometimes called a marketing consultant, sales representative, or salesperson. The station AEs target and call on local businesses to sell them advertising. This level of sales is typically referred to as direct or retail sales. The AEs may also call on local advertising agencies hired by businesses to handle their advertising. In addition to selling radio commercials, AEs often sell advertising on the station's web stream, social media, station promotions, and nontraditional revenue-generating events. The general manager, general sales manager, business manager, traffic director, and account executives comprise the "business side of the house."

The Programming Side

Program Director. The "programming side of the house" is headed up by the program director, or PD. The PD is responsible for everything that ends up on the air. PDs direct and shape the station's overall sound and make all of the decisions regarding programming, with the final approval resting with the general manager. The PD's responsibilities are expanding as stations add additional high-definition (HD) channels to the main station signal. Likewise, programming the station's web stream and social media content can add extra responsibilities to the PD's job description. Program directors do not generally make decisions based on hunches or gut feelings. There is a great deal of audience research, skill, and judgment involved on the part of the PD in interpreting the data and making programming decisions. That said, you should also be aware that in small markets sometimes the programming *is* based on hunches by the owner or their family.

PDs have three responsibilities with respect to programming. First, they must maintain the listener base the station currently has. Second, they must attract new listeners. Third, they must show consistent audience growth and achieve audience ratings goals. To accomplish these three objectives, PDs use a number of research tools to track the station's progress. These include both **quantitative** (ratings numbers) and **qualitative** (lifestyle) **research.** Such research includes audience ratings services, focus groups, surveys, web stream metrics (ratings), and **call-out research** conducted by the station to judge listeners' feelings about the station's music.

The PD uses research to help develop an overall programming strategy for the station. This includes music selection, music flow and rotation of songs, air personalities and presentation style, the number of commercials that can run (**commercial inventory load**) on an hour-by-hour basis, the style of **imaging** used to promote the station, when and if there is news, traffic reports, and weather, and the overall sound and attitude of the station, among other things. These are just a few of the elements that make up the station's programming.

Another aspect of the program director's job is music licensing. This includes being familiar with the specific copyright contracts that a station has and monitoring the station's music use. Regular reports are required by **American Society of Composers, Authors, and Publishers** (**ASCAP**), **Broadcast Music, Inc.** (**BMI**), **Society of European Stage Authors and Composers** (**SESAC**), and **SoundExchange,** which are the major organizations that represent artists' interests with regard to copyright payments and the performance of music on the radio and Internet.

Great program directors are talent coaches. They are involved in each personality's show, reviewing and helping the on-air staff to develop and increase the quality of the product (see figure 1.6). They teach their air personalities to communicate with the station's listeners while routinely doing an on-air show (air shift) themselves. Program directors are deeply involved with the promotion and marketing of the station's sound. Program directors are also required to control a budget with regard to talent and promotion expenses, and they help initiate new revenue-generating promotional opportunities.

Like many of the jobs in a radio station, the PD's position is constantly evolving and changing with the industry. In a market where there are several stations under one roof, there may be one PD who works with individual program coordinators for each station. In this situation, the PD is elevated to looking at the big picture, and the program coordinators handle the day-to-day details for the individual stations that consume a lot of a regular PD's time.

Figure 1.6 Program directors are involved in each personality's show, reviewing and helping the on-air staff to develop and increase the quality of the product. They teach announcers and anyone else working on air to communicate effectively with the station's listeners.

Music Director. In addition to the PD, many stations have a music director. A music director works closely with the program director to research and select the music that ends up on the air. The music director develops the lists of music (playlists) for the various times of the day (dayparts) and creates the music rotation order so that the music always appears fresh and forward-moving on the radio station. Usually, the music director consults with the program director on the overall sound of the station's music product and for final music approval, including songs that are being added or dropped from the station's playlist.

Promotion Director. Closely aligned with the program director and the general sales manager is the promotion director. A promotion director works with the program director, the general sales manager, and the account executive to create sales-driven, format-centered promotions for the radio station. These are revenue-generating events that the radio station hosts or cohosts with a sponsor (see figure 1.7). The promotion director is responsible for both the internal and external marketing of the radio station to listeners and sometimes involves overseeing the station's social media presence.

Internal marketing involves using all of the resources of the radio station to promote the station to existing listeners. This includes on-air contests and giveaways, promotions of special station events such as movie premieres, and just about

Figure 1.7 Remote events/broadcasts are an important part of a station's public visibility. The promotion director coordinates sales-driven, format-centered promotions for the radio station.

anything else sales and programming can think of to be competitive in the market. **External marketing**, involves using outside media such as the Internet, billboards, newspapers, and television to promote the station to new potential listeners.

Public Service/Community Affairs Director. Station promotions often involve a community group or organization as a public service function of the radio station. The public service or community affairs director of the radio station is the person who is responsible for interfacing with the community. When a community leader or group comes to the station and asks assistance in promoting an event such as a live broadcast from a pet adoption day at the Humane Society, the community affairs director serves as a liaison between the community group and the station. It is not unusual for the duties of the community affairs director to be combined with another position.

Production Director. The production director is another individual responsible in part for the sound of the radio station. Production directors work closely with the program director so that the creative product that the department turns out fits the station's format, style, and personality. The production director guides the production staff in producing commercials, station promotional announcements, and station imaging. The production director also maintains a close relationship with the general sales manager and account executives so that they are aware of the production department's talents and creative abilities. Occasionally, if a client's idea for a commercial is a little too creative and one that may not fit the station format or style, the production director may consult with the GSM and program director to clear the client's commercial idea.

The production director also works closely with the radio station's promotion department. The promotion department comes up with and directs all of the creative ways to involve the station with the listeners and the community. There is an old saying in radio: "If it's worth doing, it's worth promoting." Each event the promotion department dreams up has to have support from the production department to sell it to the listener in a creative fashion—the more creative, the better.

The production director, like many others in radio management, has a staff to lead. It is the production director's job to put his or her production people to use in the most cost-effective manner possible. The production director must know each staff member's creative strengths and weaknesses. Production directors also serve as coaches for staff members, guiding and helping them to produce the best creative product possible. By developing the staff, production directors are able to schedule production more effectively and give those clients that require the most attention to the most creative and productive people.

The production director is often asked to assign production people to collaborate with the news department on the production of public affairs programs. Such a partnership reflects the most efficient use of the newsperson's time. Although most newspeople are quite competent at production, it makes more sense for them to serve as talent on long-form public affairs shows and to allow the production person to perform the routine editing and final on-air preparation of the show.

Radio stations are required by the FCC to produce some kind of public affairs programming on a regular basis to serve their community.

The production director has to be able to work with deadlines and ensure that every commercial is produced on time and is ready to go on the air as scheduled. An ad not ready to air on time is an expensive error—the station is left with an ad that did not run and an unsold time slot, which means a loss of money. Further, the station will likely have to offer the client some "make good" ads in order to compensate for the error, which will be free of charge. It is very important that you understand that each minute of inventory has a real cash value. Finally, the production director is also required to work with budgets for new equipment, software upgrades and support packages, production music, and the production department payroll, among other things.

Production Person. Although you are called a production person and you work in the production department, you play a vital role in the sales department. In addition to the other station staff a production person works with, a good production person seeks to develop a synergy with each of the AEs, so that he or she more clearly understands the styles and types of commercials the AEs want for their clients. By working with the AEs to develop effective commercials that produce results for the clients, the production person is putting the financial success of the radio station first and foremost. When you cooperate in this way, you will be noticed by the AEs and the GSM because you are helping them reach the station's projected sales goals. There is nothing more exciting than having an AE return to the radio station and tell everyone that the $35,000 contract she just landed with a new car dealership was because the dealer loved the commercial *you* created. This is how the production person contributes to the station's profitability. Production people get pay raises and bonuses based on their contribution to the success of the sales department and the station.

Production people also play a vital role on the air through entertainment and news production. Production people help produce the comedy bits, the ongoing elements, imaging, promotional announcements, and a lot of the things critical to being on the air that gives the station its unique sound and personality. The on-air morning team may give a production person a concept or idea for a show feature that makes the station leap out of the radio at the listener with vivid sound imagery. Although their job is not as flashy as that of the morning team, and no one ever asks for an autographed picture of a production person, their creativity and skill are what helps the morning team and the other on-air staff sound great and achieve higher ratings. Higher audience ratings translate into higher advertising rates, which means that your radio station makes its budget goal.

Continuity Director. A production person works closely with the station's continuity director. The continuity director is the person who is responsible for writing all the commercials that turn listeners into customers for your station's clients. Continuity directors are an interface between the sales department and the production department. Not only do they prepare the copy, they work with the AE to get the copy approved by the client before it goes to production. This allows a production person to focus his or her time and energy on the creative presentation of the ad, as opposed to having to create the script from scratch.

The continuity director knows the station's commercial policies and is familiar with federal, state, and local laws that advertising copy must comply with. The continuity director also works with the program director to determine what kinds of ads are suitable for what dayparts. Would your radio station accept an ad for a gentlemen's club (expensive strip bar) during the daytime, or does your station only allow such ads to run between 11:00 PM and 3:00 AM, if at all?

In larger markets the continuity director may have a small staff of copywriters. In smaller markets, though, don't be surprised to find that there is no continuity director and that all of the duties of the continuity director fall to the production person.

News Director. Another important individual in the radio station is the news director. The news director reports either to the general manager or to the program director, depending on the station.

Generally, the news director meets on a regular basis with the program director to make sure that the style of news that is being presented fits the format (see figure 1.8). A rock station might want news written for 18- to 34-year-olds with a bit of an edge or attitude in the presentation. Even the length of a newscast is determined somewhat by the format and the age group or **demographic** (demos)

Figure 1.8 Because of industry consolidation, program directors typically work with smaller news and sports staffs. This newsroom handles the news and sports for six radio stations in one market.

the station is trying to attract. People from younger demographic groups tend to want short-and-to-the-point newscasts; people from older demographic groups tend to want more in-depth stories and longer-format newscasts. News directors must also be sensitive to the station's web stream and social media. To drive traffic to the station's website reporters are now toting everything from a smart phone to 35 mm digital SLR cameras that shoot stills and high-definition video to enhance the news presence on the station's web stream and social media. Of course, this description is based on a radio station having a commitment to news. Many radio stations have a "news" department in name only. This generally consists of one member of the morning team being designated the "news director" and preparing short-format (two minutes or less) newscasts using material prepared by a service for the morning show.

Sports Director. In addition to a news director, a station may have a sports director. The sports director is responsible for the sports programming that is on the station. This might include daily sportscasts, any long-form sports coverage (such as college football), and doing play-by-play for local sporting events (see figure 1.9). Like news, sports reporters look for photos and video of local sports events to increase the traffic to the station's website and social media. As with the news director, not every radio station employs a sports director, and in many cases the station uses the services of one of the top-rated television sports anchors from

Figure 1.9 In addition to daily sportscasts, a sportscaster might be actively involved in play-by-play coverage of local sports.

the area for their sportscasts. This is both a cost-saving measure and a promotional opportunity to be associated with a popular television station.

Engineer. Engineers are the people who keep your studios in excellent shape and keep your radio station on the air. Most engineers have excellent technical skills and understand that their job is to make sure everything is working 24/7/365. A good relationship with your station engineer is vital to a production person. Regular communication with the engineer can help you maintain audio quality and get good technical advice and assistance when needed.

This is not an easy job, and you may rarely see the engineers, as they often work late at night when your production studios are not in use. Most stations have specific policies about the repair of equipment, and generally your role as a production person is confined to reporting any problem accurately. Do not try to play junior engineer and make repairs yourself. This is especially important in a station in which the engineers and station staff are members of trade unions.

The Radio Team in Action

From a practical standpoint, not every radio station has all of the staff described in the preceding section. Often several jobs are combined into one position in smaller markets. For example, the general manager may also be the general sales manager, and the business manager may also be the traffic director. But, for the moment, assume you are in a larger market and let's walk a commercial order through fictional WBSP-FM so you can have some idea of how things come together. The process of getting a commercial on the air is truly a team effort, with a lot of communication going on among the various staff members at each stage of the ad's creation. At all times during the process an eye is kept on the deadlines for copywriting, ad production, client approval, and the first day the commercial is to be aired.

From a Budget to a Commercial

Earlier in the year, fictional radio station WBSP's general manager created an annual budget for the radio station based on the specific return on investment that the owner seeks. With the help of the general sales manager, the GM establishes the sales budget for each quarter of the year. Because it is the GSM's job to carry out this budget and reach the projected station income in each quarter, the GSM assigns the account executives businesses to call on and to consult with about using radio to increase their retail sales. After meeting with a new business, each AE develops a plan to meet the client's needs and desires. The plan ensures that the client uses radio effectively and includes a schedule of ads to run on WBSP-FM and its website and social media.

An AE often checks with the business department before proposing a schedule to a client to see if the client can afford to be on the station and that the client has acceptable credit. Some clients by their very nature are required to pay cash in advance. Bars, concerts, restaurants, and political candidates most often are required to pay cash in advance. Nor is it unusual for a new start-up business to be required to pay in advance until it establishes a credit history with the radio station.

The AE then meets with the client again, and because the AE has done an excellent job of researching and meeting all of the client's needs, the client agrees to use WBSP-FM in her media mix of advertising. The AE returns to WBSP-FM and submits the advertising plan and sales order to the GSM for review and approval. With the GSM's approval, the sales order travels to the business department where a credit check is run on the business, a credit limit is established, and the client is cleared to go on the air.

From the business department, the sales order moves to the traffic department. Here the available commercial time (inventory) is scheduled according to the sales order. Most good AEs check with the traffic department before they propose a schedule to a client; that way they know the inventory they are proposing (pitching) to the client is available. Once the order clears traffic, it heads to the continuity director, where the ad is written. The AE has already had a meeting with the client to determine what will be advertised and how the ad should be produced. After some discussion, the client simply asked the AE to produce a great sounding ad. The continuity director formulates an ad and then works through the AE to get client approval of the written script. At this point, the script is given a unique client name and number to identify the ad, and a production order is created and sent to production.

When the script and production order arrive, the production director assigns the job to a production person, and the job is given a specific deadline for production. At this point, the production person who has been assigned the job meets briefly with the AE to discuss any special requirements or desires the client might have. After the ad is produced, the production person contacts the AE and assists with the technical aspects of getting client approval. Such assistance might include uploading the ad to a secure location on the station's website so the client can listen to the ad. An **MP3** of the ad might be emailed to the client for her to listen to or it might be as simple as assisting the AE in calling the client and playing the ad on the phone to get the client's approval. The production person always works with or through the AE; he or she never contacts a client directly unless specifically told to do so by the AE. Once the ad is approved, the production person places the ad on the radio station server used for commercial playback on the air. The production order is completed and filed, and traffic is notified that the ad is ready to air.

Traffic generates the daily logs that tell the on-air staff what ads get played when. Traffic also generates the ad schedules that will get played on the station's web stream and social media. After the ad plays at the predetermined times, the on-air announcer verifies this by signing the program log. In some cases this will be an actual signature on a piece of paper, in others it will be an electronic signature in the form of a computerized confirmation report of what actually ran on the radio station. From the completed and verified logs the business department generates the billing invoices and sends them out to the clients.

While the process of moving a commercial project through a radio station sounds complicated, it is typically handled by computer and occurs simultaneously in the various departments.

The basic organizational structure of radio stations is quite similar from one station to the next, with the business side and the programming side working together to ensure the station's financial success.

Suggested Activities

1. Arrange a tour of a local radio station and explain your interest in radio production and how a radio station operates.
2. While there, observe the number of studios present and how the station and each studio is laid out.
3. Ask questions about how the radio station is structured and who performs the different jobs in the radio station.

Websites for More Information

For more information about:

- *how the FCC regulates broadcasting,* visit the FCC at www.fcc.gov
- *how radio stations operate as a business,* visit the Radio Advertising Bureau at www.rab.com
- *issues facing the broadcast industry,* visit the National Association of Broadcasters at www.nab.org

Pro Speak

You should be able to use these terms in your everyday conversations with other professionals.

American Society of Composers, Authors, and Publishers (**ASCAP**)—A membership-based performing rights organization that protects the copyrights of members' works by licensing and distributing royalties for the performances of their copyrighted works.

Broadcast Music, Inc. (**BMI**)—A membership-based performing rights organization that protects the copyrights of members' works by licensing and distributing royalties for the performances of their copyrighted works.

call-out research—Research conducted by a radio station or a research firm to ask listeners questions about how they perceive the station's music format; often used to keep a station's music format on track from week to week.

commercial inventory—The number of commercials that a radio station has to offer for sale. Commercial inventory can be broken down into weekly, daily, and hourly inventory.

commercial inventory load—How many commercials are actually scheduled to run in a given time period, such as an hour, half hour, or quarter hour.

demographic—A market or segment of the population that is identified by demographics, which includes statistical data such as age, average income, marital status, and geographic area.

external marketing—Any medium or marketing tool outside the radio station (such as billboards, newspaper ads, and television commercials) used to attract new listeners to the radio station.

imaging—Radio station imaging consists of the short live and recorded elements that are used between segments of a station's programming that tie everything together, creating a station's signature sound. For example, "Asheville's home for classic rock, Rock 107.9."

integrated studio audio system—A digital audio system that links together all of the audio equipment in a radio station from the microphone to the transmitter using Audio over Internet Protocols (AoIP). Stand-alone control boards and audio wiring are replaced with digital audio over a local area network (LAN) linking all of the audio consoles and equipment together. The integrated system provides a seamless digital audio workflow to the transmitter.

internal marketing—Any marketing or promotion tool used on a radio station to retain the station's current listeners, such as contests, T-shirts, bumper stickers, etc.

MP3—Moving Picture Experts Group Layer-3 audio compression compresses, or shrinks, CD-quality audio files by a factor of 12 or more, maintaining the original perceived CD-quality sound so that audio files can be transferred via the World Wide Web.

national rep—A national representative is a person or a company that represents a radio station to large national advertising agencies handling regional or national products.

qualitative research—Research centered on a person or group's lifestyle choices. For example, a radio station targets 18- to 34-year-old single males living in Jackson County who rent an apartment, make monthly car payments on a 2016 or newer vehicle, eat out three times a week, drink domestic beer, and go to the movies twice a month.

quantitative research—Research centered on numerical or statistical data; most commonly used is audience ratings research to determine the number of listeners a radio station has.

Society of European Stage Authors and Composers (**SESAC**)—A membership-based performing rights organization that protects the copyrights of members' works by licensing and distributing royalties for their performance.

SoundExchange—The nonprofit performance rights organization recognized by the US Copyright Royalty Board that collects statutory royalties from satellite radio, Internet radio, cable TV music channels, and similar platforms for digitally streaming sound recordings.

spectrum use fees—Fees charged by the FCC to broadcasters for the use of the public airwaves.

Basic Science

Analog and Digital Audio

HIGHLIGHTS

Introduction

A production person works with sound every day of his or her life. Before you can efficiently use the equipment in the radio station to perform your job, you need to understand what is moving through the equipment and how it reacts with the environment around you.

It's difficult to use a microphone properly without an understanding of sound and basic acoustics. Likewise, it's difficult to master the hardware and software used to record, store, and manipulate audio on without some understanding of the raw material (sound). Although you live in a digital age, digital audio is not sound's natural state. To understand digital audio you need to understand nature-based analog sound first, including how your own ears work.

It's an Analog World

Deep in the forests of the Great Smoky Mountains of western North Carolina, far from any four-lane highway or city, there is a 13,000-acre tract of 400-year-old old-growth trees that reach over 100 feet into the air. Some of the oldest trees occasionally lose limbs, or in some cases, collapse due to age or weather. This brings up an age-old question: If a tree falls in the forest and there is no one there to hear it, does it make a sound?

On a windy day a tree can give way and send tons of lumber crashing to the forest floor. As the winds blow, the tree is pushed forward, forcing the molecules of air surrounding it from their original resting places and shoving them into other air molecules. Energy is transferred from one air molecule to another, creating a chain reaction, or wave, moving away from the tree.

The chain reaction of one molecule being pushed into another creates an area of high pressure between the molecules, known as compression. The high-pressure compression pushes up a crest in the wave (see figure 2.1). However, the molecules really don't move very far from their original position or average resting space. The wave is very similar to what happens in a sports stadium when the crowd does "the wave." Someone gives a signal, and rows of people begin standing up and sitting down, creating crests in the wave of people. And while the wave moves all the way around the stadium, the nonactive participants stay in their seats, or their average resting place.

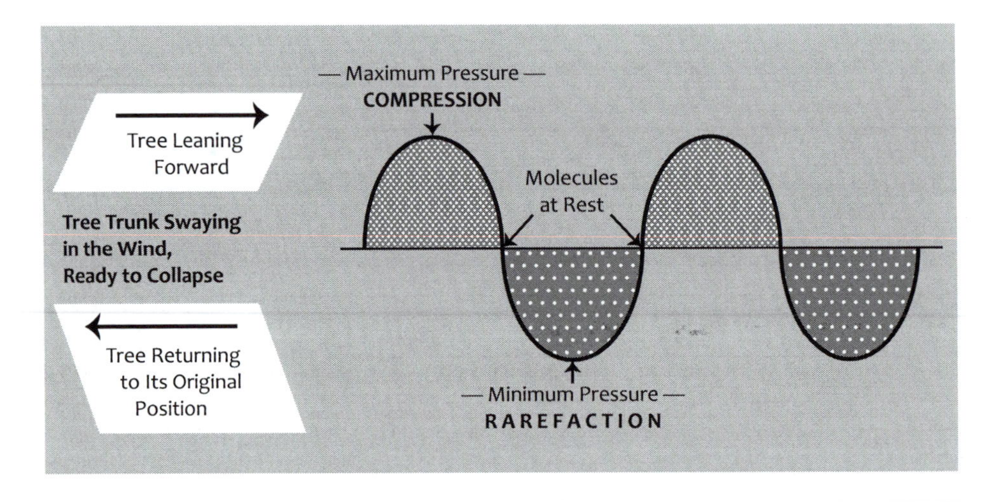

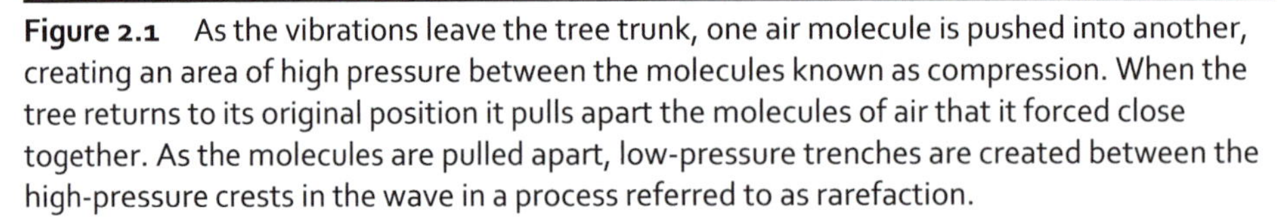

Figure 2.1 As the vibrations leave the tree trunk, one air molecule is pushed into another, creating an area of high pressure between the molecules known as compression. When the tree returns to its original position it pulls apart the molecules of air that it forced close together. As the molecules are pulled apart, low-pressure trenches are created between the high-pressure crests in the wave in a process referred to as rarefaction.

As the winds let up and the tree returns to its original position, it is no longer pushing the molecules out of the way in that particular direction. In fact, as the tree returns to its original position, it pulls apart the molecules that it forced close together. As the molecules are pulled apart, low-pressure trenches are created between the high-pressure crests in the wave in a process referred to as rarefaction.

The sequence of a single compression crest, or high-pressure point, and a rarefaction trench, or low-pressure point, is one complete cycle and is represented by the sine wave (see figure 2.2). The cycles repeat themselves over and over, traveling away from the object that originated the vibrations until the wave runs out of energy. The continuous wave is called a sound wave.

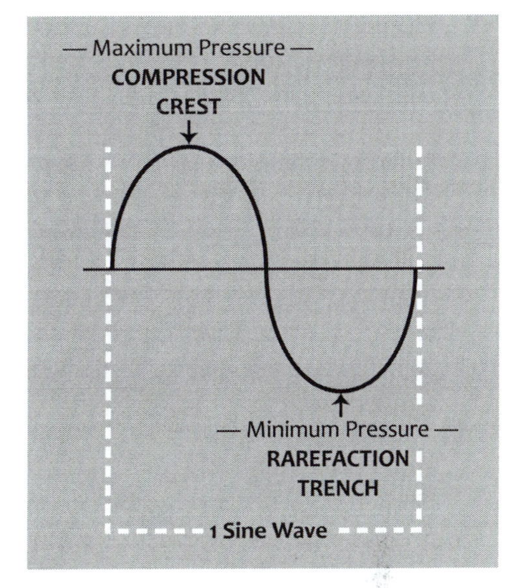

Figure 2.2 A sine wave is one complete cycle of a compression crest and a rarefaction trench. The cycles repeat themselves over and over, traveling away from the object that started the original vibrations until the wave runs out of energy. The continuous wave is called a sound wave.

Identifying Sounds

The primary way that sound waves are identified from one another is by using electronic equipment to count the number of cycles in the sound wave in one second of time, or the cycles per second. The number of cycles per second is designated as the **frequency** of the sound wave. Interestingly, humans perceive the frequency of a sound as pitch. As a general rule, the fewer the cycles per second, the sound is perceived to be lower in pitch. The generic term bass describes a broad range of low-pitched frequencies. The higher the number of cycles per second, the higher in pitch the sound appears. The generic term treble describes a broad range of these high-pitched frequencies. Every sound source has a unique frequency that can be identified.

Heinrich Hertz was a nineteenth-century researcher who did pioneering work with radio waves and theorized about cycles per second. In his honor, the term hertz is used in place of the term cycles per second and is abbreviated as Hz. If a sound has 180 cycles per second it becomes 180 hertz, or 180 Hz. As the number of hertz increases, metric terms are used to simplify the terminology. For example, instead of using the term 1,000 Hz, the metric term kilo, which means 1,000, is substituted. Thus, 1,000 Hz becomes 1 kilohertz, or 1 kHz. After kilo, the next metric step is mega, meaning one million (one million Hz becomes 1 mHz). An electrical signal with a frequency of 100 kHz or more is classed as a radio frequency. As an example, rather than saying a radio station's frequency is 90,500,000 Hz on the FM dial, we say the station is at 90.5 mHz.

A human being with excellent hearing can hear a range of sound waves from near 20 Hz, at the low end of the scale, to near 20 kHz, at the upper end of the scale. The range of sound each person hears varies depending on the person's physical condition. The complete range of audible frequencies is referred to as the sound frequency spectrum (see audio demo, cut 2-1).

There are many sounds that are beyond the range of human hearing. Even though we cannot hear these sounds we can sometimes feel them. Sounds below 20 Hz are called **infrasonic** and are sometimes referred to as sub bass. Sounds above 20 kHz are called **ultrasonic**. Not all people are able to sense the presence of these sounds. When you listen to the radio you are not listening to the radio frequency, such as 90.5 mHz. Humans cannot perceive radio frequencies. You are listening to the audible sounds between 20 Hz and 20 kHz that were piggybacked onto the radio frequency to deliver them to your radio. In other words, the radio frequency serves as a carrier to send the audio to your radio.

To this point, all of the frequencies discussed have been pure tones or frequencies, as represented by the sine wave. A pure tone is referred to as a fundamental frequency, or first **harmonic**. Harmonics are sounds that are created by multiplying the original fundamental frequency. Not all sounds in life are pure tones; most are made up of a combination of tones. Consider the case of the doomed tree in the forest; the tree trunk makes a low rumbling sound as it bends back and forth at its base, sending vibrations into the ground around it. The wind whistles through the leaves, the leaves are hitting one another as they thrash about in the wind, and there is the occasional creaking sound as the tree trunk becomes weak; in all, a multitude of fundamental frequencies and other sounds are coming from the tree.

Each fundamental frequency has harmonics, or tones, that are exact multiples of the original frequency. As an example, starting with a fundamental frequency of 400 Hz as the first harmonic, the second harmonic is 800 Hz, the third harmonic is 1,200 Hz, and so on. Applying this to a real-life situation, 400 Hz is about the top end of the primary range of a trombone. In addition to the harmonics, each fundamental frequency also has tones that are *not* exact multiples of the fundamental frequency, called **overtones**. Overtones are the frequencies that occur between the exact harmonic frequencies. In the case of the trombone, the harmonics and overtones extend from the primary sound of about 400 Hz to nearly 10 kHz.

Most sounds that you perceive are a combination of the fundamental frequency, the harmonics, and the overtones. Timbre, or the tone quality, is the term used to describe this unique combination of tones that distinguishes one sound from another. Musical instruments, when played properly, create many tonal combinations that are quite pleasing to the ear. Typically, engineers talk about bass, midrange, and treble. These broad definitions can be further defined and divided.

Sub Bass. The sub bass is typically 16 Hz to 50 Hz. These are the sounds that pipe organs do such an incredible job of reproducing, down to 32 Hz, with some large pipe organs capable of 16 Hz. These are the sounds that you don't really hear as much as you feel them. Such frequencies are associated with the power and fullness of a sound. However, if these sounds occur continuously or too often, or are too loud, the sound becomes muffled. The very nontechnical term used to describe this condition of the sound is "down in the mud" or "muddy."

Low Bass. The low-bass region of 50 Hz to 80 Hz is where some of the distinctive qualities of musical instruments begin to appear. In this frequency range are instruments like the lower frequencies of a kick drum, tuba, and the lower notes of a bass, among others. This range is well into the hearing range of most people.

One of the interesting qualities of the sub-bass and low-bass regions of sound is that humans have a very difficult time sensing the direction these sounds originate from. Sounds at these frequencies lose their directionality to us. This is why a surround sound speaker system only has one sub-bass speaker or **subwoofer**. It can be placed just about anywhere in the room, since you are not capable of sensing which direction the sub-bass sounds are coming from.

Upper Bass. The upper-bass frequencies include the sounds from 80 Hz to 315 Hz. This is an area that includes some of the lower ranges of the human voice and frequencies that provide warmth and body to sound. The sounds in this frequency range are very satisfying to the human ear and include instruments such as the piano, drums, and stringed instruments like the bass and cello. Some horns are also in this range.

Lower Midrange. In the lower midrange are the frequencies between 315 Hz and 2.5 kHz. The human voice extends into this range, and it is these frequencies that provide us with our basic sense of speech intelligibility. Many of the sounds in this frequency range are harmonics and overtones of lower frequencies. Sounds in this range add a rich quality and intensity to music.

Midrange. The midrange frequencies include the band between 2.5 kHz and 5 kHz. Of all the frequency bands, this is the one that humans are most sensitive to, for several reasons. Many of the frequencies from 2.5 kHz through 3.5 kHz contribute to speech intelligibility, particularly sounds like *m*, *b*, and *v*. Starting at 3.5 kHz, and extending into the bottom end of the lower treble (to about 6.5 kHz), are frequencies that add definition, realism, and clarity. In fact, it is the sounds in this area that give us a sense of realism and help us to pinpoint direction. People perceive sounds in this range as being nearby, which is a quality called "presence."

Another interesting acoustic quality of this range is what happens around 5 kHz, or about dead center of the presence range. If an engineer boosts the sound level at 5 kHz, the listener perceives that there has been a significant boost over the entire midrange, making vocals sound tight and very nearby.

Lower Treble. The lower-treble range includes the frequencies from 5 kHz to 12.5 kHz. This includes vocal sibilance, or the sizzle produced as some people use words containing the letters *s*, *c*, *x*, and *z*. Amazingly, female sibilance can range as high as 11 kHz to 12 kHz. Natural sibilance is fine and adds an intimate quality or presence to vocals. However, this is the frequency range in which a person who has a sibilance problem will stand out.

Upper Treble. The upper-treble frequencies include those from 12.5 kHz to 20 kHz. Earlier it was stated that a person with excellent hearing could hear sounds as high as 20 kHz. However, the reality is that most people cannot hear much past

16 kHz. In the upper-treble range are those sounds that provide a sense of airiness, sizzle, crispness, and brilliance. One of the downsides to this frequency range is that any electronic noise or hiss appears here if present in the signal.

Loudness

Sound is created from vibration and from the compression and rarefaction of molecules that follow. Specific sounds are identified using frequency, harmonics, and overtones. Because you are going to be dealing with sound on a daily basis, it is important for you to be able to determine how loud a sound is, how loudness is measured, and the effect loudness has on people.

A vibration sets off a cycle of compression and rarefaction of air molecules, creating a sound wave. The stronger the vibration and the more molecules that are forced into the cycles, the greater the height of the crest and depth of the trench of the sound wave. In other words, the taller the sound wave is physically, the louder the sound. The size of the sound wave is measured from the bottom of the trench to the top of the crest and is called **amplitude**. Subjectively, the amplitude represents the loudness or softness of a sound wave (see figure 2.3).

The amplitude of a sound wave is measured in **decibels (dB)**. The decibel is a unit of measurement developed by Alexander Graham Bell and represents one tenth of a "bel." A decibel is a mathematical calculation based on ratios of two quantities. The quantities can be acoustic sound pressure or electrical energy, such as voltage. Unless you are going into engineering, it is not necessary for you to know the formulas, as they involve advanced engineering and mathematics beyond the scope of this book.

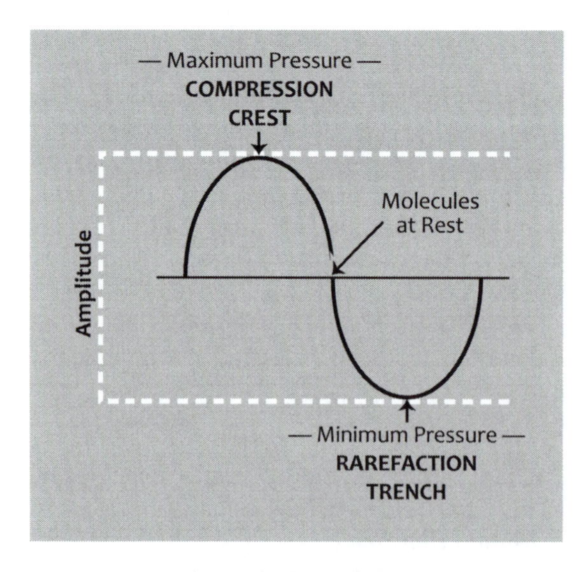

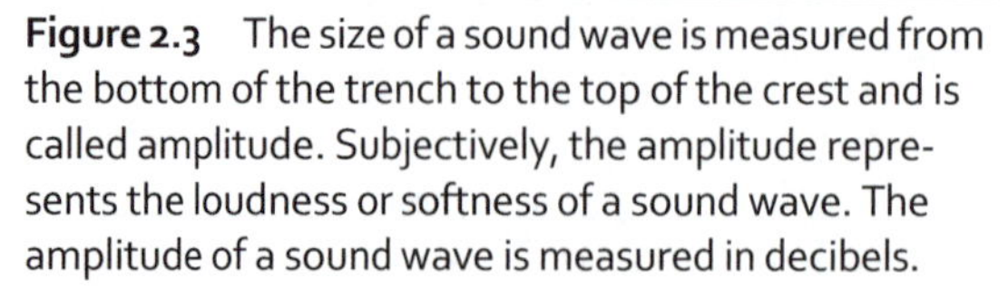
Figure 2.3 The size of a sound wave is measured from the bottom of the trench to the top of the crest and is called amplitude. Subjectively, the amplitude represents the loudness or softness of a sound wave. The amplitude of a sound wave is measured in decibels.

A sound wave passing through the air causes variations in the air pressure rising above or dropping below the **barometric air pressure**. The variations in air pressure accurately track up and down with the compression and rarefaction of the sound wave. The variation in air pressure is called the sound-pressure level (SPL). Sound, or acoustic energy, is measured in decibels of sound-pressure level, or dB-SPL, on a sound-pressure meter. A sound-pressure meter includes a microphone and a calibrated electronic filter so that the microphone and meter react to sound just like your ears do.

Scientifically, microphones are transducers, or devices that change one form of energy into another form of energy. In this case, a microphone converts acoustic energy, or sound pressure, into electrical energy. Microphones, by their very nature, measure the compression and rarefaction of a sound wave and convert it into electrical energy. When a microphone converts acoustic energy to electrical energy, the sound wave is converted into an audio signal. Audio is a Latin word meaning "I hear"; it usually refers to a signal or electrical waveform. Audio depends on wires, fiber optics, or radio frequencies to carry it from place to place.

When a microphone converts a sound wave to an audio signal, the audio signal is an identical electrical copy of the sound wave. That is, it is analogous to the sine wave or acoustic waveform that it represents. In analog audio, the voltage of the audio signal varies in amplitude the same way that the sound pressure varies in amplitude before it enters the microphone.

Humans can hear sounds from 0 dB-SPL, the threshold of hearing, to sounds that exceed 120 dB-SPL, the threshold of pain (see figure 2.4). The difference between the softest sound and the loudest sound created by something is called its **dynamic range**.

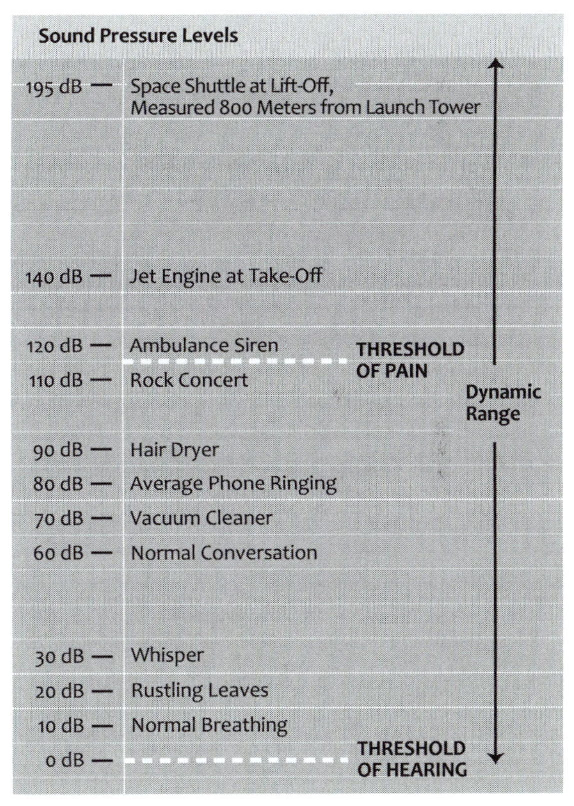

Figure 2.4 Sound pressure is measured in decibels of sound-pressure level (dB-SPL). Humans can hear sounds from 0 dB-SPL, the threshold of hearing, to sounds that exceed 120 dB-SPL, the threshold of pain. The difference between the softest sound and the loudest sound created by something is called its dynamic range.

You need use caution when it comes to your hearing. Extended exposure to high sound-pressure levels causes a temporary loss of hearing that can last several hours and in some cases becomes permanent. At a sound-pressure level of 120 dB-SPL, humans begin to experience pain. At extremely high sound-pressure levels, excruciating pain and serious physical damage to your hearing occurs.

The National Institute for Occupational Safety and Health (NIOSH) has established safe levels of noise exposure for workers in industry. The maximum continuous safe sound-pressure exposure limit for an 8-hour workday is 85 dB-SPL. However, even at 85 dB-SPL and less, many employers require workers to wear hearing protection. Safe-exposure limits get shorter as the sound pressure increases. A worker exposed to 94 dB-SPL has a safe-exposure time limit of 1 hour. At 106 dB-SPL, the safe-exposure time limit is 3 minutes and 45 seconds!

To give you some meaningful comparisons, normal conversational speech is at about 60 dB-SPL, a ringing telephone has about 80 dB-SPL, and a hair dryer 90 dB-SPL. At the upper end of the spectrum, an ambulance siren creates 120 dB-SPL, and a jet engine at takeoff is about 140 dB-SPL on the runway.

If you have ever attended a rock concert, where sound pressures can reach 100 to 110+ dB-SPL, you now know why everyone was yelling at each other when they went home. They were suffering temporary hearing loss. At extreme sound-pressure levels, your eardrums can be blown in or imploded. One of the side effects of military service is that many service personnel suffer permanent hearing loss due to exploding ordnance.

Figure 2.5 Sound-pressure meters are a valuable studio tool. Many radio stations have preset maximum-volume controls for headphones and monitor speakers so that you cannot exceed safe limits of exposure. Long production sessions can take their toll on a producer's ears, causing a temporary loss of hearing and poor judgments when it comes to the final mix.

Recording studios typically have a sound-pressure meter sitting on the main console to observe music-playback levels (see figure 2.5). Long recording or mixing sessions can take their toll on the producer's and recording engineer's ears, causing a temporary loss of hearing and causing them to make poor judgments when it comes to the recording mix. Likewise, many radio stations have preset maximum-volume controls for headphones and monitor speakers so that you cannot exceed safe limits of exposure.

Frequency and Loudness

Frequency and loudness of a sound are independent of one another and not related. Your ears, though, react differently to sounds depending on their frequency. If the frequency of a sound is varied, the human perception of its loudness also varies, even though the frequency remains at a constant sound level (see audio demo, cut 2-2). Likewise, if the amplitude or loudness of a frequency is varied, our perception of pitch is affected, even though the pitch remains at a constant frequency (see audio demo, cut 2-3).

This phenomenon occurs because your ears do not hear every sound equally. We do not hear the sub-bass and bass frequencies as well as we hear the midrange frequencies. In fact, your ears are likely not very sensitive to low frequencies at low levels. You also do not hear the lower-treble and upper-treble frequencies as well as you hear the midrange frequencies. Your ears are most sensitive to the middle frequencies. Remember the engineer's trick of boosting the frequencies at 5 kHz? That is about the frequency that your ears are most sensitive to. The name of this phenomenon, in which you hear different frequencies at different levels, is the equal-loudness principle. (Which really is backward; it should be called the unequal-loudness principle.)

The equal-loudness principle is the reason why sound-pressure meters have calibrated electronic filters to accurately measure sounds the way humans hear them.

Velocity

From the moment the tree began to wobble back and forth in the forest at the start of this chapter, the emphasis has been on movement: the molecules of air displaced by the tree and moving away from it. If there is movement, then the speed, or velocity, of the movement can be measured. Unlike the relationship you just discovered in the equal-loudness principle, the speed of a sound wave has little or no impact on the perceived frequency or loudness of the sound wave. In fact, the speed of sound is fairly constant. At sea level, when the temperature is 70 degrees Fahrenheit, sound travels through air at 1,130 feet per second. Sound does change velocity in cooler or hotter air temperatures, moving faster as the temperature rises and more slowly as the temperature drops. Each degree Fahrenheit of temperature variation produces a change of 1.1 feet per second up or down.

The speed of sound also varies with the medium it is passing through. Sound passing through water travels at around 4,900 feet per second, in wood about 11,700 feet per second, and in steel, sound is transmitted at a screaming 19,000 feet per second.

A practical use of this knowledge is estimating the distance to an event or sound source. For example, if you cover a launch at the Kennedy Space Center, the press area is several miles from the launch pad for safety reasons. Reporters see the spacecraft's engines ignite and the launch vehicle clear the tower and head downrange before the first sound ever reaches the press site. In fact, it takes about 15 seconds for the sound to travel the distance from the launch pad to the press site. From the time you see any event until you hear its sound, count the seconds and then multiply by 1,130 feet per second to determine the distance. In the case of the Kennedy Space Center and the press area, this comes to about 16,350 feet, or right at 3.1 miles (see audio demo, cut 2-4). (The same thing works for figuring out how far away lightning is. See the flash, count the seconds until you hear the roar of the thunder, and then multiply by about 1,130 feet per second to determine your distance from the lightning.)

Phase

Acoustical phase is the term used to describe the difference in time between two or more sound waves as they travel together from one place to another. In a concert hall, the sounds of an orchestra may all leave the stage at the same time, but because of physical obstacles, reflections, and other acoustic conditions, the same sounds may arrive at the listener's ears at various times, altering the original sound. Phase is essentially a difference in time between two sine waves. The difference can range from a few microseconds (micro = millionths) to an extreme of several seconds, depending on the acoustical conditions present. To measure this difference in time, sound waves are divided into segments for comparison to other sound waves.

The length of a sine wave is measured in degrees. One cycle is divided into 360 degrees (see figure 2.6). The wave begins at zero amplitude, zero degrees, and after one complete cycle of compression and rarefaction ends at zero amplitude, 360 degrees. In addition to the starting and stopping points, other major points in a sound wave can be identified. The crest of maximum positive amplitude occurs at 90 degrees into the sine wave, or about one-fourth of the way through the 360-degree cycle. As the pressure decreases, the sound wave passes back through zero on its way to the lowest negative pressure in the sound wave. The point at which the sine wave passes through zero is 180 degrees, or half of the way through the cycle. The sound wave continues to lose sound pressure until it bottoms out in the area of maximum rarefaction or negative pressure, called the trench. This point is at 270 degrees, or three-quarters of the way through the sound wave, just before the sound wave begins its upward movement before coming to the end of one cycle, or 360 degrees. Phase is a difference in time between two sine waves, measured in degrees.

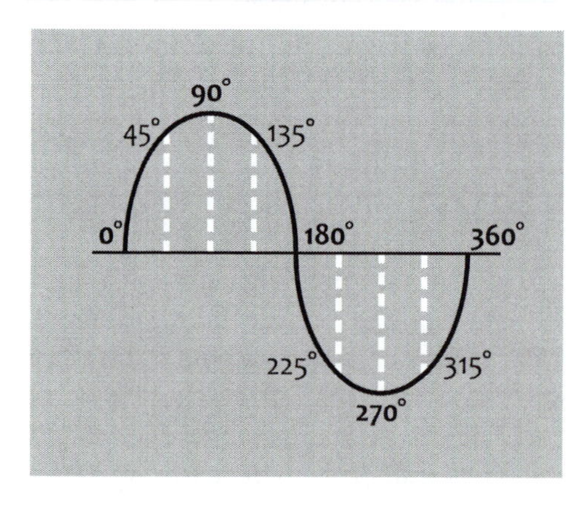

Figure 2.6 One cycle is divided into 360 degrees. The sound wave begins with 0 amplitude, 0 degrees, and after one complete cycle of compression and rarefaction ends at 0 amplitude, 360 degrees. The crest of maximum positive amplitude occurs at 90 degrees. The point at which the sine wave passes through 0 again is 180 degrees, and as the sound wave continues to lose sound pressure it bottoms out at maximum rarefaction, or 270 degrees.

Phase causes some interesting acoustical conditions to occur. When two sound waves start together and their degree interval remains the same, the waves are said to be in phase. Waves that are in phase reinforce one another, and the amplitude or loudness of the sound is increased (see figure 2.7). For example, if two trumpeters are side by side and start playing the same note at the identical time, then the sound is in phase and is amplified or made louder. In a perfect world, when two identical sound waves, of the same frequency and perfectly in phase, are combined, they double in amplitude or loudness.

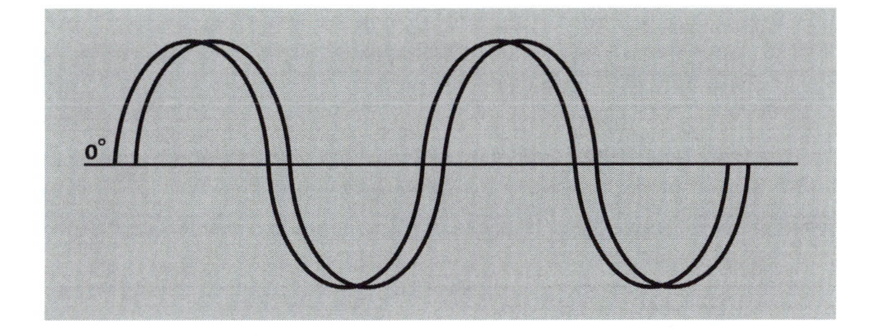

Figure 2.7 Waves that are in phase reinforce one another, and the amplitude or loudness of the sound is increased. For example, if two trumpeters are side by side and start playing the same note at an identical time, then the sound is in phase and is amplified or made louder. In a perfect world, when two identical sound waves, of the same frequency and perfectly in phase, are combined, they double in amplitude or loudness.

When two sound waves start at different times and their degree interval is not the same, however, the waves are said to be out of phase. Waves that are out of phase weaken each other and decrease the amplitude or loudness. Using a variation on our previous example, if the two trumpeters are playing the same note but start at slightly different times, the sound is out of phase and is weaker. Again, in a perfect world, two identical sound waves of the same frequency that are exactly 180 degrees out of phase cancel each other out (see figure 2.8).

These two conditions explain why a theater or concert hall has sweet spots, where the sound is in phase and is just perfect, and dead spots, where the sound is

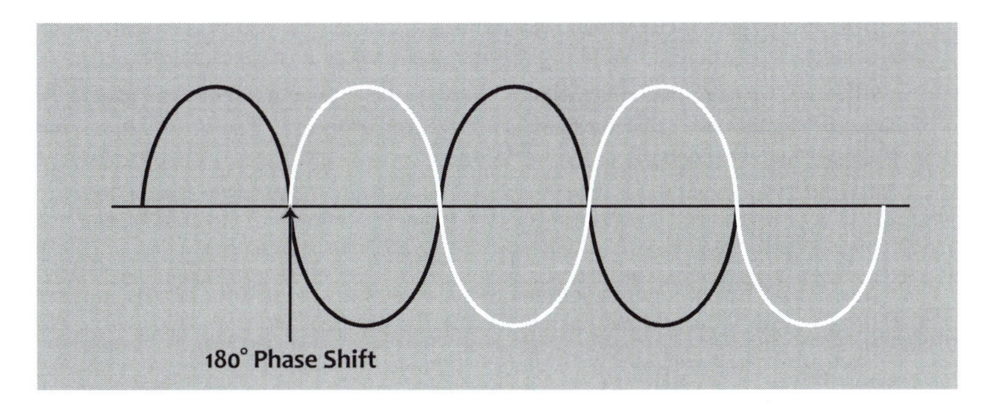

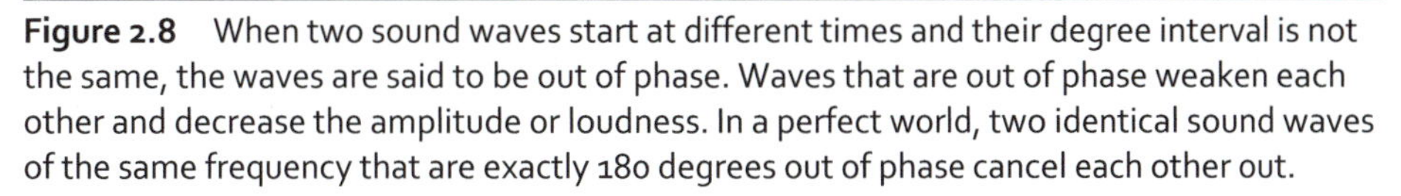
Figure 2.8 When two sound waves start at different times and their degree interval is not the same, the waves are said to be out of phase. Waves that are out of phase weaken each other and decrease the amplitude or loudness. In a perfect world, two identical sound waves of the same frequency that are exactly 180 degrees out of phase cancel each other out.

out of phase and certain sounds cancel each other out and are lost. Unfortunately for performers and listeners alike, there is no such thing as the perfect concert hall.

In reality, sound waves get started at different times, even in a great orchestra. If the sounds are together in compression and rarefaction and just slightly out of phase, it is called **constructive interference**. Constructive interference tends to increase the amplitude, reinforcing the sound waves. However, if the sounds have gotten off to such a poor start that the compression and rarefaction are occurring at different times, it is called **destructive interference** (see audio demo, cut 2-5). Destructive interference tends to decrease the amplitude and weaken the sound waves.

A practical application of phase relationships might be encountered when placing a pair of stereo microphones to record a musical group. Place the microphones in the sweet spot of the room and you get a great recording; place them in a field of destructive interference and some of the audio frequencies may be weak and cancel themselves out. Even when the sound is in phase, it is possible to improperly place two microphones and cause phase problems as the same sound enters one microphone before the other. Microphone placement is discussed in detail in chapter 3.

Your Ears: How They Work and How to Take Care of Them

To this point in the chapter the focus has been on sound: its frequencies, harmonics, overtones, velocity, and phase. Sound waves passing through the air eventually impact a human ear (if someone is present). The sound wave, consisting of compression and rarefaction, enters through the outer ear and travels through the external ear canal until reaching the eardrum (see figure 2.9).

The compression and rarefaction of the sound wave push and pull the eardrum back and forth. Attached to the back of the eardrum are three bones: the hammer, anvil, and stirrup. These three interconnected bones act as a mechanical lever, amplifying the sound pressure and transferring it to the cochlea. The cochlea is a small, snail-shaped organ that contains a microscopic amount of fluid, in which about 30,000 hair cells are located. Each hair cell is attached to the auditory nerve; they generate the electrical signals sent to the brain that you perceive as sound. When the electrical signals from each of your two ears reach the brain stem, the two sets are combined and the signal eventually ends up in the auditory cortex of the brain. Once in the auditory cortex, the electrical signals are interpreted by the brain and, finally, perceived by you as sound.

Constant exposure to high-pressure sound levels damages the microscopic hair cells in the cochlea, in severe cases breaking the hairs off. As the hair cells are damaged or lost, your hearing is affected proportionately.

In addition to this method of sound arriving at the brain, there is another way you perceive sound. Sound can also be transmitted to the brain through the central nervous system. Exposure to high sound-pressure levels such as those produced by a gunshot, or sitting front row at a concert in front of a speaker tower, can cause unexplained excitability, tension, fatigue, headaches, and nausea. When your body has to absorb powerful sound-pressure levels, it reacts negatively. Your heart rate

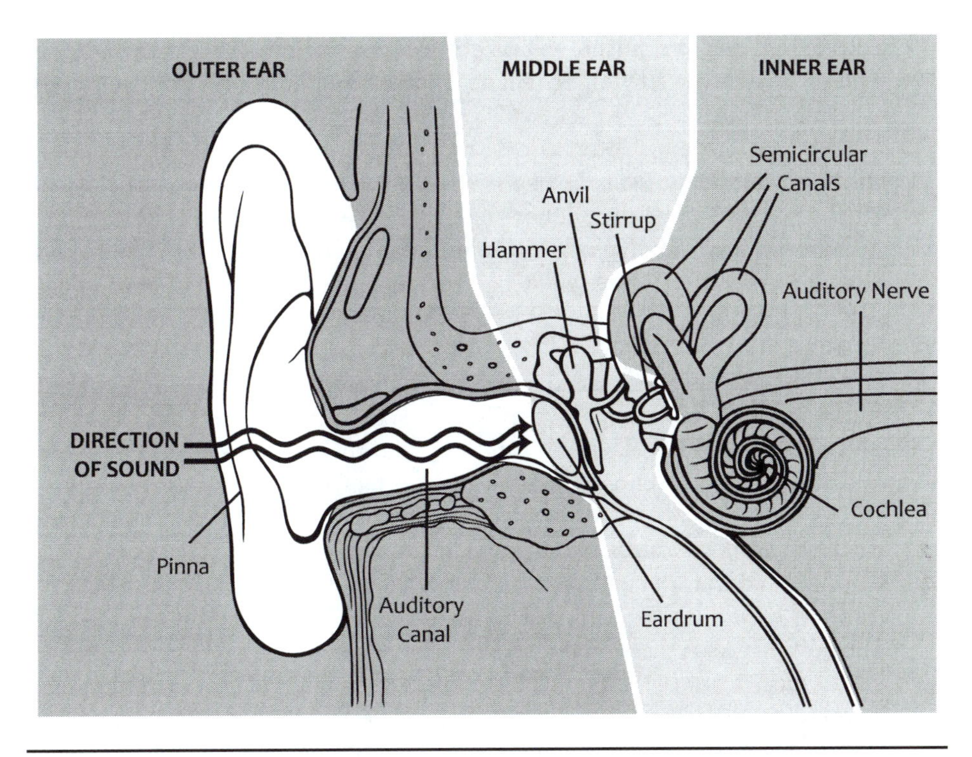

Figure 2.9 The outer, middle, and inner ear convert sound waves into electrical impulses that our brain interprets and identifies as sound.

jumps first, signaling alarm. Your blood pressure increases, muscles begin to contract, and as a result of the instinctive fear you are experiencing, adrenaline is released. In addition, your pupils typically dilate, your mouth becomes dry, and your stomach becomes upset.

You Only Get One Set of Ears. As you progress in this text you are going to be given tips on how to listen to or monitor sound. The speakers, headphones, or ear buds you use on a daily basis can, and will, damage your hearing if not properly used. When you combine them with all of the other sounds that you are exposed to on a daily basis, including live music concerts and car stereos, your hearing can be impaired over time.

In many ways, production people can be the victims of their own industry's success. By using the latest advances in studio equipment, production people work with gear that produces incredibly clean, distortion-free sound at higher sound-pressure levels. According to the world-renowned House Ear Institute of Los Angeles, although the audio is clean and sounds great, the downside is that production people are listening at levels of 10 to 15 dB-SPL higher than in the past. According to the National Institute on Deafness and Other Communication Disorders about 26 million people age 20 to 69 in the United States have high-frequency hearing loss due to exposure to noise at work or during leisure activities. Based on other estimates, by 2025 there will be some 40 million hearing-impaired people in the United States.

You can preserve your hearing by listening to your work in the studio at a maximum sound-pressure level of 85 dB-SPL. This applies both to the speakers in the studio and to the headphones you may be using. Smart producers in recording studios monitor sound levels with a sound-pressure meter near the mixing position. You can purchase an inexpensive sound-pressure meter at many electronics stores. The 85 dB-SPL rule of thumb also applies to portable music players and ear buds.

If you are in a band, attend concerts frequently, or like to watch or participate in motor sports, use inexpensive (less than a dollar) foam earplugs that can reduce the ambient sound level up to 30 dB-SPL. You still get to hear the concert or see the race, just not at earsplitting levels. Other, often-overlooked sources of loud noise are gas-powered lawn mowers, string trimmers, and lawn blowers. Again, inexpensive foam earplugs can prevent hearing loss. Although loud sounds can be a thrill, keep in mind that the end result could be impaired or permanent hearing loss or constant ringing in the ears, just to name a couple of common hearing disorders. The bottom line is that you only get one set of ears.

Ironically, one of the critical listening tests to judge if a commercial or program is mixed properly is to turn the monitors *down* to a low level and see if everything is still clear and intelligible. There is nothing wrong with mixing your projects at lower levels; this is how many pros prefer to mix, particularly in long sessions.

With your new knowledge of analog sound, go back to the first part of the chapter and ask yourself the original question again. If a tree falls in the forest and there is no one there to hear the event, does it make a sound? Well . . . does it?

It's a Digital World

The first analog sound recording and playback device was Thomas Edison's phonograph. It operated with a hand-cranked, spring-driven motor that, when wound tightly, provided several minutes of cylinder playing time. (Edison's original design used a cylinder to record and play back sound; it was Emile Berliner who developed the flat record that could be easily mass-produced.) Whereas analog audio served us well throughout its long history, it was plagued with many problems inherent in the mechanical, magnetic, and vinyl media used for recording and playback. Additionally, analog recordings have a limited dynamic range and when copies are made, each generation degrades slightly from the last.

Digital audio eliminates the problems associated with recording and storing analog audio; in addition, digital audio provides an incredible dynamic range of up to 144 dB. The most obvious advantage of digital recording is that once sound or audio is converted to a digital format, it can be stored, manipulated, and played back on a variety of devices. Digital audio can be transported and sent anywhere that computer data can be transmitted, even attached to a text message. A further advantage of digital audio, although some in the music industry consider it a distinct disadvantage, is that each copy is identical to the original, with no loss of quality or degradation of the audio.

The Foundation of Digital Audio

The origins of digital audio technology can be traced to the development of the binary number system, or base-2 number system (1 and 0), in the late 1700s. In 1840, Samuel Morse invented the telegraph and a binary system of communication based on dots and dashes, called Morse code. More recently, Harry Nyquist, working at Bell Labs in 1928, determined that it was mathematically possible to convert an analog audio signal into a digital form. Nyquist had one problem, though; he was 18 years ahead of his time. It was not until Bell Labs developed the transistor, in 1947, that the Nyquist sampling theorem could be tested!

The first electronic digital computer, using the binary numbers 1 and 0, was developed during World War II to calculate artillery range charts. ENIAC (Electronic Numerical Integrator and Computer) weighed in at a monstrous 30 tons and guzzled about 200 kilowatts of electricity. It was about 25 years later that digital audio recorders were demonstrated, in the early 1970s, and in 1982 Sony released the first compact disc (CD) player.

Whereas analog audio depends on an analogous relationship between the original sound wave and its electrical counterpart, digital audio does not. Digital sound does not occur in nature, it is created from nature. Digital audio uses numerical representations to record and play back an analog audio signal's amplitude and frequency.

Digital audio can be stored in a number of different ways. The three most popular ways are *magnetically,* on a computer hard drive; *electronically,* on some form of a flash memory card or device; and on a *server,* such as a cloud server.

Analog-to-Digital Conversion

Analog audio goes through two basic stages, ADC and DAC, to become digital audio. ADC is the analog-to-digital conversion that takes place before the audio can be recorded, stored, or manipulated by a computer or digital recorder. DAC is the digital-to-analog conversion so that the audio can be played back on speakers or headphones. From a terminology standpoint, ADC is often expressed as A/D and, likewise, DAC is expressed as D/A. A small microprocessor (computer chip) makes the A/D and D/A conversion within digital audio equipment.

The actual A/D conversion is a fairly simple process. The first step of turning an analog audio signal into a digital audio signal is sampling (see figure 2.10 on the following page). At a number of points in the analog audio signal's sine wave, the A/D converter samples, or reads, the voltage of the analog audio signal. Each one of these samples, or voltage readings, is assigned a binary number to represent that specific point in the sine wave. The process of assigning a binary number to each of the samples or voltage readings is called **quantization**.

The number of samples that need to be taken is based on the Nyquist sampling theorem. Nyquist said that the highest reproducible audio frequency in a digital

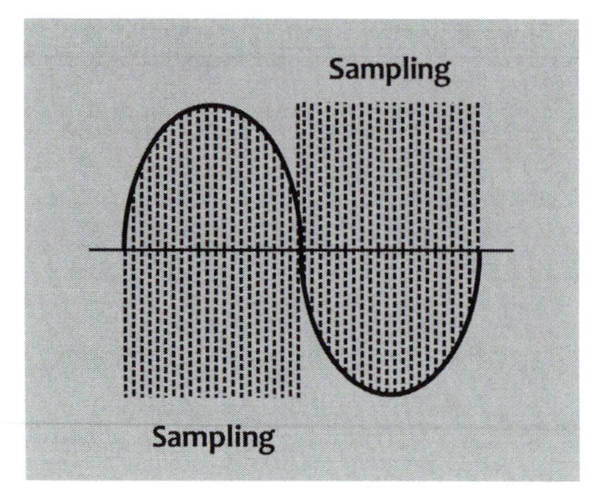

Figure 2.10 The first step of turning an analog signal into a digital audio signal is sampling. At a number of specified points in the analog audio signal's sine wave, the A/D converter samples, or reads, the voltage of the analog audio signal. Each of these samples, or voltage readings, is assigned a binary number to represent that specific point in the sine wave. The higher the sample rate, the more accurately the original frequency response can be recorded and played back.

system is one-half the sampling rate; the audio frequency is often called the Nyquist frequency.

Humans can hear up to 20 kHz. Based on the Nyquist sampling theorem, twice the number of samples of the highest frequency we want to be able to record and play back must be taken. In this case, 2 × 20 kHz, or at least 40,000 samples, need to be taken per second. However, at this sampling rate some frequencies, harmonics, and overtones that are just above 20 kHz might be lost. To ensure good-quality audio, the industry set the upper audio frequency to 22.05 kHz, resulting in a sampling rate of 44,100 Hz or 44.1 kHz (2 × 22,050 = 44,100). The sample rate determines the frequency response of the signal that is recorded and played back. The higher the sample rate, the more accurately the original frequency response can be recorded and played back.

To prevent any unwanted ultrasonic frequencies from getting into the A/D converter and causing undesirable side effects during the A/D conversion (a process called aliasing), an anti-aliasing filter is placed on the input of an A/D converter. An anti-aliasing filter blocks all frequencies above a predetermined point. With a sampling frequency of 44.1 kHz the filter might be set to block all frequencies above 22.05 kHz.

Each time a voltage sample is taken, a binary number consisting of 1s and 0s is created and assigned to the sample to represent the voltage reading. The binary digits are called bits (short for binary digits). Each binary number created and assigned to a sample is called a digital word, sample word, or just plain old "word."

Obviously, with only the digits 1 and 0 to work with, not many different digital words can be formed to represent and record the thousands of different voltage readings that are taken each second during sampling (see figure 2.11). In a two-bit system (using 1 and 0) there are only four digital words, or voltage steps, that can be formed: 1-1, 1-0, 0-1, and 0-0. The solution to the problem is to increase the length of the digital word with more bits, or more 1s and 0s. Using 8 bits results in 256 different words, or voltage steps, that can be assigned to represent each one of the samples. Although this works, the audio does not sound very good; this quality

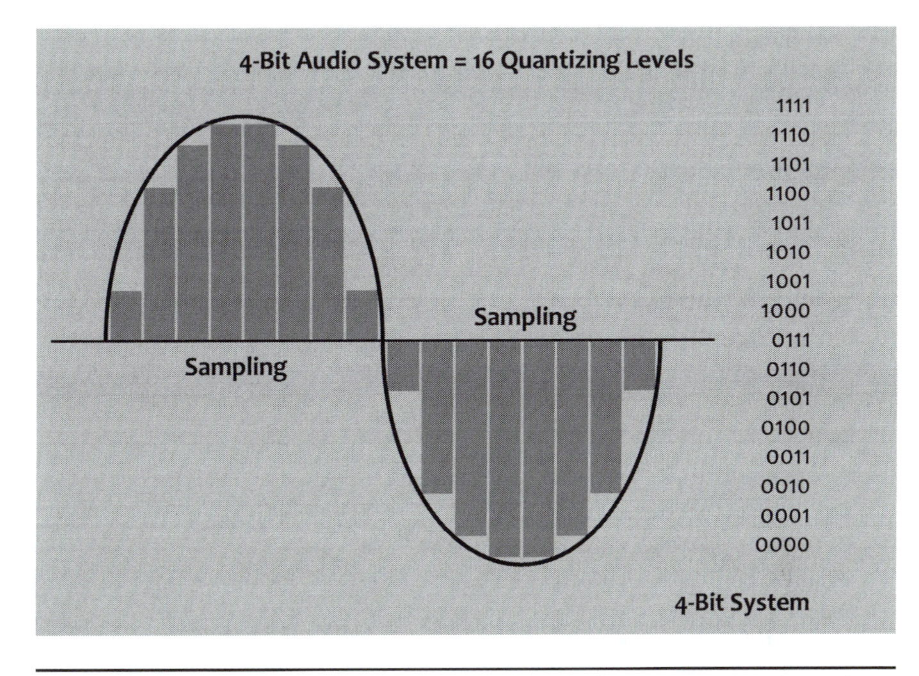

Figure 2.11 The number of samples and the number of bits in each sample word determines how accurately the original sine wave is reproduced. The 4-bit system above only produces 16 binary words, or levels of quantization, and cannot accurately reproduce the sine wave. A 16-bit system provides 65,536 binary words that can be assigned to each sample. A 24-bit system provides 16,777,216 binary words. The higher the bit rate, the more accurately the sine wave can be reproduced.

is similar to that of an inexpensive telephone answering machine. At 16 bits, however, there are 65,536 different binary words that can be assigned to represent a sample; suddenly, a magic number of sorts is reached.

The number of bits in each sample word determines the dynamic range of the audio signal. Dynamic range is the difference between the softest and the loudest sound, or in this case the signal-to-noise ratio. The ratio is a comparison of the audio signal's power to the electrical noise, or hiss, that may be inherently present in a circuit. The higher the signal-to-noise ratio, the cleaner the audio sounds. A 16-bit digital audio signal has a dynamic range of 96 decibels.

CDs use a sample rate of 44.1 kHz. For each of the 44,100 samples taken during one second of time, a 16-bit binary word is created to represent the voltage reading, or step, of the sample. The result is excellent quality sound that is the standard by which most people, particularly consumers, judge digital audio. Although this sample rate and bit rate create excellent quality audio, they still do not produce a perfect re-creation of the original analog sine wave. However, in 1982, when the CD was released, it was certainly state of the art. To create a perfect copy of an analog sine wave, more samples and binary words are needed.

Digital audio software features sample rates as high as 192 kHz, meaning that each second of audio is sliced into 192,000 individual samples. Bit rates are as high as 24 bits, creating 16,777,216 different binary words, or voltage steps, that can be used to represent the voltage reading of any one of the individual samples. The dynamic range of 24-bit audio is an incredible 144 decibels, which is easily capable of reproducing the softest sound to the loudest crescendo of an orchestra.

During playback, the digital-to-analog conversion is the reverse of the process previously described. The D/A converter reads each sample's binary word and re-creates the assigned voltage step. When all of the samples are combined, the analog audio sine wave is re-created and can be sent to headphones or monitor speakers so that our analog ears can hear the sounds.

Practical Uses of Digital Audio

Digital audio is nothing more than computer data, binary words consisting of 1s and 0s. With the digital audio tools on the market, the only real limits to your creative abilities are the limitations *you* place on yourself. Your creative opportunities extend not only to the end product you deliver to the listener, but to the hardware and software you select to record, store, and manipulate digital audio.

There was a time when broadcast-quality equipment was designed to last for 10 or 15 years. Digital audio hardware and software have changed that standard dramatically. The word digital means that the hardware and software are continuously improving and that a 10 to 15 year life span has been cut to 3 to 5 years max. There is always going to be some new and exciting hardware and software to be creative with.

In 1980, digital audio recording involved a machine the size of a kitchen stove. Today, virtually every computer on the market can provide professional-quality audio. A laptop computer can be outfitted with 192 kHz, 24-bit digital multitrack software and turned into a studio-quality recorder. Newspeople and field producers use portable recorders that fit in a shirt pocket with flash memory cards to record on. A single memory card can easily hold over three days (72 hours) of recorded material (see figure 2.12). Even your smart phone is capable of professional-quality audio recordings when combined with a professional microphone.

The manipulation of digital audio includes editing out all of the mistakes you made when you voiced your last commercial, but it offers much more than that as well. Manipulation involves altering the audio in any way. For example, digital audio processors allow radio stations to have incredible tools at their disposal for doing everything from eliminating background noise to creating rich-sounding reverb. Additionally, many audio apps, or specialty audio features, can be easily downloaded and added to a digital audio workstation or smart phone.

One of the great features of digital audio manipulation is time compression and expansion. If you create a commercial that ends up being 62 seconds, you can digitally squeeze it to 60 seconds through a time-compression program with no changes in pitch or timbre. Another feature allows you to alter voices through pitch shifting—taking the voice up or down as you desire. Again, about the only limit to the manipulation of digital audio is your imagination.

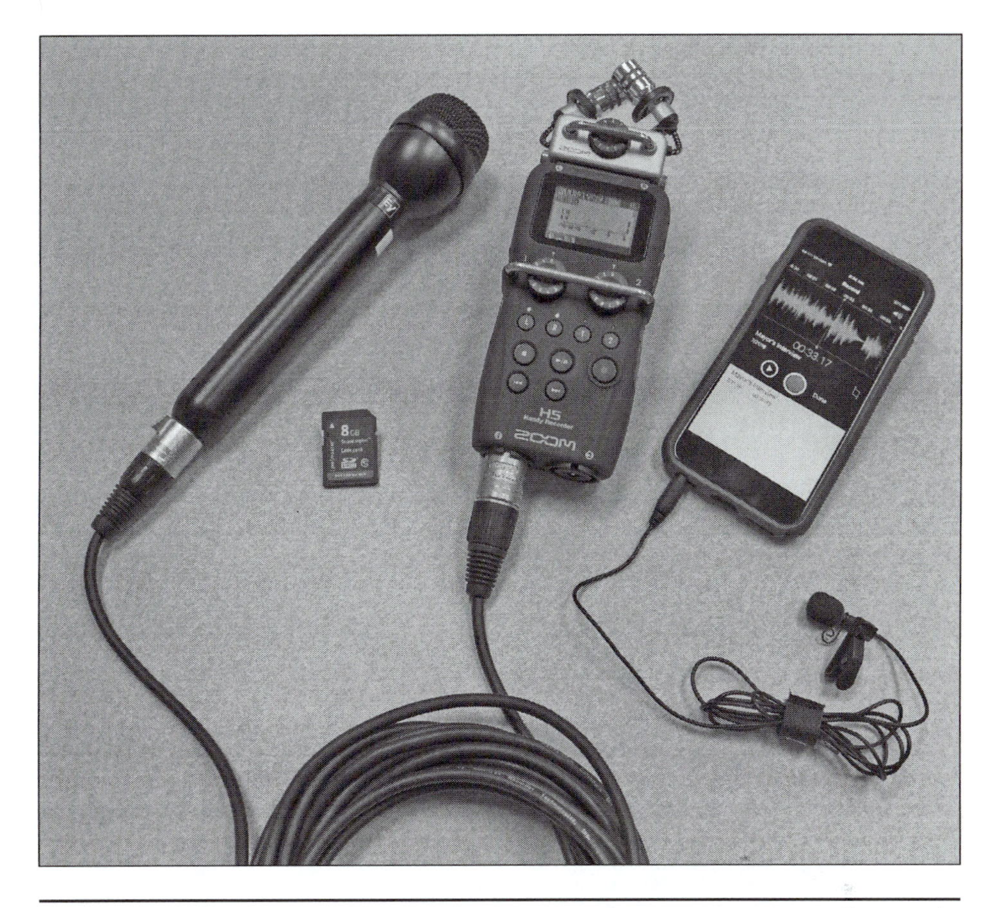

Figure 2.12 Newspeople like recorders with large-capacity memory cards. A card can easily store over 72 hours (three days) of high-quality audio. To retrieve the audio from the recorder, the memory card can be removed and inserted into a computer or you can use a USB cable to drag and drop the audio files into a computer. A smart phone along with a professional lavalier (lav) microphone can be a powerful tool. Recorded audio can be edited in the field and sent to the station via email, text message, or WiFi.

Specific production techniques and the hardware and software for the recording, editing, mixing, and mastering of creative audio products are discussed in detail later in this text.

Basic Storage Options

Radio stations use computer servers to record, store, and play back digital audio. By linking a number of broadcast facilities over a wide-area computer network, several stations can create and share the same audio files as if in one location.

There are numerous options for you to store and archive your projects and files on. For a station or group of stations there is cloud storage, main server storage, storage on individual audio workstations, and storage on portable hard drives. A 2-terabyte (1 terabyte = 1,000 gigabytes) portable hard drive can store over 3,200 hours of audio and easily fit in a shirt pocket or small purse. Even more portable for personal audio storage are small flash drives. As an example, an inexpensive 64-gigabyte (1 gigabyte = 1,024 megabytes) flash drive can hold over 6,000 1-minute commercials or over 1,000,000 typical Word document pages. Each staff member can easily wear his or her audio archive on a lanyard around their neck and take it anywhere they go (see figure 2.13).

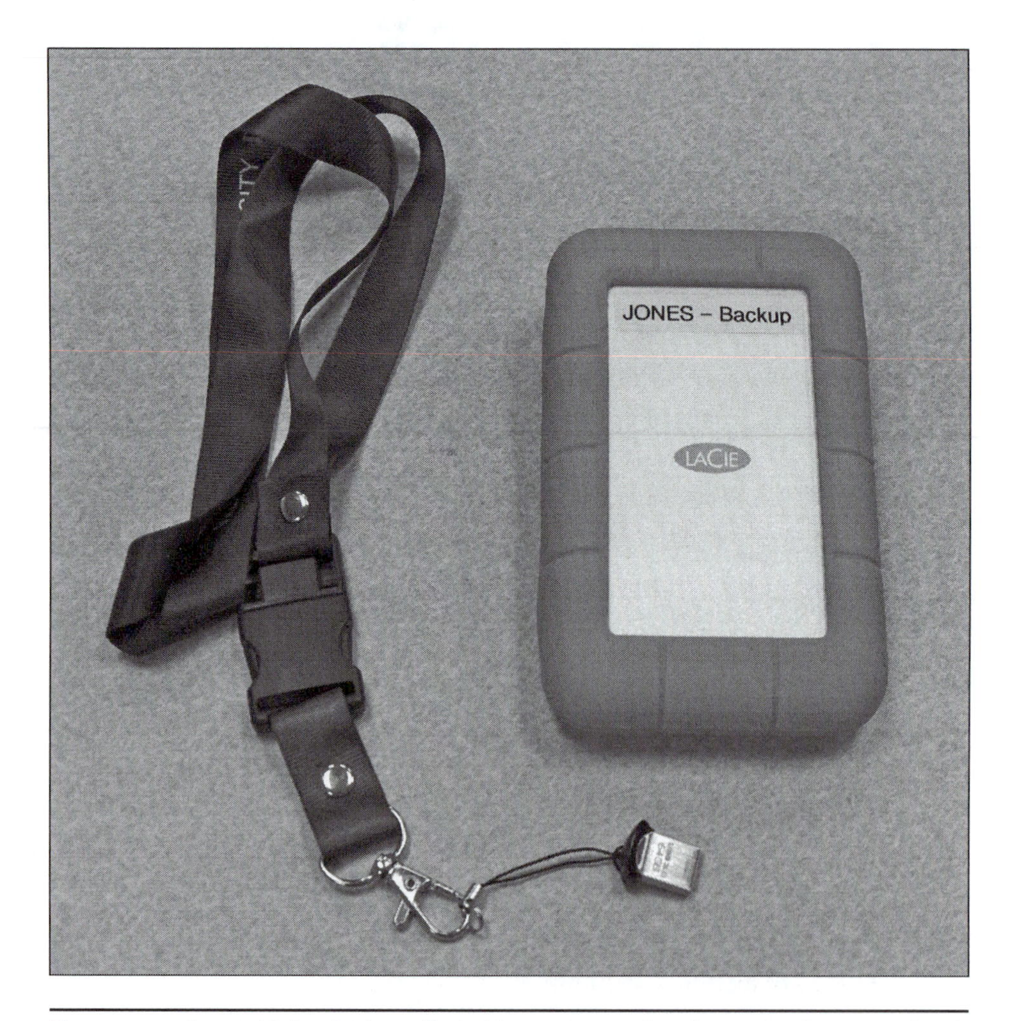

Figure 2.13 A portable hard drive or flash drive are two options for personal storage of projects. The hard drive in the illustration holds 2 terabytes of data (2,000 gigabytes) while the flash drive on the lanyard holds 64 gigabytes of data. The flash drive can hold over 6,000 1-minute commercials or over 1,000,000 typical Word document pages!

Suggested Activities

1. If your university has a speech and hearing clinic, visit the clinic and request a standard hearing test. Most universities offer such services free to students. The test can provide knowledge of your hearing strengths and make you aware of any weaknesses.
2. Save the results of your test to use as a baseline reference point for future hearing tests later in life.

Website for More Information

For more information about:

- *protecting your hearing,* visit the House Ear Institute at www.hei.org

Pro Speak

You should be able to use these terms in your everyday conversations with other professionals.

acoustical phase—A difference in time between two or more sine waves, measured in degrees.

amplitude—Amplitude represents the strength, or loudness, of a sound or audio signal without regard for its frequency. The height of the sine wave can be measured either in decibels (for sound) or voltage (for audio signals).

barometric air pressure—A measure of the weight of the atmosphere, or air, pressing down on the earth. It averages approximately 29.92 inches of mercury at mean sea level, which is the equivalent of 14.7 pounds per square inch. Barometric pressure varies with altitude, getting lighter as the altitude increases.

constructive interference—Occurs when two sine waves are just slightly out of phase with one another but are still together in their cycles of compression and rarefaction. Constructive interference tends to increase amplitude, reinforcing the sound waves.

decibels (dB)—A unit of measurement developed by Alexander Graham Bell to measure the loss of signal over one mile of telephone line. Because the bel is such a large unit, most measurements are made in tenths (deci) of a bel, thus the decibel. A decibel can be expressed as a volume of acoustic sound-pressure level or, for audio signals, as a voltage reading.

destructive interference—Occurs when two sine waves are out of phase with one another and their cycles of compression and rarefaction occur at different times. Destructive interference tends to decrease amplitude and weaken the sound waves.

dynamic range—The difference between the softest and loudest part of a sound, measured in decibels (dB). A dynamic range of 95 dB, for example, indicates that the loudest sound is 95 dB louder than the softest sound.

frequency—The number of complete cycles that occur in one second of time; expressed as cycles per second or hertz (Hz).

harmonic—A harmonic is a sound created by multiplying the fundamental frequency, or first harmonic. The second harmonic is two times the fundamental, the third harmonic is three times the fundamental, etc.

Hertz, Heinrich—Hertz (1857–1894) proved that electricity could be transmitted in electromagnetic waves, which travel at the speed of light. In 1933 his name was adopted as the international metric term used for radio and electrical frequencies (e.g., Hz, kHz, and mHz).

infrasonic—Sounds or audio signals that fall below the average person's hearing range, typically 20 Hz and lower.

overtone—A sound or audio signal that occurs in between the harmonics or multiples of a fundamental frequency.

quantization—The process of assigning a binary number to each of the samples or voltage readings taken during the analog-to-digital audio conversion.

subwoofer—A speaker designed to handle those sounds from 16 Hz to about 100 Hz.

ultrasonic—Sounds or audio signals occurring above the average person's hearing range, typically above 20 kHz.

3

Microphones and Their Role in Radio Production

HIGHLIGHTS

Microphones are one of the key tools used to create sound pictures for listeners. In this chapter you are going to learn about the different microphone types and how to creatively put them to use. One of the interesting things you will discover about microphones is that regardless of price, just about every microphone is better than another microphone at recording something. Just because one microphone is more expensive than another does not make it the best microphone. In fact, there is no such thing as the "best" microphone; it does not exist.

Microphones as Transducers

As much as it sounds like something from a science fiction movie, a microphone is a transducer, a device that transforms one form of energy into another. A microphone converts sound waves, or acoustic energy, into electrical energy. The electrical energy created is an analogous waveform (an exact replica) of the original sound wave. Of course, the quality of the microphone does have an effect on the process. A higher-quality microphone generates a more accurate replica of the original sound wave.

Prior to a sound entering the microphone, it is referred to as **sound**, or acoustic energy. Once the microphone converts sound to electrical energy, it becomes **audio**. Microphones are electro-mechanical devices and are divided into two distinct types, dynamic and electrostatic. The classification is based on how the **mic** (pronounced "Mike") converts sound to audio.

Dynamic Microphones

A dynamic microphone works using magnetic inductance, a principal of physics you probably learned in ninth-grade science. Place a coil of wire within a magnetic field and move it back and forth through the magnetic field. The result is that electricity (alternating current) is generated or induced in the coil of wire. The Western Electric Company made the first moving-coil microphones in the early 1930s.

There are two microphone types, differentiated by their physical structure, within the dynamic microphone category: moving-coil microphones and ribbon microphones. One of the quirks of the industry is that a moving-coil microphone is called a **dynamic microphone**, so from this point on in the text, moving-coil microphones are referred to as dynamic microphones.

Dynamic (Moving-Coil) Microphones

Within the dynamic microphone's body, just inside the protective metal screen at the face of the microphone, is an exceptionally thin circular plastic diaphragm (see figure 3.1). Attached on the back of the diaphragm at its center is a tiny coil of very fine wire surrounded by a permanent magnet that looks like a donut with a hole in the middle for the coil. As the sound wave hits the diaphragm and coil of wire, they move back and forth, just like your eardrum. The movement of the coil of wire back and forth through the magnetic field generates a very low-voltage electrical signal, an analogous waveform, which travels down the wires leading away from the microphone (again, see figure 3.1).

Generally, dynamic microphones accurately reproduce a wide frequency range and have a transparent sound, meaning they do not color or change the original sound.

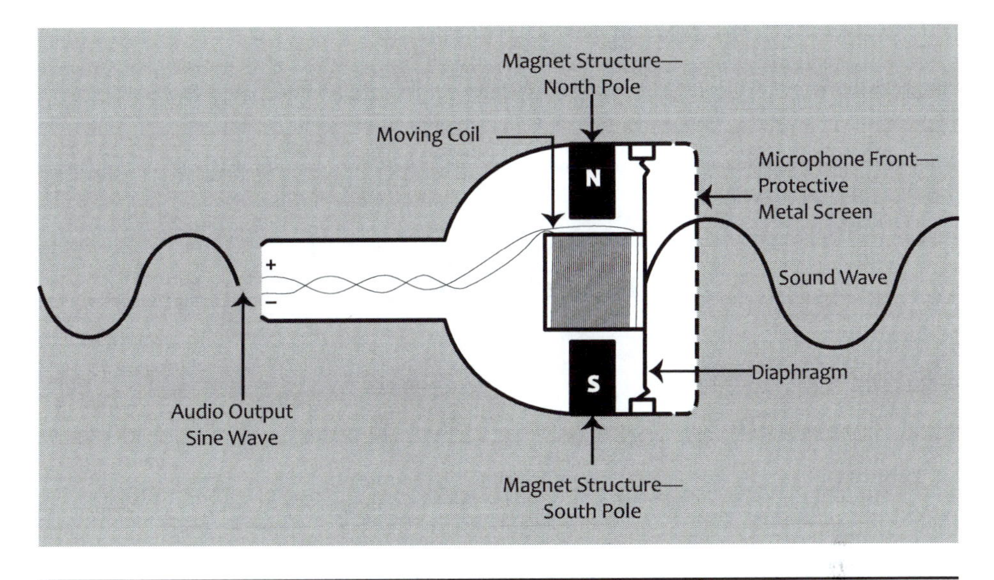

Figure 3.1 The elements of a dynamic microphone. As the sound wave hits the diaphragm it moves back and forth, just like your eardrum. The movement of the coil of wire back and forth through the magnetic field generates a very low-voltage electrical signal, an analogous waveform, that travels down the wires leading away from the microphone.

Dynamic Microphone Characteristics. Radio announcers like directional dynamic microphones because of the deep, rich qualities they can add to a good announcer's voice. This is due to the **proximity effect** (see figure 3.2 on the following page; see audio demo, cut 3-1). The closer you move to the face of the microphone, the lower your voice appears to go, taking on a richer, bassy quality. Announcers experiment to find the distance from the mic that makes their voice sound the best; this is called working the mic. Likewise, the proximity effect occurs when a directional dynamic microphone is placed closer to a sound source like a drum or other instrument rich in bass. Nondirectional dynamic microphones do not exhibit the proximity effect. Directional characteristics of microphones are discussed later in the chapter.

Dynamic Microphone Applications. Generally, dynamic microphones are very rugged and can withstand a number of different adverse environments. They operate over a wide temperature range and are not generally affected by inclement weather conditions. Dynamic microphones are used extensively in radio. Because of their design, they are also excellent for use in high sound-pressure level environments, like on stage at a rock concert. Dynamic microphones generate very little internal electrical **noise** or hiss. Electrical noise is an undesirable audio quality in any audio device.

Dynamic microphones are available in a variety of models, quality levels, and price ranges. Some are designed for very specific purposes, such as for studio use, for

Figure 3.2 The closer you move to the face of a directional dynamic microphone, the lower your voice appears to go, taking on a richer, bassy quality. Announcers experiment to find the distance from the mic that makes their voice sound the best; this is called working the mic. This announcer is about 2 inches from the microphone windscreen.

news reporting, or for use at live remotes with public-address systems. Retail prices for professional dynamic microphones range from a low of about $125 to a high of nearly $1,000, with excellent microphones available in the $300 to $500 range.

Later in this chapter you will learn how to properly place and use dynamic microphones.

Ribbon Microphones

The **ribbon microphone**, although quite different from a dynamic in its design, also works using magnetic inductance. Harry Olson, working for the Radio Corporation of America (RCA) research laboratories, developed the ribbon microphone in 1931. To this day, radio stations use photos and drawings of classic RCA ribbon microphones as logos and trademarks.

Internally, a very thin corrugated metal ribbon is suspended vertically between two poles of a large permanent magnet. At the top and bottom of the metal ribbon are the signal leads, or wires, coming off the ribbon and going to the microphone cable (see figure 3.3).

As the sound wave hits the metal ribbon, it is moved back and forth, just like the diaphragm in the dynamic microphone. This movement causes the metal ribbon to cut back and forth through the magnetic field, and a very low-voltage signal is generated in the ribbon and travels down the microphone cable.

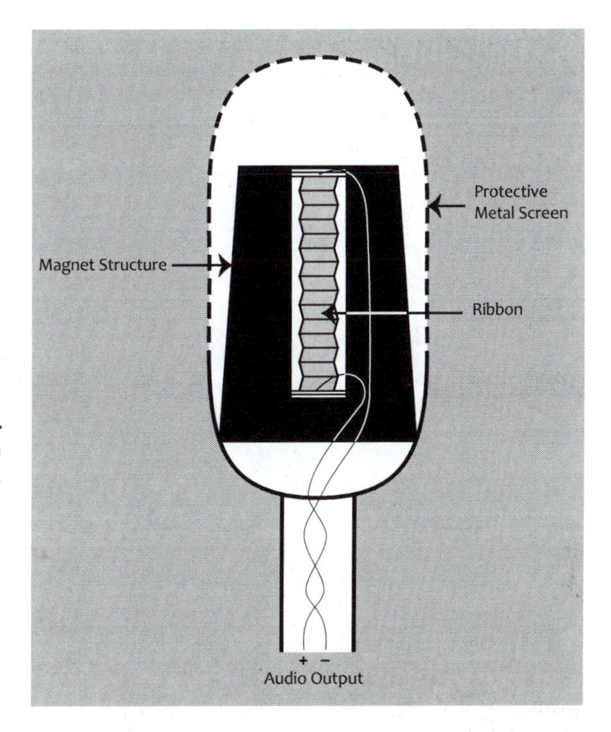

Figure 3.3 The elements of a ribbon microphone. As the sound wave hits the ribbon, it moves back and forth, just like your eardrum. The movement of the ribbon back and forth through the magnetic field generates a very low-voltage electrical signal, an analogous waveform, that travels down the wires leading away from the microphone.

Generally, ribbon microphones are known for their accurate reproduction over a wider frequency range than dynamic mics and for having a very natural, warm sound. Modern ribbon microphones have a very good extended high-frequency response that engineers refer to as open or very natural sounding (see audio demo, cut 3-2).

Ribbon Microphone Characteristics. There are a couple of characteristics of the ribbon microphone you need to know about, or you'll kill one before you ever get a chance to use it! The ribbon in older, vintage ribbon microphones is very fragile due to the fact that it is only 1.8 **microns** (one millionth of a meter) thick. Newer ribbon microphones have ribbons of up to 2.5 microns thick. You should never blow into any microphone to test it; in the case of a ribbon microphone you could stretch or tear the ribbon in two. Likewise, never take a ribbon microphone outside unless it is in a case; a breeze can stretch or tear the ribbon. Ribbon microphones should be encased in a protective cover when not in use and should be stored vertically.

Ribbon microphones have acoustic qualities that producers and voice talent like. Ribbons tend to have a very smooth, warm, natural sound at the lower frequencies, with an open high-frequency response. Ribbons make great voice microphones and are also used for a number of musical instruments in recording studios. The **transient response** (ability to handle those fast-attack, slow-to-trail-off sounds like consonants in speech) of a ribbon microphone is generally better than that of the dynamic microphone.

Figure 3.4 The classic RCA series 77 DX ribbon microphones were made starting in the mid-1930s and are prized for the unique, warm, rich, natural sound they produce on vocals. Additionally, many of these microphones have historic value based on the radio station or studio in which they were used. This particular microphone is over 60 years old.

Announcers like directional ribbon microphones because they exhibit the proximity effect just like directional dynamic microphones, while featuring the smoother, warmer, and richer acoustic quality of ribbons. A unique feature of a ribbon is that the proximity effect can start as much as 3 to 5 feet from the microphone (see audio demo, cut 3-3).

Ribbon Microphone Applications. Vintage ribbon microphones are found in recording and production studios and are rarely used as on-air radio microphones, due to their cost and fragility. Ribbons are used by many top music producers. The classic RCA series 44 and 77 ribbon microphones that were made from the mid-1930s through the early 1950s are prized for the unique, warm, rich, natural sound they produce on vocals. Additionally, many of these older microphones have historic value (see figure 3.4).

Depending on the history of the microphone and its condition, a vintage ribbon microphone can easily top the $2,500 mark. Currently, manufactured ribbon microphones range in price from $650 to $3,500.

Electrostatic Microphones

The electrostatic mic was invented shortly after Lee DeForest invented the vacuum tube amplifier in 1906. The theory behind the design is based on an early electrical component called a condenser. The name of the condenser was later changed to capacitor to reflect more clearly its function within an electrical circuit.

Although the condenser microphone operates on electrostatic principles and is technically a capacitor, the industry has stuck with the original name of condenser. From this point on in the text this microphone type is referred to as a **condenser microphone**. Also included within this microphone type are electret-condenser and digital condenser microphones.

Condenser Microphones

At the front of the microphone there is a very thin circular diaphragm (see figure 3.5). This diaphragm is made of a flexible plastic material coated with a very thin layer of gold for conductivity. One impressive feature of the diaphragm is that it is only a couple of microns thick.

Right behind the diaphragm, and parallel to it, is a rigid back plate. The diaphragm and the back plate are separated by an air space of just a few hundredths of an inch. The assembly created by the back plate, the air space, and the diaphragm is a capacitor (also known as a condenser). For a capacitor to work it must have a polarizing voltage.

Positive (+) voltage is applied to the back plate, and the diaphragm serves as the negative (–) side of the circuit. The voltage passing through the back plate and diaphragm sets up an electrostatic field of energy in the air space between the two. As long as the microphone remains in an absolutely silent room the electrostatic field and the voltage passing through the capacitor remain constant.

When a sound wave hits the diaphragm, it moves back and forth in relation to the back plate. As the movement occurs, the stable electrostatic field is upset, or changed, causing the voltage passing through the mic to vary and creating an analogous waveform (an exact replica) of the original sound wave. The output of a condenser microphone is very weak. For the microphone to be functional, a **preamplifier** is placed very near the microphone element to amplify the signal to a usable level. Depending on the microphone model, the preamplifier may be located inside the microphone or be a separate unit located near the microphone.

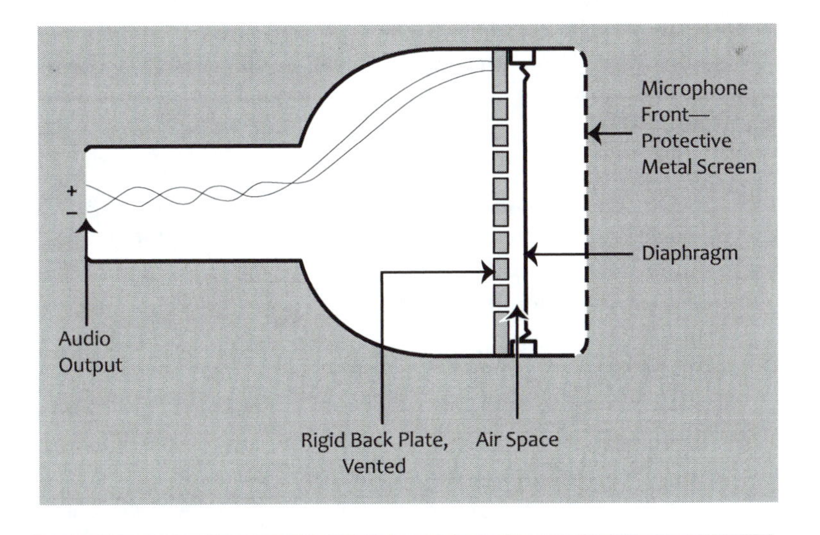

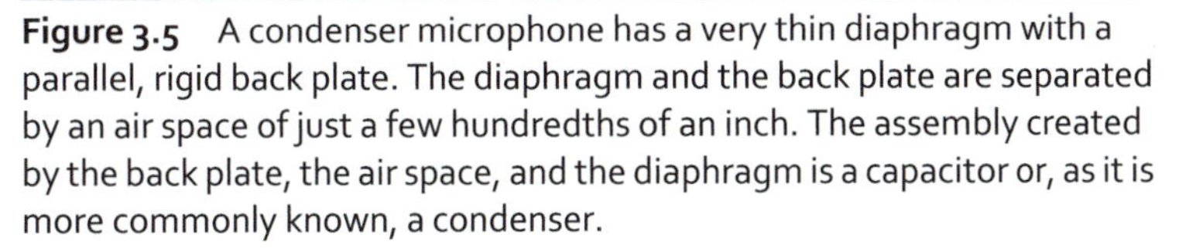
Figure 3.5 A condenser microphone has a very thin diaphragm with a parallel, rigid back plate. The diaphragm and the back plate are separated by an air space of just a few hundredths of an inch. The assembly created by the back plate, the air space, and the diaphragm is a capacitor or, as it is more commonly known, a condenser.

Power for the preamplifier can be provided with a separate power supply, a battery power supply located within the microphone, or with **phantom power** provided by a control board or portable mixer. Phantom power is a system of sending the necessary voltage for the condenser microphone preamplifier down the microphone cable. Phantom power in professional equipment is 48 volts, direct current. Phantom power supplies of between 18 and 24 volts are common in **prosumer** equipment.

Condenser Microphone Characteristics. Condensers are generally very sensitive, high-performance microphones that deliver incredibly accurate sound reproduction and are used for the most critical applications. A unique characteristic of the condenser microphone is its sensitivity. In fact, some condensers are sensitive enough to pick up instantaneous variations in barometric pressure (atmospheric pressure) within a room.

Because condensers are so sensitive, they offer excellent high-frequency and transient response (fast-attack, slow-to-trail-off sounds). A condenser microphone typically has a slight peak in the high frequencies that gives the microphone a very bright, crisp sound that helps to create a very open, natural quality.

Whereas sometimes you see on-air talent working close to the face of a dynamic microphone to get a **bass boost**, that is usually not the case with a condenser. Talent typically back off a condenser microphone as much as 1 to 2 feet to achieve a very natural, open presence in their voice, rather than the proximity effect common to dynamic and ribbon microphones (see audio demo, cut 3-4).

Although the microphone's sensitivity is an advantage in a production studio, it presents some really unusual problems for on-air use. For example, a high-quality condenser is capable of picking up the sound of air rushing out of an air conditioning vent, the sounds of control board switches, and any other stray noise in an on-air studio. A condenser microphone mounted directly on a microphone stand easily picks up stray building vibrations, such as the thud of a person walking across a studio floor or the rumble of traffic that is passing by the studio, and so forth. To solve vibration problems, condenser microphones are suspended in shock mounts, isolating them from any type of vibration. Microphone shock mounts are discussed later in this chapter.

Electret-Condenser Microphones

An **electret-condenser microphone** differs from a condenser microphone in that the diaphragm or the back plate, depending on the particular design, holds a permanent electrical charge. Because there is a permanent electrical charge present, there is no need for the separate polarizing power supply. The only power an electret-condenser microphone requires is a small battery to operate the preamplifier, which is usually included inside the microphone.

Generally speaking, electret-condenser microphones can be made much smaller, with a good example being the very tiny lavalier microphones used in television and motion picture production. Lavaliers, or **lavs**, are sometimes referred to as body microphones. Many of these microphones are so tiny there appears to be nothing on the end of the slim 3-millimeter diameter microphone cable (see figure 3.6). Lavs are discussed in detail later in the chapter.

Retail prices for condenser microphones extend from just under $200 for an inexpensive electret-condenser microphone to $5,000 for a high-end studio condenser microphone (see figure 3.7). There are a number of superior condenser microphones available in the $400 to $600 range.

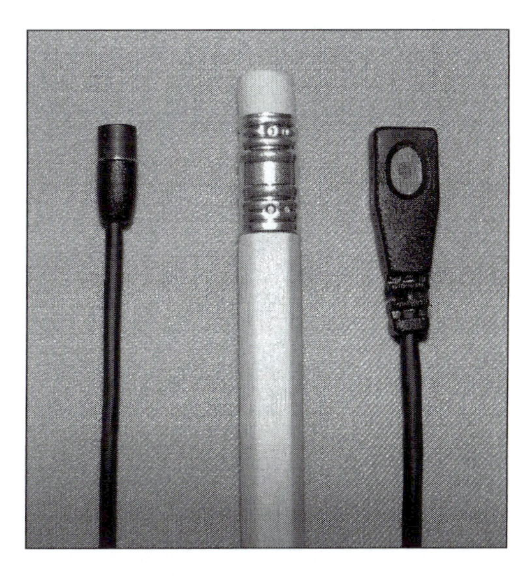

Figure 3.6 A lavalier microphone, or lav, is a specialty class of electret-condenser microphones. Sometimes lavs are referred to as body microphones, as they can easily be hidden on clothing due to their small size.

Digital Condenser Microphones

Several condenser microphone models are equipped with preamplifiers that include A/D (analog-to-digital) converters. The output of these professional microphones consists of binary numbers rather than analog waveforms. **Digital condenser microphones** offer the advantage of being able to be

Figure 3.7 From left to right, the Sennheiser shotgun microphone, Neumann U-87, and AKG 414 all represent outstanding condenser microphones. The Sennheiser is often used in television and motion picture production; the Neumann U-87 and AKG 414 are typically studio microphones. Prices range from about $1,000 to over $3,000.

Figure 3.8 This professional microphone cable to digital USB converter makes it possible to turn any broadcast microphone into a digital output microphone. This makes it possible to plug the microphone into a computer USB port for direct recording. (Copyright Shure Incorporated, used with permission.)

plugged directly into the digital inputs of a control board, or digital recorder, for higher audio quality.

A number of condenser microphones are available that feature a **Universal Serial Bus (USB)** output and are intended to plug directly into a personal computer. An advantage of this type of condenser microphone is that no control board or stand-alone preamplifier is required. The computer provides the power the microphone needs through the USB cable. USB condenser microphones are good choices for news reporting with a laptop computer from the field or a home studio.

At least one major microphone manufacturer offers a USB preamplifier/adapter that plugs into any professional microphone and converts it to a USB output (see figure 3.8).

Condenser Microphone Applications

Condenser microphones are used in the studio or in the field, provided the particular model is matched to the application. Generally, condensers are fairly rugged and can withstand a lot of adverse environments. Condenser microphones often see duty on news crews and on entertainment television and motion picture sets. Condenser microphones come in a variety of models, quality levels, and price ranges.

Microphone Pickup or Directional Patterns

Learning to use microphones includes not only selecting the right type of microphone for the job but also placing it in the right position to capture the best sound. Pros talk about being on mic or in the sweet spot. They are referring to having the microphone positioned so that the direction it is most sensitive to is aimed in exactly the right spot to capture the best sound.

Each microphone, regardless of whether it is a dynamic, ribbon, or condenser, has a specific pickup **pattern** or direction from which it picks up sound and is most sensitive. Usually, this is straight into the front of the microphone. Another term for the pickup pattern is the polar response of the microphone. The inherent pattern of a microphone type can be changed or modified by designing a special case in which to house the microphone element.

Usually, such microphones have slots or openings in the microphone body behind the head of the microphone. The openings create an acoustical environment within the microphone that reinforces some frequencies or sounds coming into the front of the microphone and cancels out others coming from the rear of the microphone. By changing the design of the microphone body, the pickup pattern can be altered to change the direction in which the microphone is sensitive.

The Five Basic Microphone Directional Patterns

Omnidirectional. The **omnidirectional**, or nondirectional, microphone pattern is just what the name implies. A microphone with this type of pattern is sensitive to sound from all directions, regardless of where the microphone is pointed. In figure 3.9, the polar pattern illustration shows the 360-degree sensitivity. An omnidirectional microphone has no dead spot and picks up everything. Because the microphone is sensitive in all directions, you have no control over what the microphone picks up other than to turn it on or off.

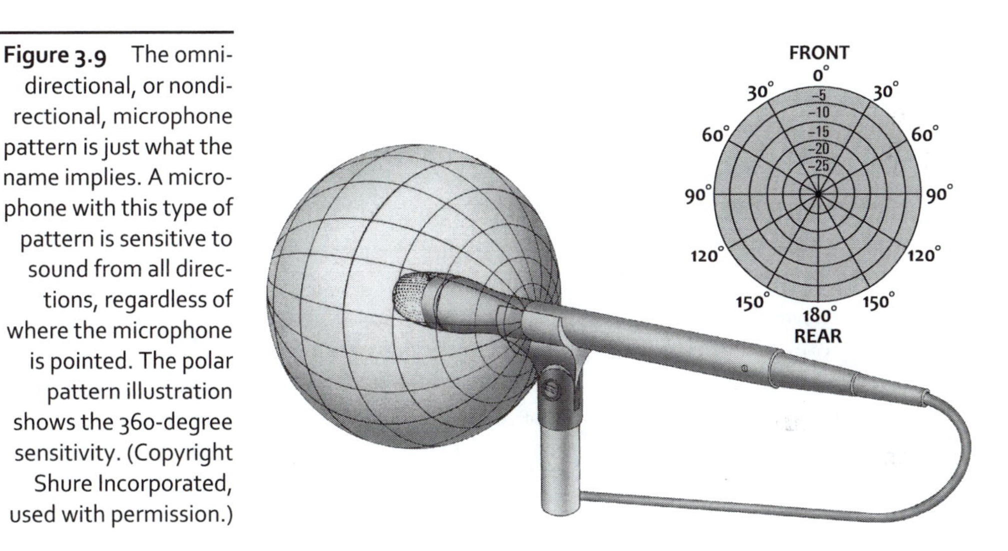

Figure 3.9 The omnidirectional, or nondirectional, microphone pattern is just what the name implies. A microphone with this type of pattern is sensitive to sound from all directions, regardless of where the microphone is pointed. The polar pattern illustration shows the 360-degree sensitivity. (Copyright Shure Incorporated, used with permission.)

Omnidirectional microphones are often issued to radio news crews as handheld interview microphones because even in confusing situations, such as covering breaking news, if the microphone is turned on something will be recorded to go on air (see audio demo, cut 3-5).

Bidirectional. A **bidirectional** microphone pattern provides the user a great deal of control over what sounds the microphone picks up, since it is most sensitive to the front and rear. The pickup pattern looks like a figure eight and is ideal for interviewing people with one person on the front side and the other person on the back side of the microphone. In a recording studio or at a live remote, careful placement between two people provides great pickup on each person and blocks other unwanted sounds from the sides (see audio demo, cut 3-6). The polar pattern illustration for the bidirectional microphone shows that it is most sensitive at 0 and 180 degrees (front and back). At 90 and 270 degrees (left and right) the microphone is least sensitive (see figure 3.10).

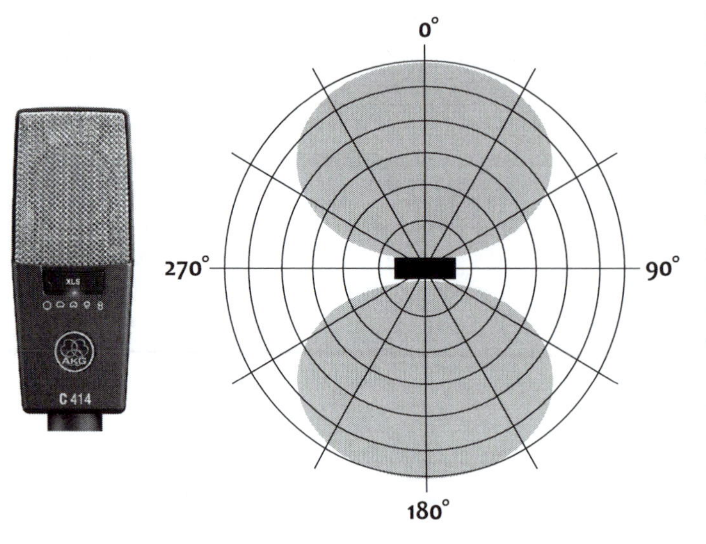

Figure 3.10 A bidirectional microphone provides the user a great deal of control over what sounds the microphone picks up. This polar pattern shows the microphone is most sensitive to the front (0 degrees) and to the rear (180 degrees). A bidirectional microphone is least sensitive to the sides at 90 degrees and 270 degrees. A bidirectional microphone is ideal for conducting interviews.

Cardioid. A **cardioid** microphone pattern is shaped like a heart and provides a greater level of control over unwanted sounds, such as room reverberation or noise (see figure 3.11). Cardioid pattern microphones are often referred to as **unidirectional** microphones, meaning that they pick up sound from one particular direction. In fact, other than the omnidirectional and bidirectional patterns, all other microphone patterns are referred to as unidirectional or single-direction microphones.

The ratio of direct sound to room reverberation that the microphone picks up is called the direct-to-reverberant sound ratio. Because the cardioid microphone pattern has a higher direct-to-reverberant sound ratio, cardioid pattern micro-

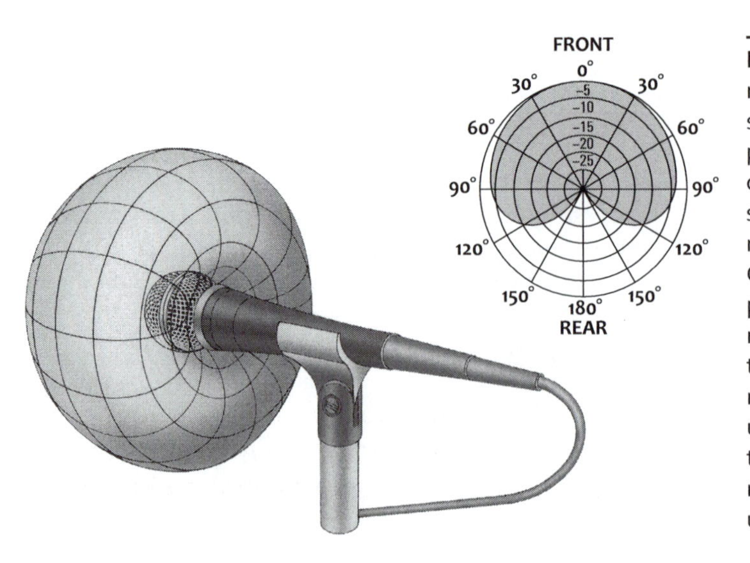

Figure 3.11 A cardioid microphone pattern is shaped like a heart and provides a greater level of control over unwanted sounds, such as room reverberation or noise. Cardioid pattern microphones are often referred to as unidirectional microphones, meaning that they pick up sound from one particular direction. (Copyright Shure Incorporated, used with permission.)

phones can be placed about one-and-a-half times farther away from a sound source than an omnidirectional pattern microphone and still maintain the sweet spot.

The polar pattern for a cardioid microphone shows it is most sensitive at 0 degrees, or straight into the face of the microphone. This is termed as being on axis. The cardioid pattern is least sensitive at 180 degrees, or off axis (see audio demo, cut 3-7). From the head of the microphone forward is where it is most sensitive to sound, and from the head to the rear of the microphone it is least sensitive.

A cardioid pattern is truly directional and is one of the most popular patterns for this reason. For example, if an announcer uses a cardioid pattern microphone in a radio studio, the face of the microphone is aimed at the announcer and the dead spot might be aimed at the door to the studio. People coming and going during the show cannot be heard because they are on the dead side of the microphone. Or the dead spot could be aimed at equipment or other staff in the studio, such as a telephone call screener or producer, whose sounds you don't want going on air.

In a talk-radio studio, the use of super-cardioid pattern microphones allows a producer to isolate and control the level of each guest individually. By arranging the guests around a table and carefully placing the microphones, each guest is on mic, but at the same time each guest is in the other microphones' dead spots (see figure 3.12).

Figure 3.12 Super-cardioid microphones are used around the desk in this talk-radio studio to allow individual-level control of the guests. Each guest is on mic, but at the same time each guest is in the other microphones' dead spots. The computer monitors display caller information to the show hosts.

Overall, a cardioid pattern gives you a great deal of control over what sounds are picked up and what sounds are not picked up, depending on microphone placement.

Super-Cardioid. A **super-cardioid** microphone pattern is a refinement of the cardioid pattern. The super-cardioid pattern shows that it is most sensitive at 0 degrees, or on axis, and least sensitive at 180 degrees, or off axis (see figure 3.13). However, notice that the sides of the super-cardioid pattern are narrower than those of the cardioid pattern, giving the user more directional control. Because of the improved direct-to-reverberant sound ratio, super-cardioid microphones can be placed nearly twice as far from a sound source as the omnidirectional microphone and still maintain the sweet spot. The cardioid and super-cardioid patterns are the two most popular patterns used in radio.

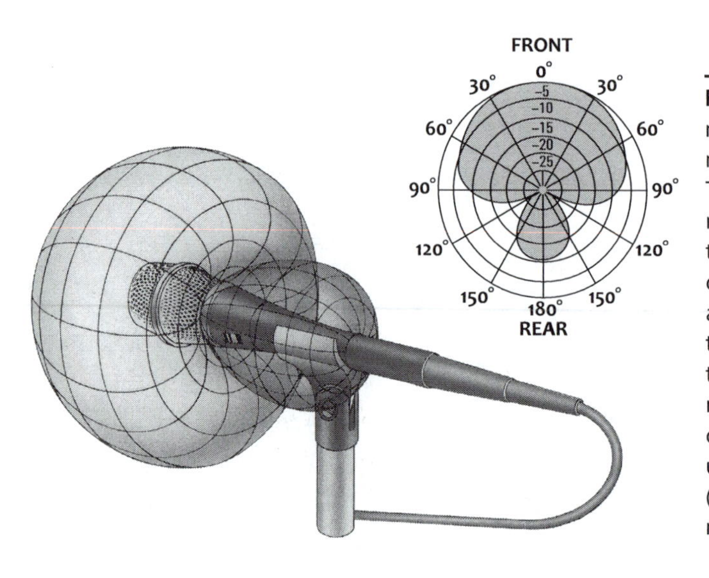

Figure 3.13 A super-cardioid microphone pattern is a refinement of the cardioid pattern. The super-cardioid pattern is most sensitive to the front of the microphone at 0 degrees, or on axis, and least sensitive at 180 degrees, or the rear of the microphone. The sides of the super-cardioid pattern are narrower than those of the cardioid pattern, giving the user more directional control. (Copyright Shure Incorporated, used with permission.)

Hyper-Cardioid. A **hyper-cardioid** microphone pattern is a further refinement of the cardioid pattern. The design goal is to be able to move the microphone farther from the sound source without increasing the apparent room reverberation. A hyper-cardioid microphone can pick up sound at slightly more than twice the distance from a sound source as can the omnidirectional pattern microphone, while still maintaining the same direct-to-reverberant sound ratio (see figure 3.14; see audio demo, cut 3-8).

A hyper-cardioid pattern is a very narrow and highly directional pattern typically found in **shotgun microphones** used by radio and television news crews, in television studios, and for location motion picture shoots. Voice-over talents sometimes use shotgun microphones aiming the microphone slightly below the chin to pick up both the direct voice from their mouth and the resonance of their upper chest and throat for a much warmer natural sound.

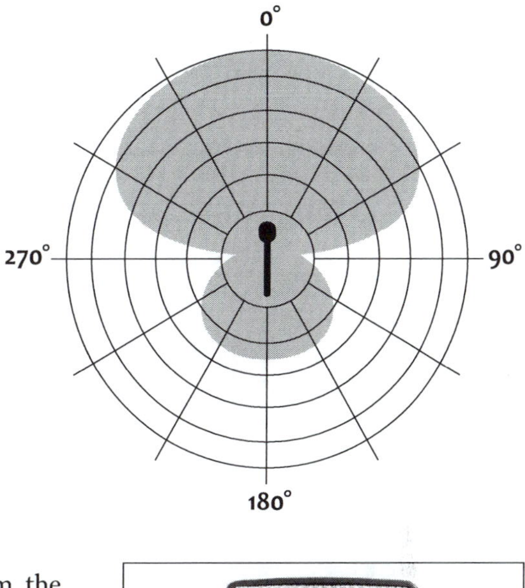

Figure 3.14 The polar pattern for a hyper-cardioid microphone pattern shows that it is most sensitive to the front of the microphone at 0 degrees. Notice how the pattern is similar to a bidirectional microphone in that it has extreme rejection to the sides at 90 degrees and 270 degrees. The combined side and rear rejection of sound allows the hyper-cardioid microphone to pick up sound clearly at slightly more than twice the distance from a source than an omnidirectional microphone and still maintains the same direct-to-reverberant sound ratio. (Copyright Shure Incorporated, used with permission.)

The name shotgun comes from the early hyper-cardioid designs that required a large microphone structure to accomplish the task. At the time, the microphones employed long tubes on the front of the mic element to attain the desired pattern and some did look like a shotgun. Today, hyper-cardioid microphones are usually less than 18 inches in length and about an inch in diameter. Shotgun microphones and their radio applications are discussed in detail later in this chapter.

Multipattern Microphones

Some microphone models have one or more elements, or pickups, within the microphone. These are known as multidirectional or poly-directional microphones, meaning that their patterns can be switched or altered. Generally, multidirectional microphones are either ribbons or condensers. Such microphones are more expensive but offer the user up to four or five different pickup patterns from which to choose, making the microphone more versatile for production (see figure 3.15).

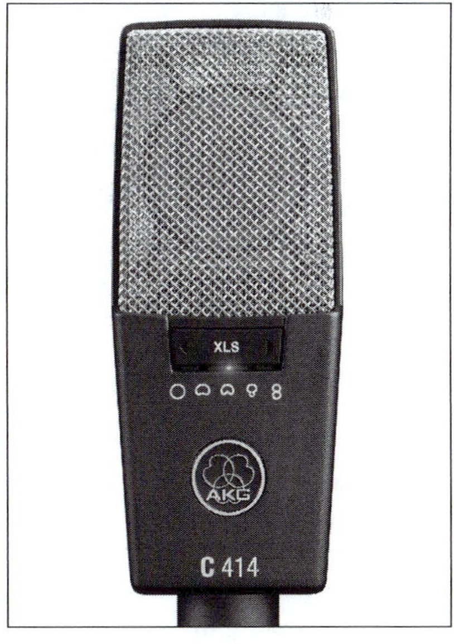

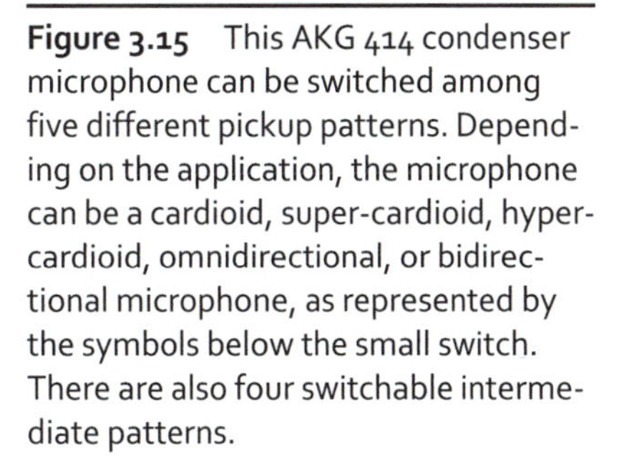

Figure 3.15 This AKG 414 condenser microphone can be switched among five different pickup patterns. Depending on the application, the microphone can be a cardioid, super-cardioid, hyper-cardioid, omnidirectional, or bidirectional microphone, as represented by the symbols below the small switch. There are also four switchable intermediate patterns.

Another popular multidirectional microphone type is based on separate components or capsules. The microphone element, or capsule, is designed to detach easily from the microphone body. By purchasing different capsules, one microphone body can have any one of a number of different microphone patterns, depending on which capsule is attached to the microphone body at the time. Modular types of microphones are often called microphone systems or system mics and offer a variety of options when budgets are limited (see figure 3.16).

Figure 3.16 Modular types of microphones are often called microphone systems or system mics and offer a variety of options. By selecting different capsules, the user can have any one of a number of different microphone patterns on one microphone body. Shown from left to right are a bidirectional capsule, mic body, omnidirectional capsule, and cardioid capsule. Below is an assembled microphone body and capsule.

Stereo Microphones

Several companies manufacture stereo microphones that have two elements within the microphone. Stereo microphones offer an easy way for a production person to capture a stereo music or sound image without setting up two microphones and running two cables (see figure 3.17). Such microphones can provide a unique sound picture for a remote or sports broadcast when used for background or ambient sound. As tempting as it might seem, stereo radio stations do not use stereo microphones for announcers. The reason is that any movement of the announcer's head to the left or right causes the announcer's voice to shift left or right in the stereo speakers. Due to the acoustics in a car, this would cause the announcer's voice to appear to fade in and out around the driver.

Factors to Consider in Selecting a Microphone

Although selecting the proper microphone type and directional pattern is important, there are three other qualities that you should consider when selecting a microphone: frequency response, sensitivity, and overload and maximum sound-pressure level.

Frequency Response

Just like your ears, a microphone has a range of frequencies that it can "hear," called the frequency response. Ideally, a microphone should pick up all frequencies at the same level or within a narrow decibel range. For professional microphones, the range is plus or minus three decibels. This means that no sound picked up by the microphone is three decibels louder or three decibels softer than any other sound the microphone is capable of picking up. A microphone's frequency range is called the **frequency-response curve**.

A microphone's frequency-response curve is graphically charted to document the microphone's performance. Most charts begin at 20 Hz and track to 20 kHz along a horizontal line, with the various frequencies listed in between (see figure 3.18 on the next page). A vertical scale lists the decibels of amplitude or loudness. In theory, the frequency-response curve of a perfect microphone is a flat line from 20 Hz to 20 kHz, indicating that the microphone does not favor one frequency over another or has little sound coloration.

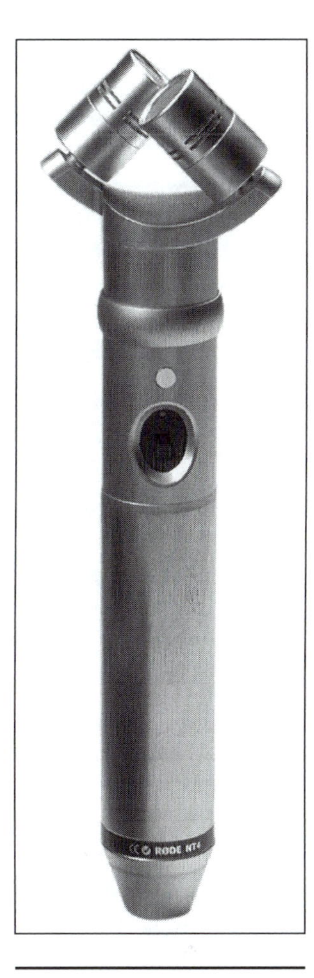

Figure 3.17 This Rode microphone has two capsules arranged as a coincident stereo pair to pick up sound. Stereo microphones offer an easy way for a production person to capture a stereo music image without setting up two microphones and running two cables. Such microphones can provide a unique sound picture for a remote or sports broadcast when used for background or ambient sound.

The Polar Frequency Response. Another aspect of the frequency-response curve of a microphone is how the microphone perceives sounds that are not coming directly into the face of the microphone. This is called the microphone's **polar frequency response**. In theory, when a sound enters the front of a microphone on axis, or at 0 degrees, the microphone is the most sensitive and treats all sounds equally, reproducing the best sound quality. This is sometimes referred to as the **centerline** of the microphone. As you may have noticed, when you turn your head from side to side, the way you hear a particular sound changes. The same thing is true for a microphone.

As a directional microphone is turned away from a sound source, off centerline, the sensitivity and the frequency response of the microphone

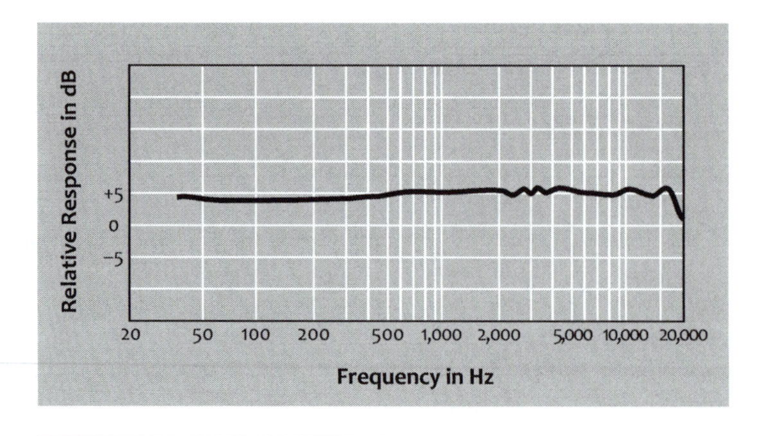

Figure 3.18 A microphone's frequency response is graphically charted to document the microphone's performance. Most charts begin at 20 Hz and track to 20 kHz along a horizontal line with the various frequencies listed in between. A vertical scale lists the decibels of amplitude or loudness. (Copyright Shure Incorporated, used with permission.)

degrades and changes (see audio demo, cut 3-9). This is called **off-axis coloration** of the sound, and it gets progressively worse as the angle to the sound source increases. High frequencies are lost first, making the microphone sound muddy the farther it is moved **off axis** or is turned away from the sound source.

Sensitivity

Like people's ears, some microphones are more sensitive to sound than others. The **sensitivity** of a microphone is connected to its output. Generally, a sensitive microphone has a higher output voltage and does not have to be placed as close to the sound source to do its job. Likewise, a sensitive microphone does not have to be amplified as much when its signal gets to the control board or mixer. Condenser microphones are generally the most sensitive microphones with the highest output levels.

Overload and Maximum Sound-Pressure Level

Microphones distort and in some cases just plain stop functioning under extremely high instantaneous sound-pressure levels, which is very similar to what your ears do. Such a failure is called **overload**.

Typically, dynamic and condenser microphones reproduce the highest sound-pressure levels without distorting the sound. Under some circumstances, the element in a ribbon microphone can be stretched or torn if placed too close to a sound source like an electric bass guitar amplifier.

As with your ears, the sound pressure a microphone can handle before distorting and failing is expressed in dB-SPL (decibels of sound-pressure level) and is listed in its specifications as its **maximum sound-pressure level**. For example, dynamic microphones are capable of handling sound pressures in excess of 125 dB-SPL; condenser microphones can handle up to 150 dB-SPL. Again, humans start to experience pain and hearing loss at 120 dB-SPL.

If you remember, the dynamic microphone is basically a small generator with a coil of wire moving back and forth in a magnetic field. So it stands to reason that the louder the sound, the more the diaphragm moves back and forth, generating a higher microphone output voltage. At some point the microphone may generate a high enough voltage to overload the input of whatever it is plugged into. For example, if a microphone that generates one volt of signal output is plugged into a control board that can only handle three-quarters of a volt as the maximum input, that extra quarter of a volt from the microphone will cause the control board input to become overloaded. The overload results in the sound being distorted. Often this condition is misdiagnosed as a bad microphone.

The maximum output of a microphone can be determined by looking at the particular model's specification sheet. Likewise, the maximum input voltage that your recorder or control board can handle can also be determined by looking at its specification sheet.

Gain Staging

Gain staging is the term applied to matching the maximum microphone output voltage to the maximum input voltage of the device the microphone is plugged into. The maximum voltage ratings can be determined by reading the equipment specifications.

To prevent overloading of a recorder or control board, a number of microphones have built-in switchable pads. A pad is the term used for a preset, or fixed, volume control. For example, a –10 dB pad lowers the audio output of a microphone 10 dB below whatever is coming into the microphone. The user can select between two or more pads to reduce the output of the microphone (see figure 3.19). In addition to microphones, many recorders, consoles, and other devices also have built-in switchable pads to lower the input or output level so as not to overload the electronic circuits.

By matching the output of the microphone to the input of the control board or recorder (gain staging), you are assured of a clean audio signal with no distortion. As simple as this sounds, failing to gain stage the microphone and equipment is one of the most common mistakes made when using microphones.

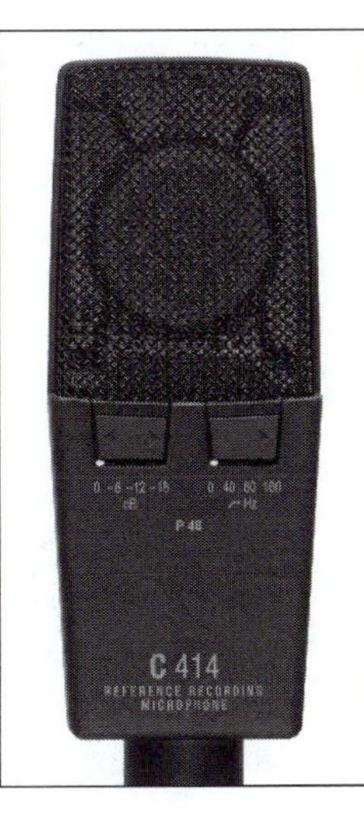

Figure 3.19 This AKG 414 condenser microphone has a built-in button on the left that can be switched to lower the audio output from 6 to 18 decibels. The button on the right is a frequency roll-off switch used to adjust the amount of low-frequency bass the microphone picks up.

Microphone Mounts

There are a number of ways to mount a microphone properly, ranging from a factory mount designed for a particular microphone model to the most unique microphone stand—you.

People

Often, people become microphone stands. As silly as it sounds, people are quite well designed for the purpose. Radio and TV newspeople hold microphones for field interviews. Many times, to get just the right sound, a person is required to hold a microphone to get into tight spaces or close enough to the sound source. Since humans are pretty flexible they can easily position a microphone just about anywhere. However, a person should grasp the microphone firmly and not change hand positions. Any movement of your hand on a handheld microphone can create a thumping noise through the body of the microphone. Likewise, clothing needs to be kept clear of the microphone, as any movement may cause a rustling sound of the clothing that the microphone can pick up.

Floor Stands

Microphone floor stands come in a variety of styles and materials. Stands usually consist of one of two types, a permanent base or a folding-leg base, both with a vertically telescoping pole to mount the microphone on.

A permanent base stand has a round or triangular base that is made of cast iron and is fairly heavy—approximately 20 to 30 pounds, and sometimes more. A folding-leg stand has three legs that fold neatly to allow the stand to be easily transported for location and concert sound work. The telescoping pole usually extends from 3 feet to a little over 5 feet (see figure 3.20).

A useful addition to a floor stand is a baby boom, or an arm about 3 feet long with a sliding counterbalance weight that mounts to the top of the mic stand. It gives the stand additional reach either horizontally or vertically. The mic is mounted on one end of the boom and the counterbalance weight is adjusted to balance the boom at whatever angle it is placed.

Floor Booms

A floor boom is a large stand with an extended arm to hold a microphone over the sound source. Typically, they are on casters or wheels so they can be easily moved, as most have a very heavy iron base. A "small" floor boom extends vertically over 6 feet, and the boom arm extends out 5 to 6 feet (see figure 3.21). Occasionally, you may find one of these in a radio station.

Desk Stands

Desk stands are usually miniature versions of a floor stand. Some have short, telescoping mounts for the microphone, and some do not (see figure 3.22 on p. 72). A disadvantage of a desk stand is that any vibration in the table is transmitted

Figure 3.20 A folding-leg stand has three legs that fold neatly to allow the stand to be easily transported for location and concert sound work. A useful addition to any floor stand is a baby boom, or an arm about 3 feet long with a sliding counterbalance weight that mounts to the top of the mic stand.

Figure 3.21 Solid-base floor stands come in a variety of styles and materials from the half stand with a short boom (center) to large, rolling floor booms. Finishes vary from bright chrome to nonreflective black.

Figure 3.22 A desk stand can easily position a microphone when an announcer is sitting in a fixed position. A disadvantage of a desk stand is that any vibration in the table is transmitted directly through the stand to the microphone, producing a thumping sound. This often occurs when someone is nervously tapping his or her fingers on the table or tapping a foot against a table leg.

directly through the stand to the microphone, producing a thumping sound. This often occurs when someone is nervously tapping his or her fingers on the table or tapping a foot against a table leg.

Desk-Mounted Microphone Arms

For radio, it's best to mount a microphone so that it is adjustable for height and reach, since no two people who work at a radio control board are the same size and shape. A desk-mounted mic arm is a cantilever arm mounted on a swivel. The arm is spring-loaded to counterbalance the weight of the microphone and to allow the arm to be positioned at any height (see figure 3.23).

The microphone is also mounted on a swivel, so it can be turned in any direction. The arm allows the microphone to be positioned anywhere within a range and "hang" there for the announcer to use. Often, desk arms are mounted on a 12-inch riser to raise the arm above any equipment on the desk.

Clips, Clamps, and Goosenecks

Necessity is the mother of invention when it comes to mounting a microphone in an unusual circumstance. There are a number of very creative products on the market to help solve the problem (see figure 3.24). There are generic microphone clips that look like large clothespin-type devices, which are basically spring-loaded clamps. Just about any size microphone can be placed in them.

There are microphone quick-release adapters, swivel adapters, and microphone thread adapters to match microphones to microphone stands. There are also a number of different screw-type C-clamps designed to clamp down on something (table, desk, the pole of a mic stand, etc.) and hold a microphone.

Figure 3.23 A desk-mounted mic arm is spring-loaded to counterbalance the weight of the microphone and to allow the arm to be positioned anywhere within its range and "hang" there for the announcer to use. This arm is mounted on a 12-inch tall base to place the arm above the control board.

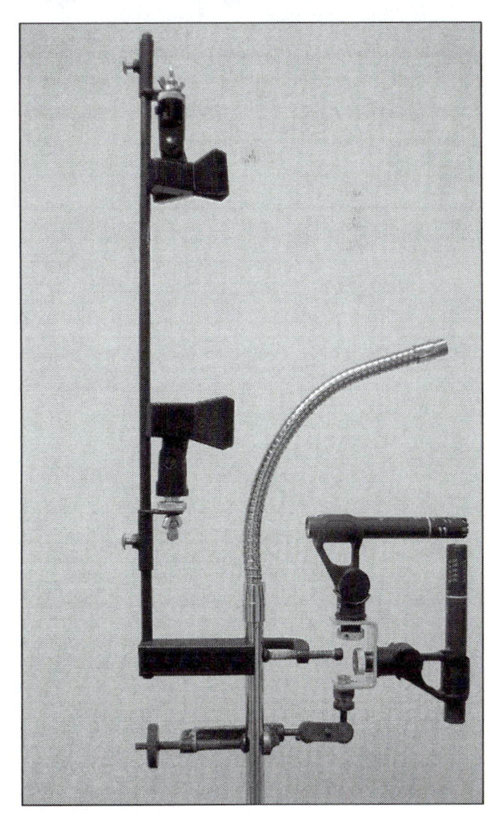

Figure 3.24 When a situation calls for something out of the ordinary, many technicians open a bag of clips, clamps, and an occasional gooseneck (center). There are no hard and fast rules here, except that the microphone must be mounted securely so it can pick up the sound. There are eight mic mounts in this picture that can be arranged in a number of combinations.

One of the more unusual microphone mounts is a **gooseneck**. A gooseneck is a flexible piece of metal conduit about 16 inches long that can be twisted, turned, and bent in just about any direction you can imagine (see figure 3.24). Whatever you do, though, don't bend or flex a gooseneck with the microphone turned on because a gooseneck makes a loud creaking noise that is picked up by the mic.

Of course, there's always **gaffer's tape**. Tape the microphone to something, keeping in mind not to cover any vents on the microphone body. And keep in mind that any vibration in whatever the microphone is taped to will be transferred through the body of the mic as a thumping sound. Although gaffer's tape looks like duct tape, the two are different products with different properties and adhesives. Gaffer's tape is designed to have strong holding abilities and to be removed without leaving a sticky residue. Duct tape is intended to seal permanently and leaves a sticky residue all over everything.

Shock Mounts

All microphone mounts have one common fault, regardless of whether they are made of steel, plastic, or carbon fiber. The mount transfers mechanical noise and vibration from the stand into the microphone body, producing unwanted rumble and thumping sounds.

Figure 3.25 A shock mount is a mic mount situated within a larger supporting mount. The inner mount holds the microphone at its base and is suspended on what looks like tiny bungee cords. The suspension system isolates and absorbs building vibrations so they can't reach the mic and be picked up.

A **shock mount** is a mic mount situated within a larger supporting mount (see figure 3.25). The inner mount holds the microphone and is suspended on what looks like expensive rubber bands or tiny bungee cords. The rubber suspension system isolates and absorbs the vibrations so they can't reach the mic.

Why You Need a Shock Mount

A microphone mounted on a floor stand is essentially connected to the floor. Any mechanical vibration in the floor travels up the microphone stand into the body of the microphone. Such sounds include vibrations from air conditioning and heating systems, a person walking across the floor, and general building noise originating anywhere within a building (see audio demo, cut 3-10). Vehicle traffic passing by a building can introduce ground vibrations into the building foundation. The foundation passes the rumble into the floor, into the microphone stand, and finally to the microphone.

Desk stands and microphone arms are also notorious for passing any vibration on the desk into the microphone. This includes someone tapping their fingers on the desk, bumping the desk with a chair, or slapping the desk to emphasize points when speaking.

Generally, the more sensitive a mic is, the greater the need to use a shock mount. A large number of microphones come with a factory shock mount. Specialty shock mounts that are more expensive are available.

Microphone Placement

Using a microphone is part science and part art. Part science from the standpoint that there are some basic rules to use as guides; part art from the standpoint of capturing the exact sound *you* want from an event, musical instrument, or voice talent. Once you learn the basic science, you can start practicing your art.

It would be nice to have a chart that told you where to place the microphone on radio talent and each different instrument or sound source you might encounter, but no such chart exists. The art of sound recording is just too subjective and open to creativity. Although there are certainly guidelines to follow, much of microphone placement is based in science with a lot of additional experimentation. The experimentation is challenging and fun as you try different microphone placements to capture the sound you want.

Microphone Myths and the Science behind Microphone Placement

There are at least three myths with regard to how microphones operate and, thus, their proper placement. First, microphones do not reach out to capture sound, nor is there such a thing as a working distance or "microphone reach" measured in inches and feet. The fact is that sound waves have to travel to the microphone and impact the diaphragm in order to start the process of converting acoustic energy to electrical energy.

How effective a microphone is at picking up sound at a particular distance from the sound source depends on a number of factors, including the initial sound-pressure level of the source you want to record and the background, or ambient, noise level present. For example, if you hike into the boonies in one of the national parks with a recorder and a microphone, you might be amazed to hear sounds that are hundreds of feet away, or even the sound of a stream cascading down a mountainside a quarter-mile away (see audio demo, cut 3-11). Does this mean the microphone has a quarter-mile reach? Absolutely not. What it means is that there is almost no background noise present to interfere with the sound, allowing it to reach your location.

A second microphone myth is that directional microphones somehow pull in or enhance sound waves coming from the front of the microphone. The fact is that directional microphones do not enhance or pull in sound from any particular direction, just as microphones do not reach out for sound. Directional microphones are designed to *reject* sound from specific directions. In the case of a cardioid microphone, sound is slightly rejected from the sides and entirely rejected from the rear of the microphone. This gives the impression that the sound to the front of the microphone is enhanced, when in reality, the front is simply the only active side of the microphone left to pick up sound.

A third myth surrounding microphones is that hyper-cardioid shotgun microphones somehow focus and zoom in on sounds, pulling them in and making them appear closer. Here again, hyper-cardioid microphones do not focus and do not zoom in, and they certainly don't pull in sound. What hyper-cardioid microphones are capable of doing is *rejecting* sound from the sides and rear of the microphone so that the active side of the microphone has a very narrow pickup pattern. It is the extreme rejection of the sound from the sides and rear that gives the impression that the microphone is focused and pulling in sound.

During a nature recording session for this text, a hyper-cardioid shotgun mic was used to capture the sounds of white-water rafters as they passed through a rough section of a popular river. The goal was to record their hoots and hollers as they went crashing through the water. Their speaking voices were clearly recorded from over 75 feet away as they talked to one another before hitting the rapids, with a state highway only 30 feet behind the mic (see audio demo, cut 3-12). Listen carefully to the cut. Can you hear any vehicle traffic? The shotgun mic did what it was designed to do: it rejected the vehicle sounds, which are not in the recording. Regardless of the directional qualities of a microphone, unwanted sounds can still enter a directional microphone by being reflected off walls and objects.

Basic Microphone Placement Guidelines

Using a simple three-step process, you can properly place a microphone on just about any sound source. First, identify the source of the sound. What is creating the sound? For an announcer, it is his or her mouth. At a news event, what is the source of the sound you are trying to capture? Is it the fire engine racing to the scene of the fire, or the crackling of the fire itself? What did your producer ask for? In the case of a musical instrument, something has to vibrate to cause a sound to be generated. For example, on a violin, piano, or guitar, it is the strings. Regardless of the situation, identify the source of the sound.

Second, once you have identified the sound source, focus your attention on the sound's amplifier. Something is amplifying the sound so that you can hear it. In the case of radio talent, people's mouths are the amplifiers. On a piano, it is the sounding board that the strings are mounted to; on a tuba, it is the huge bell-shaped opening at the end of all the plumbing; on an acoustic guitar, it is the guitar body. Outdoors, the sound reverberating off buildings or objects often serves to amplify sound. Regardless of the situation, identify what is amplifying the sound.

Third, with the sound source and the amplifier in mind, determine the direction that the amplifier is aimed and the direction the sound waves are traveling. Place the microphone in the sound's direct path, facing the amplifier. The question now is, how far away from the sound source do you place the microphone? And the answer is: within the critical distance.

Critical Distance. You perceive sound as a combination of the direct sound from a source and the reverberant sounds from the space surrounding the source. Earlier we discussed the direct-to-reverberant sound ratio, which is the ratio of direct sound to the reverberant sound that you hear. At a certain distance from each sound source the direct sound and the reverberant sound are equal, and the ratio between the two is one. Everything between the sound source and this point is part of the **critical distance**.

The critical distance depends on the sound source. A source that is louder and is more directional in nature has a longer critical distance. Likewise, a softer sound that is nondirectional has a much shorter critical distance. For example, if you are standing in front of a blaring trombone, its direct sound path is several feet long before you can begin to hear the reverberant sound from the room you are in. At the opposite end of the spectrum, if you are trying to record someone whispering, you leave the direct sound path and begin to hear the reverberant sound from the room fairly quickly.

Microphone Placement within the Critical Distance. Ideally, a microphone should be placed within the critical distance from the sound source to obtain the most accurate and intelligible recording. However, there are times when, for effect, a production person intentionally places a microphone outside the critical distance to capture more room reverberation or reverberant sound.

There are questions that come into play in microphone placement that cross the line between science and art. Science says that, to get the perfect recording, you place the microphone one way. But art says that to get the effect you are seeking, you might place it another. For example, try placing a microphone very near the source and another a few feet away, recording each on a separate track. When you listen to the playback, mix the ratio of the direct mic track to the ambient mic track and see the incredible range of sounds that you can create (see audio demo, cut 3-13). The best sound is usually achieved by experimentation.

Great recording engineers and production people are not known for their ability to place a mic in just the right place according to a book or chart. They are known for their creativity in capturing sound, sometimes by some very unconventional methods. You should feel free to experiment to achieve the sound *you* or your client wants.

A Nontraditional Mic Placement. Have you ever heard an announcer on a national commercial or voice over and wondered to yourself, how did they get "that" sound? There is often a rich, resonant quality to the voice. Many people automatically assume that some sort of electronic processing was used. If you would like to use the same technique and microphone many of the best voice talents use get a Sennheiser model MKH-416 shotgun microphone. This microphone is well known for the warmth of its sound, which is often compared to the tonal qualities of a well-played woodwind instrument.

Sit facing the microphone and aim the microphone at the upper part of your chest or lower throat area so you are in essence talking just over it. Because the microphone is aimed at the lower throat or upper chest it not only picks up your voice, but the resonance of your chest. Due to the unique hyper-cardioid pattern of the microphone, the talent can be a little farther away from the mic to eliminate unwanted breath sounds but retain an intimacy that you can usually only get by working very close to the mic. Experiment with the aim and placement of the microphone and you will be surprised at the sound quality you can achieve for your voice.

Stereo Microphone Placement Guidelines

The goal in recording or broadcasting an event in stereo is to re-create a broad, life-like sound field between two speakers during playback. A stereo recording requires at least two microphones, one for left channel, one for right channel. How these microphones are placed determines the stereo image you capture. There are four possible ways to capture a stereo sound field.

The Spaced Pair or A-B Stereo Configuration. This is the simplest and most straightforward way to create a stereo image. Two identical omnidirectional microphones are placed on stands in front of the band or event to be recorded. However, if you desire less room ambiance, two cardioid microphones can be used instead. The two microphones are spaced apart, facing the band, with each microphone angled slightly outward to the left and right edges of the sound source to form the **spaced pair**.

There is one potential problem to be aware of with a spaced pair. Because the microphones are spaced apart, some sound will arrive at one microphone before it arrives at the other, and vice versa. Due to the different arrival times, the sound may or may not be in phase. As covered in chapter 2, when a sound arrives at a pair of mics out of phase it can be canceled out and disappear from the stereo image. Essentially, there is going to be both constructive and destructive interference of the sounds. To prevent this, observe the three-to-one rule.

The three-to-one rule states that when using a spaced pair of microphones, they should be placed 3 feet apart for every foot they are from the sound source. Using the three-to-one rule helps ensure that phase cancellations are kept to a minimum (see figure 3.26).

By listening to the stereo microphone placement in mono (left and right combined into a single channel), you can tell when you are getting phase cancellation and acoustic losses in the recording. The sound takes on a fuzzy, nondirectional quality, and things that should appear in the center of the stereo image seemingly disappear or become quite faint.

The X-Y Stereo Configuration. X and Y in the **X-Y stereo configuration** represent your left and right ears, respectively. There are two methods of obtaining an X-Y recording, with a **coincident pair** of microphones or a **near-coincident pair** of microphones. The object of this type of mic placement is to eliminate or reduce the possible phase cancellation that can occur with the spaced-pair mic technique.

A coincident microphone pair includes two cardioid (or other directional-type) microphones placed in a horizontal position facing the group or instrument. The two microphone *heads* (the part of each microphone that contains the sound pickup) should be nearly touching. The right microphone head should be aimed toward the left side of the sound source. The left microphone head should be aimed toward the right side of the sound source (see figure 3.27). The degree of angle at which you aim the microphones determines the size of the stereo image that is recorded. Because the microphone heads are so close together, a

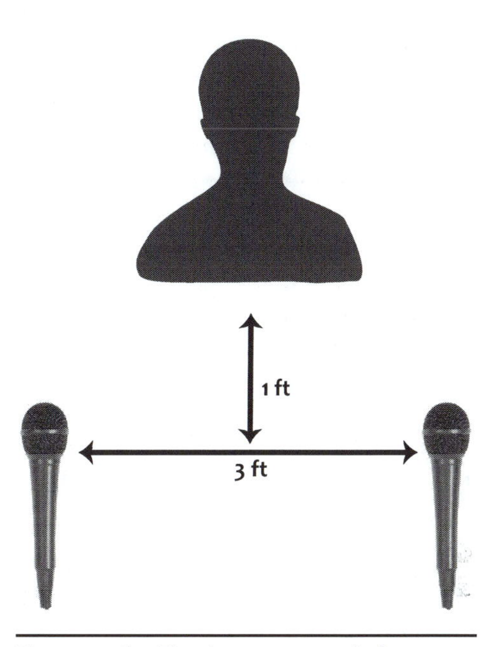

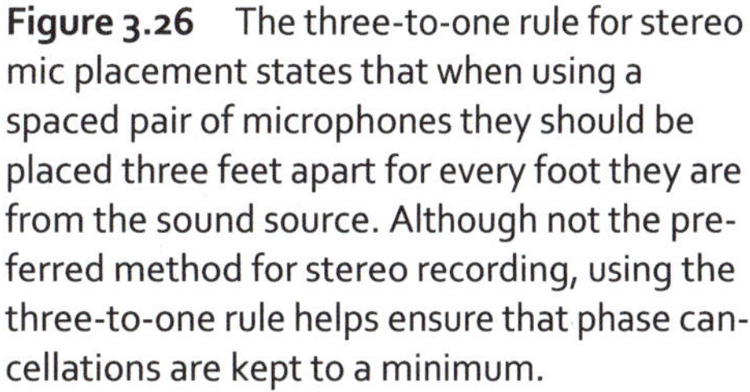

Figure 3.26 The three-to-one rule for stereo mic placement states that when using a spaced pair of microphones they should be placed three feet apart for every foot they are from the sound source. Although not the preferred method for stereo recording, using the three-to-one rule helps ensure that phase cancellations are kept to a minimum.

Figure 3.27 A coincident microphone pair includes two cardioid (or other directional-type) microphones placed in a horizontal position facing the group or instrument. The degree of angle at which you aim the microphones determines the size of the stereo image that is recorded. Because the microphone heads are so close together, a coincident microphone pair introduces very little phase shift and produces an accurate stereo picture.

coincident microphone pair introduces very little phase shift and produces an accurate stereo picture.

A near-coincident microphone pair includes two cardioid (or other directional-type) microphones placed a few inches apart in a horizontal position facing the group or instrument. The base of the two microphone *bodies* (the part of each microphone that is below the microphone head) should be nearly touching. The right microphone should be aimed toward the right side of the sound source, using the midpoint of the mic body as the rotation point. Using the same procedure, the left microphone should be aimed toward the left side of the sound source (see figure 3.28).

The two microphone heads should generally be four to seven inches apart. This positioning of the microphones allows for just a little bit of phase shift and adds an open sound with some width and depth to the stereo image. The angle of the microphones can be varied to adjust the stereo image.

Stereo Microphones. Stereo microphones usually have two elements arranged within the microphone head in a coincident X-Y pair. This is a convenient way to record stereo using only one microphone body, one microphone mount, and one microphone cable. The other advantage, or disadvantage depending on your viewpoint, is that the aiming of the X-Y pair is fixed, giving you consistency of the stereo image, especially on live remote broadcasts. Stereo microphones work well as crowd mics for sports broadcasts, adding a realistic ambiance.

Middle/Side Microphones. Another way to capture a stereo image using just one microphone is with a **middle/side microphone** (M/S). A middle/side

Figure 3.28 A near-coincident microphone pair includes two cardioid (or other directional-type) microphones placed in a horizontal position facing the group or instrument. The two microphone heads should generally be 4 to 7 inches apart. This positioning of the microphones allows for just a little bit of phase shift and adds an open sound with some width and depth to the stereo image. The angle of the microphones can be varied to adjust the stereo image.

microphone has two directional elements within a single microphone. The first of these is a cardioid element facing straight forward to pick up the direct sound and is the middle (or M) element. The second is located behind the cardioid element and is a bidirectional microphone element. The bi-directional pattern is aimed out each side of the microphone and is the side (or S) element (see figure 3.29).

By combining the sound from the cardioid element and the left half of the bidirectional pattern the left channel is formed, likewise for the right. Most mid/side microphones have adjustable patterns to vary the size of the stereo image.

Stereo microphones are offered in a wide variety of directional patterns. Prices range from $100 to $5,000 for a high-end condenser stereo mic.

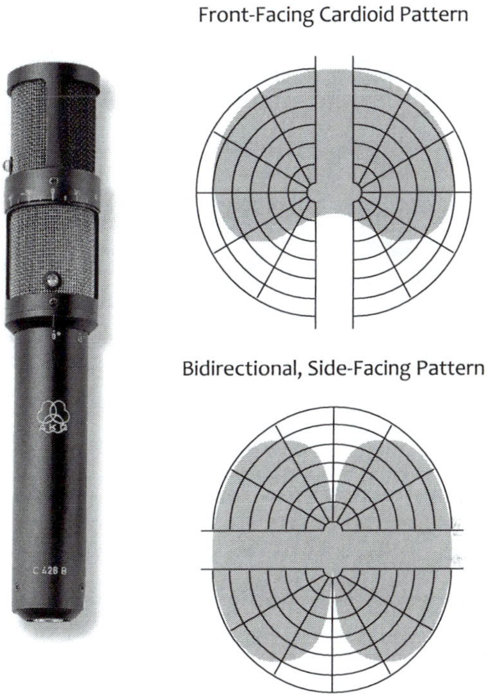

Front-Facing Cardioid Pattern

Bidirectional, Side-Facing Pattern

Figure 3.29 A middle/side microphone has two directional elements within the microphone. The middle element is a cardioid pattern facing straight forward to pick up the direct sound. The side element is located behind the cardioid element and is a bidirectional microphone pattern aimed out each side of the microphone.

Windscreens and Pop Filters

A **windscreen** or **pop filter** protects a microphone's diaphragm from blasts of wind or plosive speech sounds. A plosive speech sound is created when you obstruct the flow of air through your mouth and nose prior to making the sound and suddenly release the air when you make the sound or say the letter. The plosive sounds include the letters *b, d, g, k, t,* and the most famous plosive of all, the letter *p*. Many people deliver a significant blast of air along with the letter *p*. It is called popping your *p*'s (see audio demo, cut 3-14). The blast of air is so strong that the letter is lost, and all that is heard from the mic is what sounds like a loud thump, or small explosion. This occurs because the diaphragm bottomed out or was forced all the way to the limits of its ability to travel by the blast of air.

Good announcers learn to pull their *p*'s, or slightly restrict the plosive, so that this condition does not occur. But you are not always going to be working with experienced talent. Other sounds that can cause problems are the sibilant sounds, like "s" and "ch."

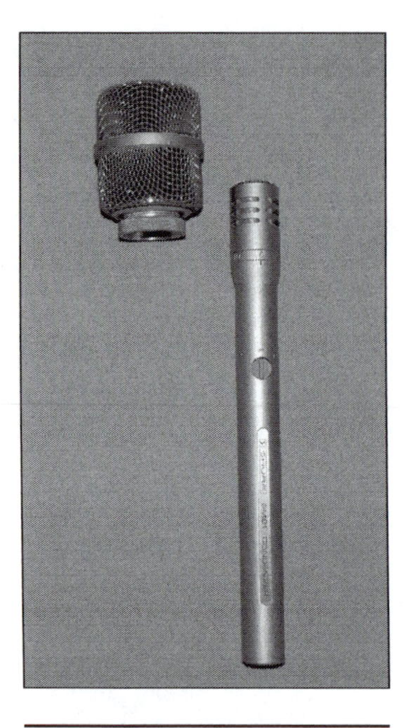

Figure 3.30 A pop filter helps prevent the nasty thumping sounds from plosives and sibilant speech. This particular condenser microphone has an easily removable pop filter that allows, depending on the application, the microphone to be used with or without the filter. (Copyright Shure Incorporated, used with permission.)

Pop Filters

A pop filter is a small, acoustic-foam windscreen built into the head of the microphone in an effort to eliminate those nasty thumping sounds from plosives and sibilant sounds (see audio demo, cut 3-15). In theory, the acoustic foam allows the sound through, and the blasts of air from the plosives get lost in the foam and never make it to the diaphragm (see figure 3.30). However, not everyone speaks the same and, in many cases, the built-in pop filter just can't handle a serious "*p* popper."

If the microphone's built-in pop filter is not stopping your popper, turn the microphone 90 degrees to the left or right and have the popper talk across the face of the microphone so that the plosive blast of air goes past the microphone instead of directly into it (see figure 3.31; see audio demo, cut 3-16). However, some people can still pop a mic, even under these circumstances.

A more aggressive approach for a pop filter is to use a **popper stopper** or external pop filter. These mount on the mic stand and are placed a few inches off the face of the microphone. They appear to be a fine mesh cloth, similar to women's stockings, stretched over a small circular frame (see figure 3.32). Popper stoppers are acoustically transparent, and the mesh stops or protects the microphone from the blasts of air created by the plosive sounds. If one is not enough, try a second placed behind the first.

Figure 3.31 If a microphone's built-in pop filter is not stopping an announcer from popping, turn the microphone 90 degrees to the left or right and have the announcer talk across the face of the microphone so that the plosive blasts of air go past the microphone instead of directly into it.

Popper stoppers are also sometimes called **spit screens** as they offer the additional benefit of stopping spit from reaching the face of the microphone and building up a film on the microphone face. As gross as it sounds, it is something to think about if you are using something like a $1,000 voice-over microphone.

Internal pop filters and popper stoppers are intended for indoor use and cannot provide adequate protection from the wind outdoors.

Figure 3.32 Popper stoppers are acoustically transparent, and the mesh stops or protects the microphone from the blasts of air created by the plosive sounds. Popper stoppers are also sometimes called spit screens, as they offer the additional benefit of stopping spit from reaching the face of the microphone.

Windscreens

A microphone can be protected outdoors from the wind with a foam windscreen. Made from special acoustic foam to fit the microphone, windscreens come in various shapes and sizes. A foam windscreen slides over the head of the microphone and is held in place by friction or a small piece of Velcro (see figure 3.33). The sound passes through the foam with only a very slight effect

Figure 3.33 Foam windscreens are popular in radio stations because they allow announcers to work microphones very closely for the proximity effect and not pop their *p*'s or sizzle their *s*'s. This foam windscreen slides over the head of the microphone and is held in place by friction. It will completely cover the head of the microphone when fully in place.

on the high frequencies. The wind gets lost in the foam, loses its power, and never makes it to the diaphragm (see audio demo, cut 3-17).

Foam windscreens are also popular inside radio stations because they allow announcers to work microphones very closely for the proximity effect and not pop their *p*'s or sizzle their *s*'s.

Microphone Cables

There are two types of microphone cables, distinguished by their physical structure and electrical properties.

Balanced Cables

Professional **balanced microphone cables** have a positive (+) wire, a negative (–) wire, and a braided, or foil, shield that encases the positive and negative wires.

A balanced microphone line is always preferred, as it is the least susceptible to electrical noise and hum or to radio-signal interference. Electrical interference occurs in a number of ways, such as laying a mic cable close to an AC power line. If you are close to a radio station, a mic cable can serve as a very good antenna and help a portable audio mixer or recorder to act as a radio. In some location situations, two-way radios can cause interference.

The shield in a balanced mic cable protects the audio traveling on the positive and negative wires by being connected to earth ground through the equipment the mic cable is plugged into. The shield acts to pick up or receive the interference and carries it harmlessly to ground. Balanced mic cables can be run up to 1,000 feet without serious audio degradation.

Unbalanced Cables

An **unbalanced microphone cable** has only one conductor, positive (+), with the shield as the negative (–). This presents two problems. First, unbalanced cables are highly susceptible to electrical noise and radio-signal interference, since the shield is also the negative (–) of the audio signal. Second, such cables need to be kept as short as possible due to the electrical noise and radio-signal interference they are capable of picking up. With very few exceptions, such as adapting a professional microphone to a prosumer-type recorder, unbalanced mic cables should be avoided if possible.

Microphone Cable Care

A microphone cable is not just a plain old piece of wire. It is a special cable designed to offer low impedance (resistance to electrical flow) so that it can more easily pass low-level microphone signals. It is also designed to be flexible under a variety of conditions.

Microphone cable should be rolled up using the natural looping of the cable as a guide. Once looped, it should be tied with a Velcro cable tie-wrap, a reusable plastic cable wrap, or a piece of gaffer's tape (see figure 3.34). Each of these makes

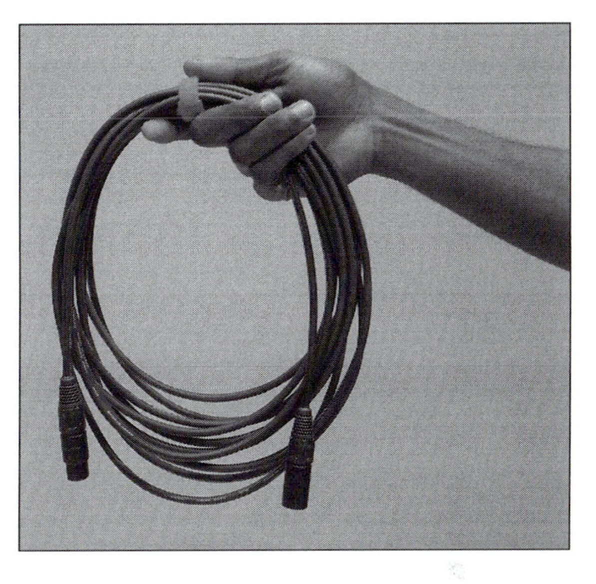

Figure 3.34 Microphone cable should be rolled up using the natural looping of the cable as a guide. Once looped, it should be tied with a Velcro cable tie-wrap, as this cable is, or with a reusable plastic cable wrap. It is also acceptable to use a small piece of gaffer's tape.

it very easy to hang the cable up or stack it. Another acceptable way to roll and store cable is with a microphone cable reel.

NEVER wrap a mic cable around your bent arm from hand to elbow or in a figure-eight hand to elbow pattern. This twists and kinks the wire within the cable and will cause it to fail. In addition to causing the wire to fail at some point, the kinks and twists formed by this method of wrapping a cable make it difficult to lie flat or hang straight from a microphone stand.

Microphone Connectors

XLR "Cannon" Connectors

Professional microphone cables use three-pin **XLR connectors**. The connector is also called a Cannon connector. The connector is named after James Cannon of the Cannon Electric Company, who developed the first three-pin connector in the late 1920s.

The XLR connector is a cylindrical metal housing with three protected connector pins or sockets. The male connector has protruding pins; the female connector has pin sockets. Typically, outputs are male and inputs are female (see figure 3.35A on the next page).

The pins and sockets are numbered 1, 2, and 3. Pin number 1 is the shield, or earth ground, and is slightly longer than the other two pins. This is so that the first thing that connects is the ground, to eliminate any electrical pop when the other two pins are mated together. Pin number 2 is the positive (+), and pin 3 is the negative (–). An XLR connector is a balanced connector, just like the mic cable that it is attached to.

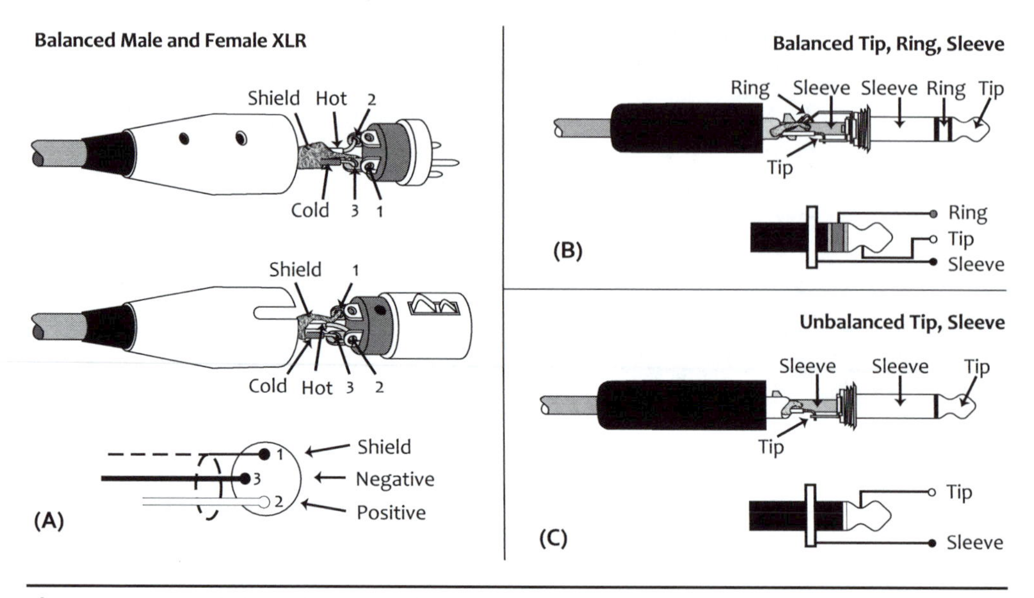

Figure 3.35 (A) A balanced male and female XLR connector. (B) A balanced male TRS (tip, ring, and sleeve) connector. (C) An unbalanced male T/S (tip and sleeve) connector.

A professional microphone has a male connector built into its base. Microphone cables have a male and a female connector; the female end plugs into the microphone. The male connector plugs into the console or recorder.

Quarter-Inch Connectors

The TRS Connector. **TRS connectors** are sometimes used in radio, but more typically they are used in the concert industry on musical instruments, mics, and portable mixers or consoles. The TRS connector is often called a quarter-inch phone plug since it is one-quarter inch in diameter and looks just like the plugs used by operators in the days when they literally connected each telephone call by patching in a cable with a plug on the end of it. TRS male connectors are called plugs, and the female connectors are called jacks.

TRS stands for tip, ring, and sleeve. Looking at the connector, the "tip" of the connector is the positive (+). A small black insulator ring separates the tip from the next part of the connector, called the "ring"; the ring is the negative (–). Another small black insulator separates the ring from the final part of the connector, called the "sleeve"; the sleeve is the shield or earth ground (see figure 3.35B).

A TRS connector is a balanced connector, just like the XLR connector. Professional microphone cables are often made with a female XLR connector on one end and a male TRS connector on the other. This type of connector is used to plug a professional microphone into a portable mixer or console.

Occasionally TRS connectors are used as "stereo" connectors. For example, stereo headphones use a TRS with the tip being positive left channel, the ring being positive right channel, and the sleeve being a common negative. In such instances the TRS is used as an unbalanced connector.

The T/S Connector. A **T/S connector** (tip/sleeve) is an unbalanced quarter-inch phone plug. The tip is positive (+) and the sleeve is the shield or negative (–) (see figure 3.35C). Unbalanced T/S connectors are generally found on public address amplifiers, speaker cables, and consumer-type microphones. Occasionally you'll encounter a microphone cable with a female XLR connector on one end and a T/S plug on the other to plug a better quality microphone into consumer gear.

Mini-TRS and T/S Connectors

The mini-plug was developed in the consumer electronics industry and found its way into the professional world through prosumer-type equipment, like the small recorders, used in many radio news operations. The mini-plug is sometimes referred to as an eighth-inch mini-plug or a 3.5-millimeter mini-plug.

The plug is constructed like the standard one-quarter-inch phone plug, just a lot smaller. It comes in TRS and T/S configurations. It is not unusual to find a female XLR connector on one end of a microphone cable and a TRS mini-plug on the other end so that a professional microphone can be plugged into a small portable recorder or other portable gear.

Microphone Preamplifiers

Boosting Microphone Signals

A microphone preamplifier, in its most basic form, is a voltage multiplier. It takes the low-voltage signal from the microphone and amplifies it to a line-level signal. Without getting too involved in electronics theory, professional audio is usually either at mic level or line level. Mic level is a very low-level signal measured in millivolts (one millivolt = one thousandth of a volt). Line level is typically 1 volt but can be as high as nearly 2 volts.

A preamplifier's design is critical, since a microphone's signal is very weak. Preamplifiers boost whatever signal is present. If the audio contains noise and distortion from the microphone or a bad microphone cable, the noise and distortion are amplified along with the original signal. Ideally, microphone cable runs should be kept as short as practicable to help prevent this.

Broadcast and production control boards and portable mixers have internal microphone preamplifiers. Such preamplifiers are designed to work with a wide range of microphones and have a limited number of adjustments, if any, to boost and shape the sound.

In a quest for sonic perfection, a producer or voice talent typically turns to a stand-alone preamplifier known as an outboard preamp, often referred to as a voice processor. Outboard gear is rack-mounted, separate from the main console (see figure 3.36). Preamplifiers also exist in virtual form within digital audio workstations.

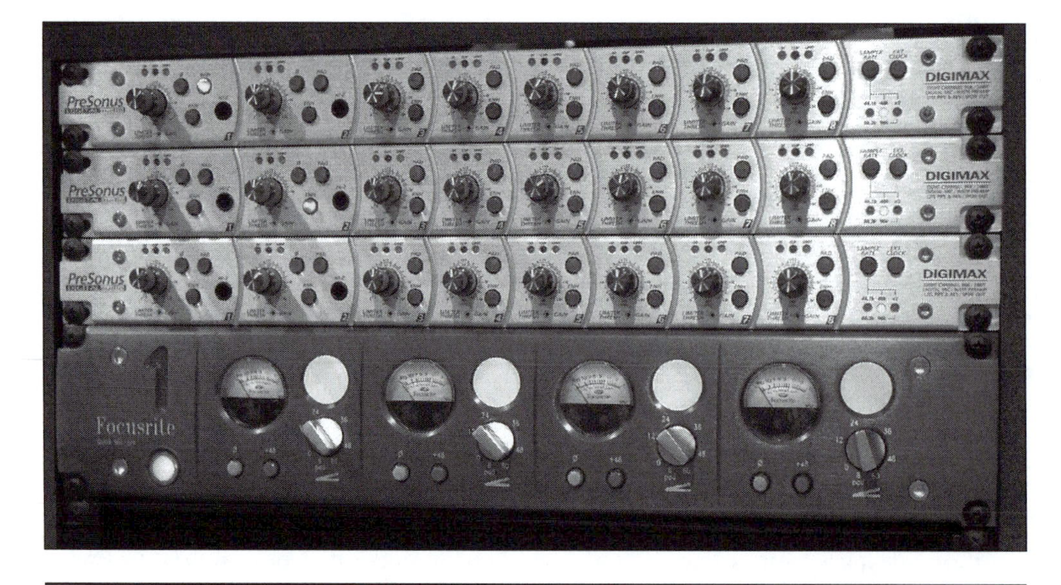

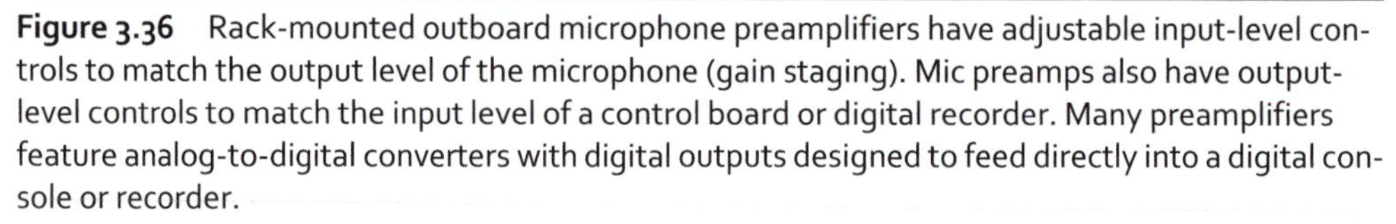

Figure 3.36 Rack-mounted outboard microphone preamplifiers have adjustable input-level controls to match the output level of the microphone (gain staging). Mic preamps also have output-level controls to match the input level of a control board or digital recorder. Many preamplifiers feature analog-to-digital converters with digital outputs designed to feed directly into a digital console or recorder.

A virtual preamplifier is software-based and emulates the same controls, functions, and effects as the hardware version of the preamplifier.

Professional preamplifiers have adjustable input-level controls to match the output level of the microphone (gain staging). Mic preamps also have output-level controls to match the input level of a control board or digital recorder. Many preamplifiers feature analog-to-digital converters with digital outputs designed to feed directly into a digital console or recorder.

Depending on the design, a preamplifier or voice processor may have additional audio processors, including compressors, limiters, equalizers, de-essers (to eliminate sibilant "*s*" sounds), and other user-selectable settings to tailor or shape the sound (again, see figure 3.36). You will learn about setting up a mic preamplifier or voice processor in chapter 6. Outboard microphone preamplifiers cost from $200 to nearly $4,000.

Selecting a Microphone Preamplifier

Golden-eared engineers and producers have written quite a bit about the perfect preamplifier. Selecting the "best" outboard microphone preamplifier is like selecting which variety of **Krispy Kreme donut** is best. You have to try them all before you know which one *you* like! Engineers and voice talent try different microphone and preamplifier combinations until they find the one that's just right for their specific purpose or application. In a typical radio station, the station engineer and the production director select a microphone preamplifier for the staff. However, if you are putting together your own voice-over studio, you may get the opportunity to select what you think sounds the best for your voice.

A professional voice-over artist experiments to find the exact microphone and preamplifier combination that makes his or her voice sound the best. Just as with microphones, there is no "best" microphone preamplifier; it does not exist. By now you should have figured out that sound is very subjective on a person-to-person basis. Regardless of published equipment specifications, more than anything else you must learn to let your ears be the final judge of what sounds good and what does not.

Special-Purpose Microphones

Lavalier or Body Microphones

Lavs, as they are called, are often referred to as body mics, since they are so small and can easily be hidden from view for television or motion picture use.

Lavs serve a very useful purpose in radio. Often news people do interviews with subjects who have no idea in the world how to use a microphone. These people turn their heads and move about, drifting in and out of the microphone pickup pattern. This kind of movement makes it very difficult to understand what they are saying. A lav comes in handy in a talk-radio studio, or on location, since it is clipped to the subject. People can move all they want to and they are always on mic.

Lavs are generally electret-condenser microphones. Some lavs require a small battery power source or phantom power for the microphone's preamplifier. A "large" lav mic is about the size of an eraser on a pencil. Tiny body mics are sometimes only slightly larger in diameter than the 3-millimeter (less than one-eighth of an inch) cable they are mounted on (see figure 3.6 earlier in the chapter).

Lavs are designed for placement below the chin, in the area of the upper chest. Because the talent does not speak directly into the microphone, most lav mics have an omnidirectional pickup pattern. The microphone is designed to be below and behind the sound source (the mouth and chin), so its designers boost the high-frequency response or roll off the low-frequency response to make the microphone sound more natural. The high-frequency boost explains why a lav sounds odd when someone picks it up and speaks directly into it (see audio demo, cut 3-18).

Although the lav is a tiny microphone, its price is not. Top-quality lav mics intended for use in the radio, television, and motion picture industry cost $300 to $400.

Shotgun Microphones

Shotgun microphones work by using a tube placed in front of the element with a series of openings along each side of the tube. As sound enters the microphone directly from the front, it also enters the side openings in succession, one after the next, in phase, adding upon itself until it reaches the element at the rear of the mic.

Sound from the rear of the microphone enters the openings in reverse order, out of phase, canceling itself out and making the microphone highly directional.

Shotguns are designed primarily for use in radio and television news and the motion picture industry. Shotguns have the ability to pick up sound at more than twice the distance of an omnidirectional microphone and retain the same direct-to-reverberant sound ratio (see audio demo, cut 3-19). Radio newsreporters use short shotguns for press conferences and on-the-run interviews to get a good clean pickup on the subject and eliminate unwanted sound or background noise. Shotgun microphones range in price from $300 to $2,000.

Headset Microphones

Headset microphones are designed to allow the talent to work with a microphone hands free, as the microphone is attached to a pair of headphones (see figure 3.37). Broadcasters use headset microphones primarily for news/sports coverage and radio talk shows where the announcer needs to have his or her hands free to keep statistics or take notes. The added benefit of a headset microphone is that no matter where the talent looks, or what the talent does, he or she is always on mic.

Headset microphones can be either moving-coil or electret condenser and are usually classed as noise-canceling microphones. They are intended to be used close to the lips and to block out unwanted background noise.

Figure 3.37 With headset microphones, these two sportscasters are always on mic as they follow the action on the court and are free to use their hands to produce the broadcast.

Recording Microphones

Recording microphones are designed to allow for tremendous freedom of movement while doing remote broadcasts, news, or sporting events. These microphones range from a simple microphone with a removable memory card to record on to sophisticated recording tools. The better-quality microphones contain dual power supplies—both manual and automatic record-level controls and the ability to power a set of headphones (see figure 3.38). The only real limitation to the recording time is the battery life. Many microphones have the ability to record for over 10 hours. The audio is stored on a standard SD memory card.

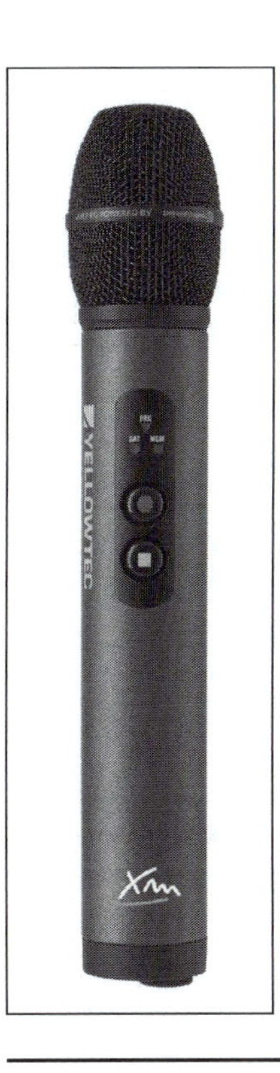

Figure 3.38
A recording microphone is handy for remotes or news and sports coverage. The device records onto a standard SD memory card that can be easily removed and replaced with another. Microphones of this type can record for up to 14 hours at a time. The primary limitation is the length of time the batteries will power the device.

Wireless Microphones

Wireless microphones provide talent the freedom to move about at a remote broadcast without being attached to a microphone cable. They also allow a production person to place a microphone at a distant location, such as on the backboard during a basketball game to capture the swish of the ball going through the hoop. Although not often thought of for radio, wireless microphones can be used to solve a lot of remote broadcast sound pickup problems. A professional wireless microphone has three separate components, a microphone, a transmitter, and a receiver. Occasionally, the microphone and transmitter are combined into a handheld microphone.

Transmitters

Professional wireless transmitters are low-power radio transmitters into which a microphone is plugged. The transmitters share frequencies in the **VHF** (very high frequency channels, 2 to 13) and **UHF** (ultrahigh frequency channels, 14 to 69) television bands. Most professional wireless transmitters are frequency-agile and can be tuned to any one of a number of channels. For example, in the UHF band (preferred by professionals), there are 256 frequencies.

Types of Transmitters. Professional wireless transmitters come in three varieties: self-contained, body-packs, and plug-ons.

A self-contained microphone/transmitter appears to be a regular handheld microphone, only slightly larger in size due to the transmitter that is contained in the microphone body (figure 3.39B).

Body-packs are smaller than a pack of cigarettes, and in some cases not much bigger than a cigarette lighter (see figure

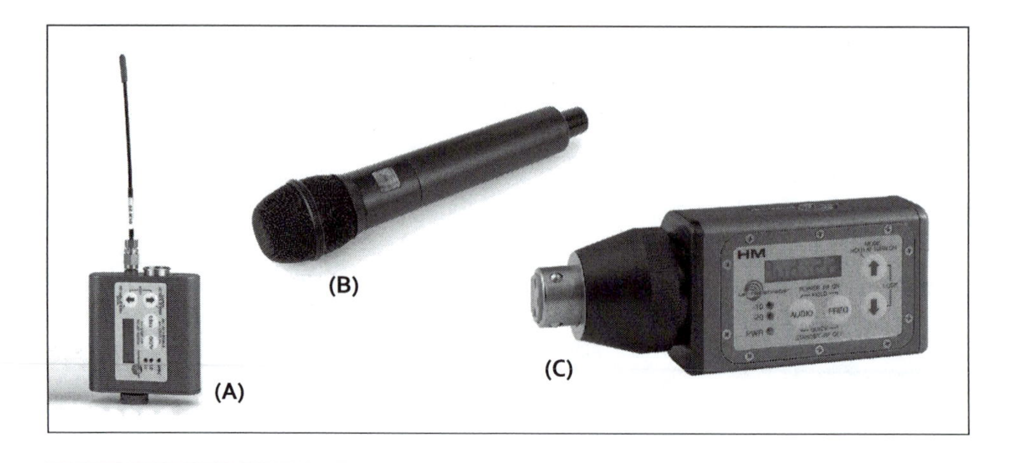

Figure 3.39 (A) Wireless transmitters known as body-packs operate on batteries, have a small rubber ducky antenna, and are designed for use with a lavalier, or body mic. Usually the transmitter is smaller than a pack of cigarettes. (B) Self-contained microphone/transmitters are popular for live remote broadcasts to free up talent so they can move around without worrying about someone tripping over a mic cable. (C) Plug-ons are small transmitters that plug into the base of the microphone and are used to turn any handheld microphone into a wireless.

3.39A). Body-packs operate on batteries, have a small **rubber ducky** flexible antenna, and are typically clipped to the back of a person's clothing with a belt clip or taped to an object. Most body-pack transmitters have been designed for use with a lavalier, or body mic.

Plug-ons are small transmitters that plug into the base of the microphone and are used for handheld microphones (see figure 3.39C). They are very popular because they can turn any microphone into a wireless by just plugging the transmitter into the microphone.

Radio stations use self-contained microphone/transmitters and plug-on transmitters effectively during live remote broadcasts to free up talent so they can move around without worrying about someone tripping over a mic cable. In many situations a radio station's remote truck is parked on the street as a portable billboard, and the talent is positioned inside the sponsoring business with a wireless mic.

A plug-on transmitter can also be attached to the microphone-level output of a portable microphone mixer at a sporting event, such as a football game, to transmit the announcers' voices from the press box to the remote truck in the parking lot, where the audio is routed back to the radio station.

However, these wireless transmitters are not usually referred to as plug-ons; they are often called by their unflattering industry nickname, **butt-plugs**, since they plug into the butt of the microphone.

Wireless transmitters have an effective line-of-sight range from 100 to 1,000 feet, depending on conditions. Professional wireless microphone transmitters are not inexpensive, with many models costing $1,000 per transmitter.

Wireless Receivers

A wireless receiver is a specially designed radio receiver that is tuned to match the wireless transmitter frequency in use. There are two basic types of receivers, single and diversity. A single receiver is one radio tuner and one antenna. Under most circumstances these work just fine, but occasionally they can experience noise, signal dropouts, and other interference, depending on the location and physical surroundings.

A **diversity receiver** has two tuners and two antennas combined to receive the signal (see figure 3.40). The two receivers are connected internally by a small microprocessor that automatically selects the better signal from the two identical receivers or, if necessary, combines the signals from the two receivers to get the best usable signal. Although diversity receivers are more expensive (near $2,000), they provide better reliability and are most often chosen by professionals who cannot afford to have poor audio quality or signal dropout.

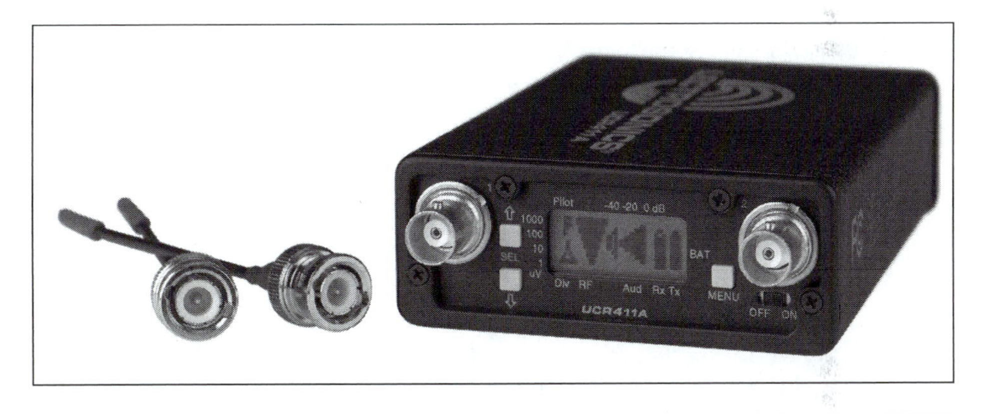

Figure 3.40 A diversity wireless microphone receiver has two built-in tuners and two antennas to receive the signal. The two receivers are connected internally by a small microprocessor that automatically selects the better signal from the two identical receivers or, if necessary, combines the signals from the two receivers to get the best usable signal.

Making the Final Selection

Selecting a microphone by type and pattern for a project is not necessarily an easy task for a beginner. What complicates things is that microphone selection and placement are not exact sciences but rather subjective judgments of technicians, engineers, and producers. In many cases, such decisions are based on accepted practice within an industry. As an example, one microphone thought to be a great voice microphone in radio is rarely used in recording studios for vocals. Likewise, there are even more different microphone preferences between television and motion picture production.

Microphone selection is, many times, a very personal decision on the part of the technician. A microphone is often chosen for the unique acoustic qualities it imparts in a particular recording situation. Sometimes the unconventional yields a very pleasing result, and many times the most expensive microphone does not yield the "best" sound. For someone starting in the industry, the best thing you can do is to observe and ask questions, just like the three-year-old who constantly asks "Why?" Also, read industry trade publications, especially interviews with top engineers and producers. If you are observant, trends and patterns of how professionals select and use microphones will become evident to you.

Suggested Activities

1. Collect a dynamic and condenser microphone, and a ribbon if one is available, and place them close together in one of the studios available at your school. Have someone stand in one spot and speak into the microphones. Simultaneously record each microphone on a separate track. Play back the recordings and see if you can find three distinct qualities that make each microphone different from the others.
2. Using the same setup as above, experiment with the proximity effect, making recordings with the talent 3 feet from the microphones, 2 feet from the microphones, 1 foot from the microphones, 6 inches from the microphones, and finally 2 inches off the face of the microphones. How was the proximity effect for each microphone?
3. Place a directional microphone in the studio and have the talent speak directly into the face of the microphone on its centerline, or 0-degree axis. Slowly rotate the microphone away from the talent, calling out the approximate position of the microphone as you go. When does the pickup pattern begin to take effect? In your judgment, how was the frequency response of the microphone affected?
4. If available, experiment with a lav mic, recording the talent with the microphone placed in a number of locations on the body as suggested in this chapter. Where is the best place to use your lav microphone?
5. Take a shotgun microphone and an omnidirectional microphone outside and record some birds or other sounds, with both microphones in the same location. Then compare the recordings. Can you name at least three different sound qualities produced by each?

Websites for More Information

For more information about:

- *condenser microphones,* visit AKG at www.akg.com
- *digital condenser microphones,* visit Neumann USA at www.neumannusa.com
- *dynamic or moving-coil microphones,* visit Electro-Voice at www.electrovoice.com
- *dynamic or moving-coil microphones,* visit Shure, Inc. at www.shure.com
- *general knowledge about microphones, patterns, and applications,* visit Shure, Inc. at www.shure.com/americas/support/downloads/publications

- *powered ribbon microphones,* visit Royer Labs at www.royerlabs.com
- *ribbon and digital microphones,* visit Beyerdynamic at www.beyerdynamic.com
- *shotgun microphones,* visit Sennheiser USA at www.sennheiserusa.com
- *wireless microphones,* visit Lectrosonics at www.lectrosonics.com

■ Pro Speak

You should be able to use these terms in your everyday conversations with other professionals.

audio—Latin, "to hear"; sound that has been converted into electrical energy.

balanced microphone cable—A microphone cable or line in which there is a positive wire, a negative wire, and a shield or earth ground surrounding the positive and negative wires.

bass boost—Another term for proximity effect.

bidirectional—A microphone pattern that has maximum sensitivity at 0 and 180 degrees and maximum rejection at 90 and 270 degrees.

butt-plug—A wireless transmitter designed for plugging into the base of a microphone.

cardioid—A heart-shaped microphone pattern with maximum sensitivity at 0 degrees and maximum rejection at 180 degrees.

centerline—An imaginary line extending straight out from the face of a microphone, or on a 0-degree axis.

coincident pair—Stereo microphone placement in which the two microphones are placed side by side facing the sound source. With the midpoint of the mic bodies serving as the rotation point, the heads of the two microphones are aimed inward nearly touching. One is aimed at the right side of the sound source, the other at the left side.

condenser microphone—A microphone that uses an electrostatic field to convert sound into audio and requires a polarizing voltage and a preamplifier.

critical distance—The distance from the sound source at which the ratio of direct-to-reverberant sound is equal, or 1; abbreviated d_c.

digital condenser microphone—A condenser microphone with a built-in preamplifier that includes an analog-to-digital converter and digital output only.

diversity receiver—A wireless microphone receiver using two tuners and two antennas to eliminate noise and dropouts.

dynamic microphone—A moving-coil microphone.

electret-condenser microphone—A condenser microphone that, although it does not require a polarizing voltage, does require a battery to run the preamplifier.

frequency-response curve—A plotted chart that shows how a microphone responds to various frequencies; called flat if the microphone treats all frequencies equally.

gaffer's tape—Available in colors, gaffer's tape has the strength of duct tape without the sticky mess afterward.

gain staging—Matching a microphone's maximum output voltage to the preamplifier's maximum input voltage to prevent overload.

gooseneck—A piece of flexible metal conduit used to support a microphone.

hyper-cardioid—An elongated, heart-shaped cardioid pattern used primarily in shotgun microphones to allow the microphones to be moved farther from the sound source without increasing the apparent room reverberation.

Krispy Kreme donut—A Southern delicacy, wonderfully light donuts slathered in mouth-watering sugar frosting, often used to entice the crew to show up in the morning; a sugar-delivery device.

lav (lavalier)—A microphone worn on the body. Lav is pronounced like lava from a volcano, just drop the last *"a."* Lavalier comes from the French word *lavaliere,* or pendant.

maximum sound-pressure level—The maximum dB-SPL that a microphone can reproduce before going into distortion and then failure.

mic—Pronounced "Mike"; industry abbreviation for microphone (mics is used for more than one mic).

micron—An outdated term still in common usage that has since been replaced with the term micrometer. A micron, or micrometer, is one millionth of a meter, or about 1/25,000th of an inch.

middle/side microphone (M/S)—A stereo microphone incorporating a cardioid element facing forward and a bidirectional element facing out the sides of the microphone. The combination of the middle element and the separate halves of the bidirectional element create the left and right channels.

near-coincident pair—Stereo microphone placement in which two microphones are placed side by side facing the sound source. With the midpoint of the mic bodies serving as the rotation point, the right mic is aimed at the right side of the sound source and the left mic is aimed at the left side.

noise—Electrical noise, which appears as hiss in an audio signal, measured in negative dB. The larger the negative figure, the lower the noise level.

off axis—Outside of the centerline of a microphone's directional pattern, usually measured in degrees.

off-axis coloration—The change in the frequency response and sensitivity of a microphone as it is turned away from a sound source and aimed in a different direction.

omnidirectional—A microphone pattern that allows the microphone to pick up sound from all directions equally.

overload—The point at which a microphone stops functioning under extremely high sound-pressure levels.

pattern—The term used to describe the directional characteristics of a microphone.

phantom power—Power supplied to a condenser microphone from a source such as a control board, portable mixing board, or other outside source through the microphone cable.

polar frequency response—The frequency-response curve of a microphone measured at various points in the microphone pattern.

pop filter—An acoustic foam filter within a microphone used to prevent popping plosive sounds from reaching the microphone element.

popper stopper—A fine mesh screen placed a few inches off the face of a microphone to prevent popping plosive sounds from reaching the microphone element.

preamplifier—A voltage multiplier used to boost the weak signal of a microphone to a usable audio level.

prosumer—Prosumer equipment is equipment that offers many professional features but is not built to rigid professional standards. It is generally priced in between consumer and professional-level equipment.

proximity effect—The closer you move to a directional microphone, the more bass your voice appears to have.

ribbon microphone—A member of the dynamic class of microphones; the ribbon uses a thin ribbon of metal as the sound pickup element.

rubber ducky—A flexible radio antenna made from silicone-based rubber products that can be bent and flexed without damage to the antenna coil contained within.

sensitivity—An electrical measurement in either dBV (decibel volts) or millivolts that tells what the audio output of the microphone will be at a given sound-pressure input level.

shock mount—A device designed to hold a microphone and isolate it from vibrations that might enter the microphone body and cause undesirable sounds.

shotgun microphone—A microphone designed with a long tube in front of the microphone element to focus the sound and to increase the microphone's direct-to-reverberant sound ratio.

sound—Acoustic energy traveling through a medium such as air or water.

spaced pair—Stereo microphone placement in which two microphones are spaced a distance apart following the three-to-one rule: three feet of separation for each foot of distance from the sound source.

spit screen—A fine mesh screen placed a few inches off the face of a microphone to prevent popping plosive sounds and spit from reaching the microphone element; same as a popper stopper.

super-cardioid—A heart-shaped microphone pattern with maximum sensitivity at 0 degrees and maximum rejection at 180 degrees. Super-cardioid microphones are even less sensitive to the sides than a cardioid microphone.

transient response—How a microphone reacts to transients, or sounds, with a fast attack that trails off slowly.

TRS connector—A one-quarter-inch diameter balanced connector with the tip as positive, the ring as negative, and the sleeve as shield or earth ground.

T/S connector—A one-quarter-inch diameter connector with the tip as positive and the sleeve as negative and shield or earth ground.

UHF—Ultrahigh frequency; a band of frequencies assigned to television channels 14 through 83; 14 through 69 are also assigned to wireless microphones.

unbalanced microphone cable—A microphone cable or line that only has one positive conductor, with the shield acting as both the negative and the shield.

unidirectional—A generic term for a directional microphone, meaning one direction.

Universal Serial Bus (USB)—The most widely used hardware interface for connecting equipment to computers. USB 2.0 can transmit up to 480 Mbps of data; USB 3.0 can transmit up to 10 Gbps of data. Virtually every portable device uses some form of USB connectivity for data transfer.

VHF—Very high frequency; a band of frequencies assigned to television channels 2 through 13 and to wireless microphones.

windscreen—An acoustic foam cover for the head of a microphone used to block wind outdoors.

XLR connector—A balanced microphone connector; sometimes referred to as a Cannon connector, after the man and the company that invented it. A connector in which there are pins or sockets for the positive, negative, and shield, or earth ground.

X-Y stereo configuration—Stereo microphone placement in which the two microphones are placed in a horizontal position facing the sound source and aimed in either a coincident or near-coincident manner.

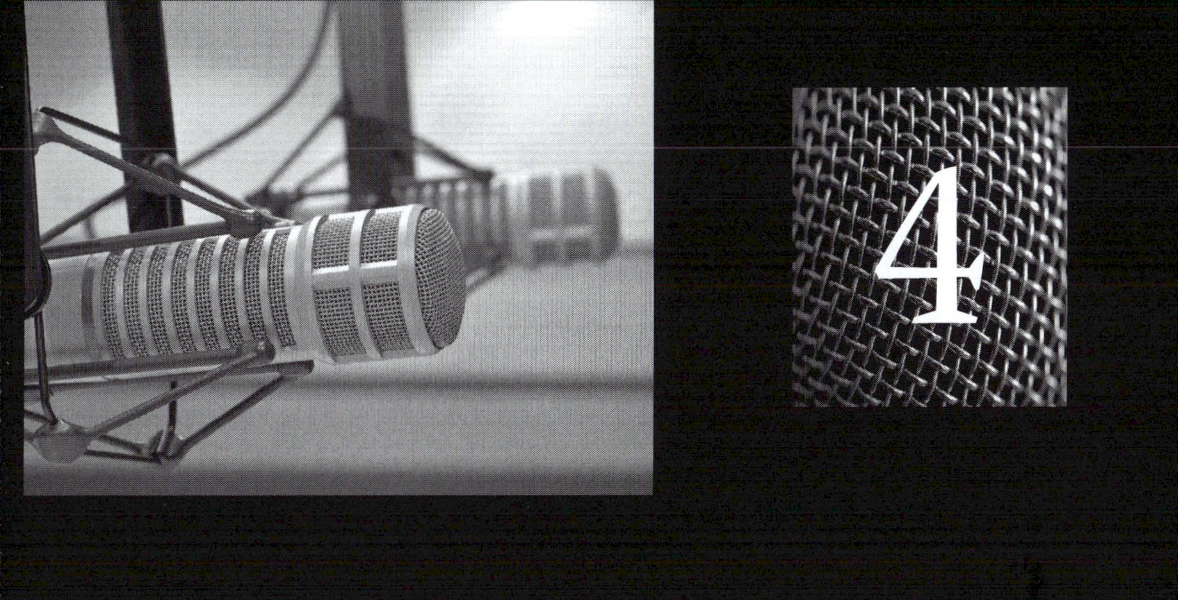

4

Control, Mixing, and Monitoring

HIGHLIGHTS

To someone new to radio, an audio control board can look pretty terrifying with all the knobs, switches, faders, and metering displays packed into a compact space. Although a control board may appear intimidating at first, it is a lot easier to learn and operate than you think. Generally, a station staff member trains new operators and takes them through the station's control board operation. With some hands-on experience you will get over your fear quickly. As one of my mentors in broadcasting said as I stood looking at my first 32-channel broadcast console, "This isn't rocket science; it's radio!" Still, operating a console requires eye-hand coordination, preplanning, and some creativity and skill.

The fact is, running an on-air control board is the simplest task a show host has. The difficult work is planning your show, following the station format, and creating a fresh, innovative relationship with listeners. The same applies to production. The operation of the hardware and software at your disposal is relatively easy. The tough part is creating and writing the script, selecting music and sound effects, and learning how to communicate with your audience. Remember, despite all of the hardware and software you have to work with, the primary goal is to communicate with the listener. The control board is simply a tool used in the creative communication process.

Broken down into their modular components, audio control consoles are easily understood since they all do basically the same thing. In fact, once you understand the basic theory of operation you will have the skills necessary to sit behind just about any control board and figure out how it operates.

Operational Theory

Understanding the Different Sections of a Console

An audio control surface performs three basic functions—amplification, routing, and mixing. Audio first enters the input section of the control board from microphones, playback devices, or other sources at either a mic or line level. The first stage of the control board *amplifies* the various audio inputs to a standardized level for use within the control board.

Using *routing*, or assignment switches, the operator determines the audio's path through the control board. The audio is eventually routed to the output section, where it is sent to digital workstations, audio processors, broadcast transmitters, or the Internet. At the same time, the audio is routed to the monitor section of the control board so that operators and talent in the control room or studio can hear the project on monitor speakers or headphones.

Mixing is the process of using the various volume controls to balance or blend the audio from the numerous sources and achieve the sound the production person or producer desires. Mixing occurs at several stages in the control board and is finalized in the master section with the master output volume controls. Mixing also involves adjusting playback levels in the monitor section, which feeds the monitor speakers, headphones, and intercoms in both the control room and the studios connected to it.

Breaking down a console's functions into amplification, routing, and mixing makes it much easier to understand. These three functions occur within the input, master, output, and monitoring sections of the console.

Types of Audio Control Consoles

There are five popular types of audio control boards or consoles. On-air and production consoles are two different types found in radio stations. Virtual consoles exist within digital audio workstations. Portable mixers are used for remote broadcasts and sporting-event coverage. Although less common in radio production, large-format mixing desks are found in recording studios and in production houses that specialize in custom audio production for the media. Regardless of the console type, they all operate on the same universal principles of amplifying, routing, and mixing across the input, master, output, and monitoring sections of the console.

A brief overview is provided here for each type of console as a starting point in order to describe the functionality of each type—a more detailed discussion will follow. A console's functionality is based on use. For example, radio station control consoles come in two variations that are as different as night and day—on-air consoles and production consoles.

On-Air Console Overview. On-air consoles are designed for controlling a number of audio sources, including microphones, audio storage systems for commercials and music, news/traffic/weather networks, phone lines, remote broadcasts, and so forth.

An on-air console's only real role is to route and control the volume levels of all the sources, eventually sending them to the radio transmitter and the station's web stream. The key design element of an on-air console is simplicity and ease of operation under pressure. In effect, the console designer tries to prevent operator errors with the ergonomic layout and design of the console. The true role of an on-air console is not creativity, but control.

Production Console Overview. A production console is just the opposite of the on-air console; it is designed for creativity. Production consoles have many more features and signal-routing abilities. These include individual channel equalization, the incorporation of outboard gear such as digital audio workstations, and the ability to mix audio in three dimensions, including width, depth, and height. Naturally, to be able to perform all of these functions a production console has more controls and metering.

Analog and Digital Consoles Overview. On-air and production consoles can be either analog or digital, the differences being that in a digital console all inputs are from digital sources; the audio remains in a digital format in the console, and the console's digital outputs are fed to digital recorders, audio processors, or transmission equipment.

Digital consoles are available in two different physical styles. The first is a console that appears to be a traditional analog console with the same simple and familiar knobs and controls. Each knob and fader has an assigned function, or knob per function (see figure 4.1). The only difference is that the console's internal components and the audio moving through the console are digital. This type of digital console requires no retraining on the part of the operator. In fact, the operator might not even know he or she is using a digital console. This type of digital console is often found as an on-air console and is simple and easy to operate.

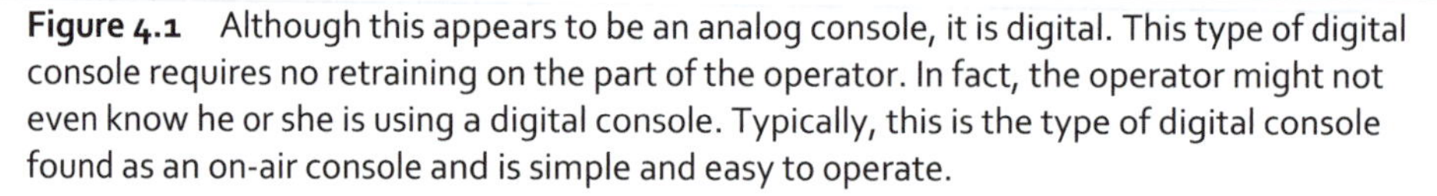

Figure 4.1 Although this appears to be an analog console, it is digital. This type of digital console requires no retraining on the part of the operator. In fact, the operator might not even know he or she is using a digital console. Typically, this is the type of digital console found as an on-air console and is simple and easy to operate.

The second type of digital console is the "any-switch-can-be-anything-any-day" console. Essentially this type of console is a part of an integrated studio audio system. In such a system stand-alone control boards and audio wiring are replaced with digital audio over a local area network (LAN) linking all of the audio consoles and equipment together. The integrated system provides a seamless digital audio workflow from the microphone to the transmitter.

The physical control board can be thought of as a giant "mouse." The central computer in the system is used to set up each console and assign the input and output functions to each of the console's switches and faders. The channel assignments, switching setup, and functions can be changed at any time, using the central computer. Literally, any input, output, or function can be assigned to any channel or control; it's like moving checkers on a checkerboard (see figure 4.2).

In some situations the station engineer sets up, or programs, the console to best serve a particular format, rearranging the console's configuration to suit the particular format's needs. The engineer saves the configuration in the console's memory under the format or show's name. At any time an operator can recall the settings and reset the console to a particular show format. For example, a sports

Figure 4.2 A digital console can be set up with a number of custom layout configurations for various formats or shows. The console can easily change from a music show layout to that of a talk show layout simply by calling up the particular console configuration from the console's memory. This type of a system provides maximum flexibility in studio use.

talk show might follow a music show in the same studio and use a different console configuration. The flexibility afforded by an integrated studio audio system can make the most efficient use of a station's studios.

On-Air Consoles

Most radio stations either have or are in the process of installing digital audio consoles. Digital consoles are needed to remain competitive in the marketplace and to meet consumer demands for high-quality audio. Additionally, an integrated studio audio system streamlines workflow allowing a station to be much more efficient. On-air consoles are generally custom-ordered to meet the station's specific control room needs or configured as a part of an integrated studio audio system. For stand-alone consoles this is easy to do since most consoles are made up of a number of different modules and individual channel strips and can be ordered in just about any configuration a station engineer desires. Prices for the more traditional stand alone on-air consoles, such as the example in this chapter, start in the $10,000 range and rise quickly, depending on the features that are ordered. Most radio stations operate in what is termed the **combo mode**, meaning that the on-air

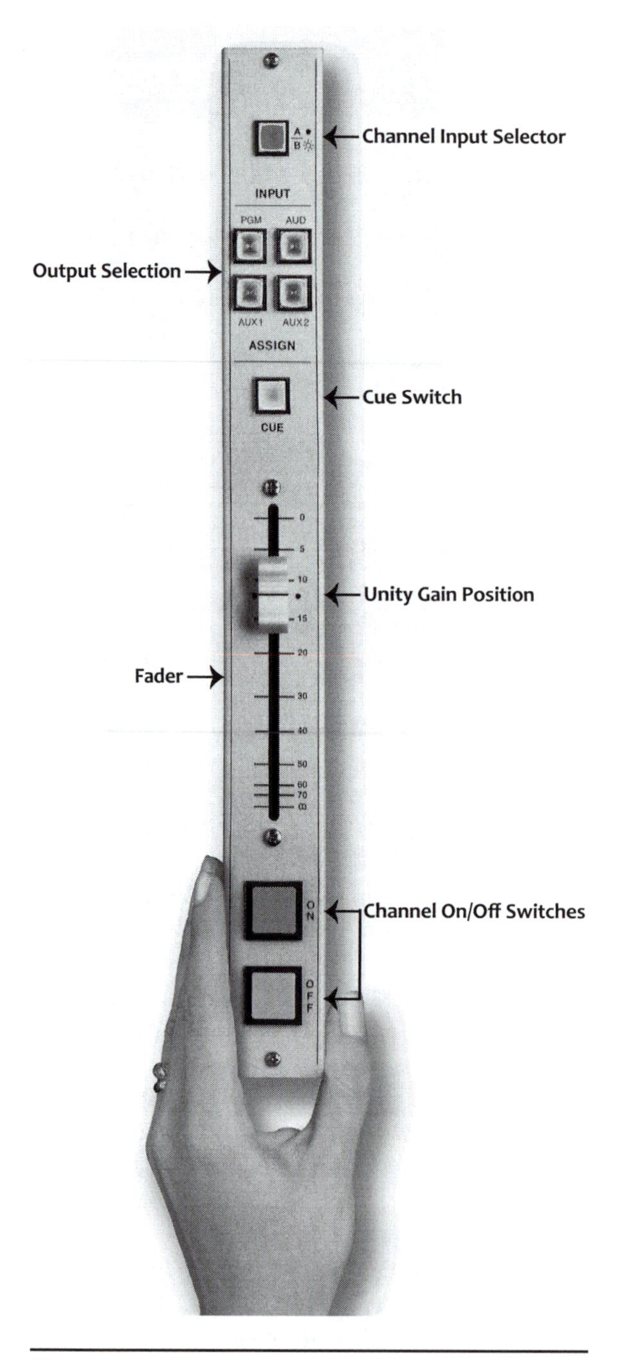

Figure 4.3 The easiest way to understand a console is to break it down into components, looking at just one channel strip at a time, top to bottom.

talent serves as their own producer/engineer, running the control board and other elements of their show.

The easiest way to understand a console is to break it down into components, looking at just one channel strip at a time, top to bottom. The digital on-air console chosen as an example for this chapter has the appearance of an analog console (see figure 4.1). It has 12 identical channel strips; if you know how to operate one, you know how to operate all 12, and if you know how to operate 12, you can operate 24.

Selecting the Channel Input

At the top of the example channel strip shown in figure 4.3 is the channel input selector (input section). This switch lets you choose between the two inputs feeding this particular channel, labeled A and B from the factory. For clarity and ease of operation, most stations have custom labels made indicating the actual audio source, such as the server for music playback. Depending on the model, some console channels have as many as four different inputs to select from. After choosing an input to the channel, the next step is to use the output section of the channel strip and assign where you want the channel's audio to go after the level has been adjusted with the fader.

The Output or Assignment Section

Traditionally, on-air consoles have two main **buses** (common outputs) to which audio is assigned or routed—program and audition. The program bus generally feeds audio to the radio station transmitter and many times to the station's web stream. Any channel with the program button depressed is routed to the transmitter. The audition bus is an entirely separate circuit and is used to preview, or audition, audio on the studio monitor speakers without putting the audio on the air. The audition bus can also feed a digital audio recorder, so that while the program bus is feeding the transmitter, the audition bus can be used to record a telephone call or other program elements.

The channel strip in figure 4.3 has two additional buses that audio can be assigned to, auxiliary 1 and 2. The outputs of these two buses can be sent anywhere the operator needs the additional audio feeds. For example, the auxiliary (aux) 1 bus might feed the telephone system so that talk show callers can hear whatever console channels are assigned to that bus. The second auxiliary bus could be used to feed a station's **interruptible foldback** (**IFB**) system so that the on-air talent can talk to, or feed, an on-air signal to the remote talent's headphones during a live remote or sports broadcast. With the example channel strip, it is possible to route the audio signal from the channel to four separate places simultaneously, if necessary.

Listening In Cue

Continuing down the channel strip, the next button is the cue button. Cue is an isolated bus routed to a separate amplifier and cue speaker so that you can listen to a channel and preview music or check microphones before you put them on the air. When a channel is placed in cue, its signal cannot go over the air; cue is a part of the monitoring section of the console. Most cue speakers are small, crummy-sounding speakers placed in an odd location in the control room, such as behind the operator or under the desk the console is sitting on. The cue speaker might also be mounted in the control console. If an operator makes a mistake and inadvertently leaves a channel in cue that is supposed to go on the air, the odd sound from the tiny, crummy cue speaker alerts him or her to the error.

Consoles do not always have a cue button; instead, they may have click-cue faders. Click-cue faders, when brought all the way down, go one more notch and snap, or click, into the cue circuit.

Faders and Level Control

Often, instead of faders or level controls, the term **pot** is used. Pot is the abbreviation for the word potentiometer, the technical term for a volume control. Other terms for the volume control include attenuator, gain control, or just plain volume control.

There are basically two types of pots, rotary and faders. Rotary pots are round volume knobs that are turned in a clockwise/counterclockwise manner to raise and lower the volume. Faders are volume controls that require more space to operate, sliding a knob along a short linear scale. They are sometimes referred to as linear faders.

Looking back at the channel strip in figure 4.3, the fader slides along a linear scale to raise and lower the channel's audio volume. At about three-quarters up the scale there is a simple dot or line with no numbers beside it. This mark is the **unity gain** position. Unity gain is an optimum audio level near 100 percent modulation and is preset by the station's engineer. This makes it very easy to run audio levels on the console. Later in the chapter, metering is discussed in detail.

Below the fader are the channel on/off switches. The switches can also be used to remotely trigger start/on and stop/off devices such as a telephone recorder or a digital playback source.

Monitoring Your Work

The monitor section controls what you hear in the control room monitor speakers and headphones. Monitoring levels should be set to a comfortable level in the control room. Something you will learn quickly is that, as an announcer, you will not have any time to stand and listen to music; you will be too busy producing a radio show. Beginning radio announcers are surprised to find out how little music they listen to or even remember hearing during their show. A simple rule of thumb is to listen to the monitors at an "office" level. You want to hear what is on the air, but you do not want it to dominate your control room since you are going to be doing other things in the control room while the music is playing, such as taking phone calls, posting things on social media, and talking to other staff coming and going from the control room.

Using our example on-air console in figure 4.1, there are multiple monitor inputs to select from on the monitor channel strips located on the right side of the console. In addition to the program, audition, and auxiliary buses 1 and 2, often there is a cue and an external input. The external input is for an off-the-air monitor, referred to as an off-air signal. Any one of these sources can be listened to in the control room monitor speakers or headphones without affecting what is going on the air. Simply select the source and turn up the monitor speaker or headphone volume.

On-air talent use headphones to monitor with when they are talking on the air. The reason for this is that the moment the on-air microphone is turned on, the monitor speakers are muted (turned off) by the control board to prevent the **feedback** (the deafening howling sound that public-address systems sometimes produce) that will occur if the sound coming out of the monitor speakers is picked up by the control room microphone, fed back through the control board and broadcast transmitter, and back out of the monitor speakers into the microphone again in a vicious, screeching, never-ending circle.

Automatic muting of the monitor speakers is a universal feature of broadcast control boards. Production control boards do not always have this feature, therefore the operator has to turn the monitor speakers down manually before using the production studio microphone.

A common beginner's error is listening to the program bus (what is coming out of the control board going to the transmitter) and not the actual air signal after it has been broadcast. When working in an on-air control room, your default monitor is always the actual air signal so that you are the first to know if there is a signal-quality problem or the station goes off the air. Many an operator has been embarrassed beyond belief when a station manager or chief engineer frantically entered the control room and the operator did not know the station was off the air. A station can go off the air for any number of reasons, including power outages, storms, lightning strikes, and so on. The point is, in an on-air control room, you always listen to the off-the-air monitor.

Lastly, on the right side of the example on-air console, just before the monitor strips, is a telephone caller section. The operator can control the audio levels and monitor multiple telephone callers at once, which is ideal for talk shows or taking requests.

An additional feature of many consoles is an assignment switch for the console metering so that the meters can display any one of the audio buses, such as program or audition.

On-Air Operational Tips

Control consoles are pretty straightforward with regard to their functionality; they are all basically the same. The key to being able to handle any console is not to learn just the knobs and faders of one console but to understand the overall operational principles that have been discussed in this chapter, which you can then apply to all consoles. As an operator you are going to have to learn to "walk and chew gum at the same time" so that you can talk on the air and operate the console without really thinking about it. Given a little experience, operating a console is like riding a bicycle; they all have pedals and handlebars. The trick to smooth operation often lies in simple things that can make the job of the operator easier.

The first rule of working in a studio is never, ever, take drinks, food, or lighted cigarettes into a control room. Sugar-saturated liquids and electricity do not mix. A soda spilled into a control board can cause thousands of dollars worth of damage and take the station off the air, costing thousands more in lost advertising revenue. Food is a real problem as crumbs find their way into equipment and greasy, sticky fingers leave fingerprints on control boards, computer keyboards, and computer touch screens. Cigarette smoke leaves a film on switch contacts and sensitive digital equipment, eventually causing equipment failure.

Although you might not get the final say in a console's layout, the operator's perspective is important. Talk to your station's operations director or engineer. From an operational standpoint, similar console inputs should be grouped together and logically arranged to assist the operator. We read left to right, so it is not unusual for a console to be laid out from left to right. For example, what is the most important input on the console? Your microphone is usually on pot 1. Any other microphones in the studio or adjoining studios follow it. From this point on it is up to the engineer and the operations staff as to how the control board is laid out.

The simplest consoles to operate are those that have been laid out ergonomically, or in relation to the studio and operator. For example, if the music playback computer monitor is located on the left side of the console, locate the music playback pots on the left side of the console, too. If the screen and computer keyboard for commercial playback is located to the right, locate the pots for these on the right side of the console as well. Make sure your console is clearly marked as to the function of each channel strip, input, and output. The marking should be clear enough that someone who has never been in your studio can read the labels and figure out what is where on the console. This makes training new operators easier and makes it easier for part-time operators who may not have learned all of the "oh yeah, that's on that pot" locations.

As a side note, never mark or doodle on a console. Consoles are expensive. If you want to see an engineer get ugly, write on his or her console. We laughed pretty hard one day when our engineer caught an operator writing on a console and returned to the studio to write his name across the operator's forehead with a permanent marker. The operator got the message.

With regard to accepted operating procedures, follow some simple suggestions to avoid problems. Leave all of the pots you are not using closed, or down. This eliminates the chance of something you don't want to go on the air making it there, and it also reduces any noise that may be present in the source. When opening a mic, turn the mic on and then raise the level on the pot. Likewise, when closing a mic, lower the level and then turn the pot off. This avoids the noise of the on/off switch getting on air and makes a smoother transition from the music to the acoustic environment of the studio.

One of the biggest mistakes new operators and production people make is relying on their ears to tell how loud or soft a signal is. Studio monitor speakers *do not* allow you to judge audio levels accurately, since the speaker's volume can be adjusted up and down to compensate for low- or high-level audio. Always rely on the console meters to judge audio levels (see figure 4.4). Metering is discussed in depth later in this chapter.

Most consoles are set up by the station engineer to run the pots at unity gain, or the three-quarters open position. At these settings you should be seeing a reading of 80 to 100 VU (volume units) on the console's meters. As an operator, it is *your* responsibility to control the audio levels from the control board. Set too low, your station sounds weak on the air; too hot, or loud, and your station is distorted on the air. Monitor your levels in accordance with your engineer's instructions. The term for watching the levels and adjusting levels or gain as necessary is called "riding the gain."

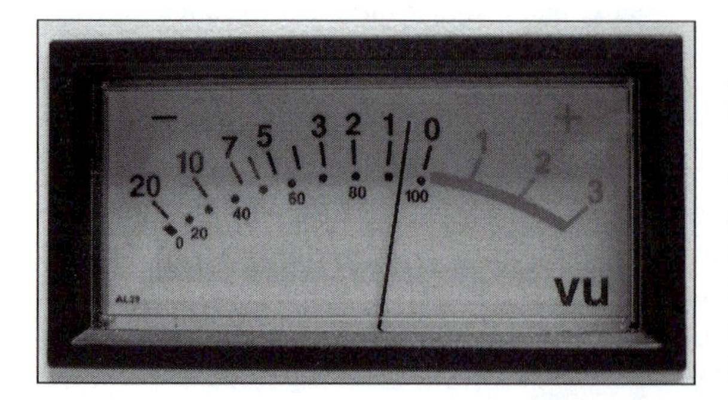

Figure 4.4 As an operator, it is *your* responsibility to control the audio levels from the control board. Set too low, your station sounds weak on the air; too hot, or loud, and the station is distorted. Monitor your levels in accordance with your engineer's instructions. Typical readings should average 80 to 100 VU.

Production Consoles

An on-air console is primarily designed for the control of playback levels. A production console is just the opposite; it is designed for creativity and manipulation of the audio in three dimensions within the stereo sound image.

The three dimensions of sound include height, width, and depth. Using frequency equalization, or EQ, creates *height,* or a position for a sound either above or below various frequencies, sometimes referred to as carving a hole (see audio demo, cut 4-1). Using the panoramic control (pan) and panning sound sources from left to right, or placing them at a given point within the sound image, creates *width* (see audio demo, cut 4-2). Reverbs and delays create a feeling of *depth* to sound, positioning it either at the forefront, close to the listener, or in the distance, depending on the timing of the effect (see audio demo, cut 4-3).

Production consoles vary greatly in price depending on the number of channels and the features offered. Consoles start at around $1,000, with prices going up proportionally for the various makes and models based on the number of faders and features.

For our purposes in this chapter, our example is a production console with 16 faders and is available off the shelf, meaning that one can be ordered from a dealer and you can get delivery in a couple of weeks or sooner. Custom-built recording consoles with over 32 faders are classed as large mixing desks and are not practical in a radio station due to their size and they can easily cost over $150,000 (see figure 4.5). Generally, such large consoles are only found in recording studios or custom production houses.

Figure 4.5 Large mixing desks are usually custom designed for a particular studio and are out of reach for most radio stations, as they can easily cost over $150,000 or more. This particular console has 32 faders and 96 digital inputs. Generally, such large consoles are only found in recording studios or custom production houses.

However, as noted before, a high price does not necessarily mean the best console for a particular job. The needs of the job must be matched to the functionality of the console. The example production console in this chapter can meet the needs of a radio station or small production/postproduction company and costs less than $1,500 (see figure 4.6).

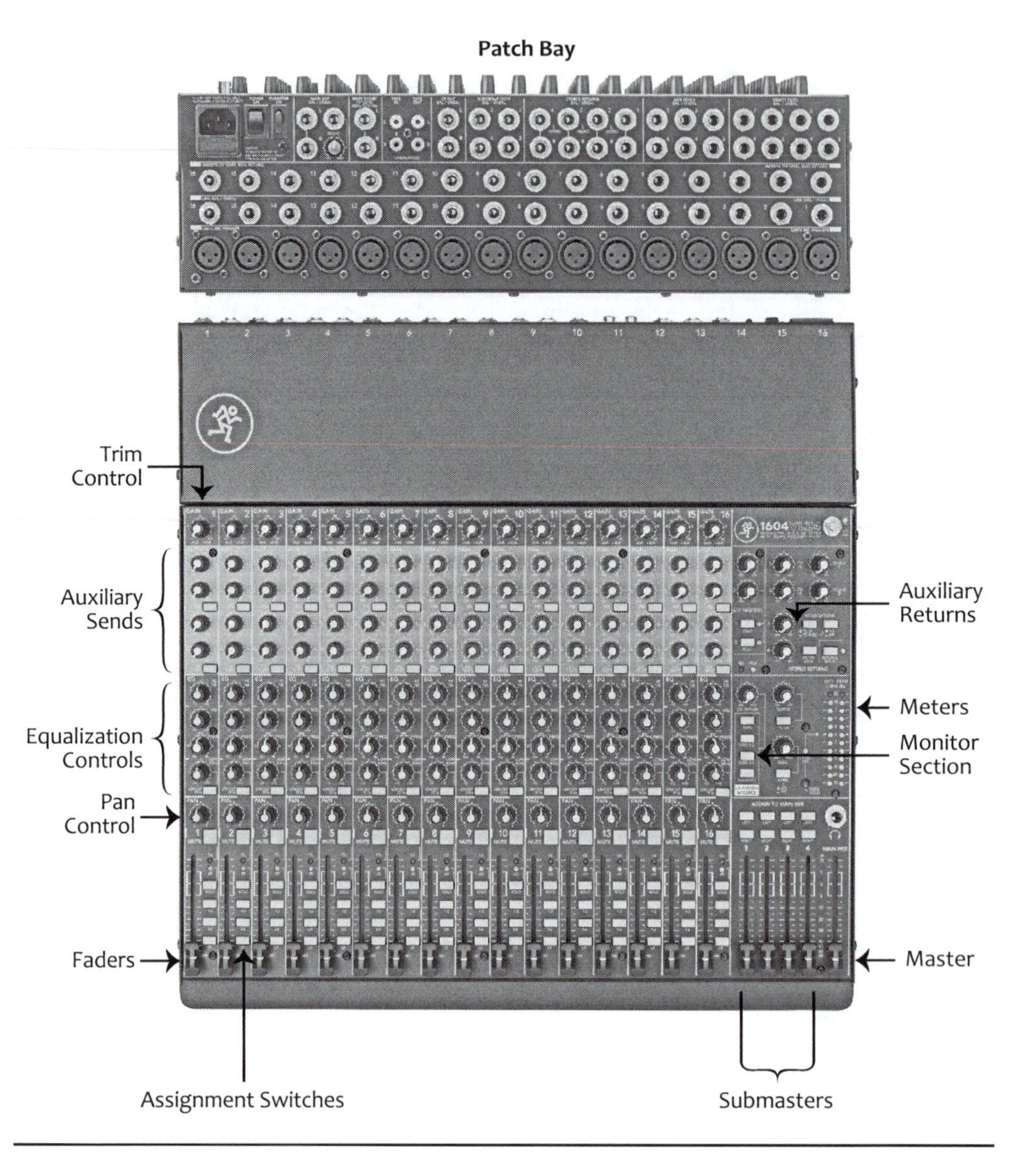

Figure 4.6 An excellent production console does not have to be expensive. This 16-channel console costs less than $1,500 and offers a number of professional features. It is quite popular with radio stations and small production/postproduction companies.

The In-Line Layout

Production consoles are typically in-line consoles. Each channel strip contains input and output routing, monitoring, equalization, and master output section controls. Each strip has individual inputs and outputs and can operate somewhat independently of the rest of the console. For example, a microphone can be plugged into a channel, and a direct output of that channel can be sent straight to a digital recorder, without the channel strip ever being routed to the master output section of the console. About the only limitation to a production console's flexibility is your imagination and creativity in routing signals.

The number of input channels, number of submasters, and number of master outputs is used to identify the production capability of a console. A 16 × 4 × 2 console has 16 input channels, 4 submasters, and 2 master outputs for stereo. Submasters are used to group together a number of input channels so you can control them with one fader or a stereo fader pair.

Although models and name descriptions vary among manufacturers, most production consoles have many of the elements used in this chapter's example.

The Input Section

Patch Bays and Routing Switchers: Getting Audio from Here to There. A patch bay is a place to patch together, or route, signals in and out of a system. Patch bays are comprised of a jack field (female connectors) with system inputs and outputs on the jacks. Using patch cords to connect the jacks together, the operator can route audio through the system easily. Often a patch bay is mounted in an equipment rack in the production room or in the engineering area of a station (see figure 4.7 on the following page).

A routing switcher is a computer-based electronic patch bay to which all of the audio inputs and outputs have been wired and in which they have been assigned an input or output number. The routing switcher is located in an equipment rack in the engineering area of the radio station, with remote terminals located at each of the on-air and production control boards (see figure 4.8 on the next page). Using the remote terminals connected to the routing switcher, you can route audio into your control board by simply entering the source's number on the remote terminal and pressing the "take" button.

In the case of the example production console, it has a built-in patch bay on the back for convenience (again, see figure 4.6). Either mic or line-level sources can be plugged into the console through XLR or TRS connectors. The console's XLR connectors have 48-volt phantom power available that can be turned on and off, as needed, for condenser microphones.

Input Gain Staging. After a source has been plugged into the console, the first step in console setup is to gain stage the input. Before a signal reaches the input of the console preamplifier it typically passes through a trim, or gain control. The gain control is usually at the top of the channel strip and is used to match the input level of the mixer to the output level of the source. Setting the gain level is

Figure 4.7 A patch bay is a place to patch together, or route, signals in and out of a system. Patch bays are comprised of a jack field (female connectors) with system inputs and outputs on the jacks. Patch cords connect the jacks together, allowing the operator to route audio through the system easily.

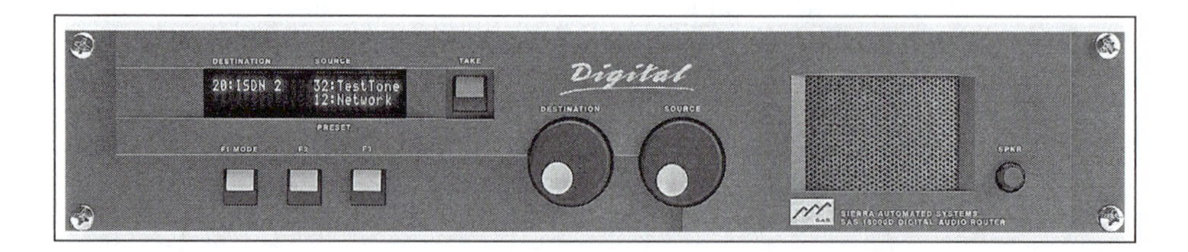

Figure 4.8 This digital routing switcher control head makes it simple to route audio in a radio station. Using the Source knob, you dial the input you want to route somewhere. Using the Destination knob, you dial the destination where you want the audio to go to. Then press the Take button and the connection is made.

very important in getting the best signal-to-noise ratio and so that you do not overload the console input and cause distortion. On the example production console, an overload light (OL) next to the channel strip's fader comes on if the input to the channel is too high and the channel begins **hard clipping**. This is an immediate indication that you need to lower the input gain level. Some consoles have a switchable pad instead of a gain control to boost or lower the audio input in fixed steps.

Channel Assignment and Routing

On the right side of each fader are the channel's routing, or assignment, switches that control where the audio from the channel goes. On the example console, the choices are to send the channel's audio to the left/right channels of the main mix, to submasters 1 and 2, or to submasters 3 and 4. For consistency in standard practices, odd-numbered submaster channels are usually left channel and even-numbered submaster channels are usually right channel in a stereo mix. For example, submasters 1 and 2 are a stereo pair of left (1) and right (2) channels. Likewise, submasters 3 and 4 are a stereo pair of left (3) and right (4) channels.

Above the channel assignment switches is another assignment switch, labeled solo. A solo assignment switch works independently of the other assignment switches. When a channel is placed in solo, all of the audio from the other channels is muted. The solo channel becomes the only channel that can be heard in the headphones or control room monitor speakers. Solo is handy for critically checking a channel or channels of audio.

Above the fader is the mute button, which is not an assignment switch per se. A mute button does the same thing as lowering the channel fader to zero: it silences the output of the channel, regardless of how it is assigned or routed.

Controlling Audio Levels

Level control is accomplished with a linear, or sliding, fader. When the fader is drawn to the bottom of the linear scale, the fader is closed and does not allow any audio to pass from the channel strip. As with on-air consoles, there is a unity gain position marked at about three-quarters of the way up the linear scale on the fader. Provided the input gain on the channel has been properly adjusted, this should indicate the optimum level to work with. If the input level is too low, you may hear noise creeping into the signal due to the poor signal-to-noise ratio. If the level is too high, there's a good chance of hard clipping and audio distortion. The overload light next to the fader on the channel strip comes on if the channel begins hard clipping. In both cases, the input gain level needs to be adjusted (either up or down). It is best to follow a specific console manufacturer's guidelines for setting optimal trim or gain levels.

Panoramic Control

The **pan (panoramic) control** permits the sonic positioning of a channel between the left and right stereo channels and works in conjunction with the left/right and other assignment switches. For example, if the left/right assignment switch is chosen to send the channel to the main stereo output, then turning the pan control to the far left or right places the audio on the left or right output channel. The same applies to the submaster 1 and 2 and submaster 3 and 4 assignment switches. Leaving the pan control in the center positions the sound equally between the left and right stereo channels, creating a mono, or single channel, feed. By carefully listening to the monitors, you can literally move a sound to any position located between the left and right speakers.

Shaping the Sound: Equalization

Usually, the EQ section of the channel strip is located above the pan pot. Equalization is the electronic process of altering a signal's frequency response by boosting or cutting certain frequencies (sophisticated bass and treble controls). Most consoles have equalization that is divided into three frequency bands: low, mid, and high. The example production console provides up to 15 decibels of boost or cut at 80 Hz for the low band, 15 decibels of boost or cut at 12 kHz for the high band, and a sweepable (adjustable) midrange. Since the human ear is most sensitive to midrange frequencies, the midrange on the example console is adjustable from 100 Hz to 8 kHz and has 15 decibels of boost or cut at the selected frequency.

Briefly, EQ is used to sweeten or correct minor problems with a mix and is covered in depth in chapter 6. Properly used, EQ can make a valuable contribution to a recording; improperly used, EQ can easily ruin a recording or project. Unless you have had specific training in equalizing audio, you would be wise to leave these controls set at their center, or neutral, positions.

Audio Filtering

In addition to EQ, many consoles have **high-pass filters** at around 75 Hz to eliminate very low frequency rumble and noise that may be present in a recording situation. High-pass filters allow only those frequencies *higher* than a specific frequency to pass the filter. Likewise, a **low-pass filter** is one at the high end of the frequency spectrum that allows only frequencies *lower* than a specific frequency to pass. On our example production console, the manufacturer has chosen to simplify these often confusing terms by calling the filter that cuts everything below a specific frequency a low-cut filter. The 75 Hz filter is located just above the pan control on the example console.

Unless you are recording a pipe organ, kick drum, bass guitar, synthesizer, or building rumble, or trying to record an earthquake, there are really not many sounds below 75 Hz that contribute anything musically. The advantage to using a high-pass filter on sources other than those just listed is that when you do want to boost the low end it sounds much cleaner and tighter. This is because the high-pass filter has already eliminated the muddy sounding low-end rumble, leaving just cleanly recorded bass to work with.

Auxiliary Sends and Returns: Connecting Outboard Gear

Above the EQ controls are auxiliary (aux) send controls. An auxiliary send is an additional output from the individual channel used to send the channel's audio to an outboard signal processor or recorder, or to be used for a separate mix. An aux send taps into the audio on the channel and feeds the like-numbered auxiliary bus output. In the case of the example, each channel has six aux-send buses and is capable of sending the channel's audio to six different locations outside the mixer.

Naturally, if you use the aux send to route audio to an outboard signal processor, such as a reverb, you need a way to get the processed signal, with the reverb, back to include in the mix. This is accomplished using an aux return input to the console and the aux return level control (located on the upper right side of the example production console) to assign the aux return to the main mix (again, see figure 4.6). An unaltered audio signal being sent to an outboard processor is called a **dry signal**, and the returned signal, with the reverb or effect added, is called a **wet signal**.

There is one other feature of the aux sends that adds to their functionality. Aux sends have the option of tapping into the channel's audio before or after the fader. In the pre-mode, an aux send taps into the audio before the fader, EQ, pan, and assignment switches. This provides a signal straight out of the preamp, which the console operation cannot affect in any way. A post-aux send taps into the audio after the fader and before the pan and assignment switches. This provides a signal over which the fader has level control. The preselect switch is located below the aux 2 send on each channel strip.

Metering: Being Able to See Audio

Why You Must Watch the Meter. It was mentioned earlier in the chapter, but it's worth bringing up again: one of the biggest mistakes new operators and production people make is to rely on their ears to tell how loud or soft a signal is. Just as pilots are taught to rely on their instruments, you have to learn to rely on the audio meters to verify actual audio levels. Otherwise, you will crash and burn like a pilot in a storm. Production studio monitor speakers *do not, cannot,* and *will not* allow you to judge audio levels accurately, since the speaker's volume can be adjusted up and down to compensate for low- or high-level audio signals.

Many a production person, with speakers blaring, has said "it sounds fine to me," only to be asked by the production director to look at the control board meters, where the audio level is barely registering. Proper audio levels are adjusted by *FIRST* observing the meters, setting the audio level, and then adjusting the monitor speakers to a comfortable listening level. Always refer to the meters to observe or set levels. Failure to rely on the audio meters has ruined many a commercial and program.

VU Meters. Metering the audio in any console is a challenge. The problem lies in developing a mechanical (or other) device to represent accurately the audio, which is constantly varying as a quantity. In 1939, during the early days of radio, Bell Labs, NBC, and CBS agreed on a standard method to measure audio levels called the VU, or volume unit. The volume unit is an electrical measurement to indicate audio level in relation to human hearing and how we perceive the loudness of a sound. A VU meter is essentially a calibrated voltage meter through which the audio passes, with a mechanical needle moving back and forth to indicate the voltage or audio level.

The meter has two scales; the upper scale shows volume units, and the lower scale is the percent of modulation (see figure 4.9). The upper scale has a range of units from –20 VU, through 0, to +3 VU. Typically, the meter's scale above or beyond 0 VU is displayed in red. The red area of the meter indicates levels that are likely to cause distortion and clipping.

The word modulation means change. Modulation is measured in percentages; it represents the percent change in the audio signal. Zero percent modulation is no change, or no audio. In theory, 100 percent modulation is the maximum change that can occur without damage to the audio signal. Beyond 100 percent modulation, the audio distorts at some point; eventually, it hard clips.

The two different scales on the VU meter are calibrated so that –20 VU is equal to 0 percent modulation and 0 VU is equal to 100 percent modulation. Signals that indicate below –15 VU, or about 20 percent modulation, are considered too weak to be of any practical value. At the opposite end of the scale, audio driven past 0 VU, or 100 percent modulation, is into the red and in danger of distorting and clipping. Audio that is driven so hard that it pins the needle to the far right, past +3 VU, is called "gone," in that it consists of nothing but distortion and hard clipping. Aside from the problems this causes in the audio itself, such abuse can make the needle stick or damage the meter. VU meters display *average* audio levels, since it is physically impossible for the meter's needle to track accurately the instantaneous audio level changes in the signal.

To combat transients and instantaneous audio peaks that can surprise anyone, manufacturers build **headroom** into the console. Headroom is the difference between the normal operating levels and the point when distortion (clipping) starts to occur in the console. Because of this safety net, it is OK to allow *occa-*

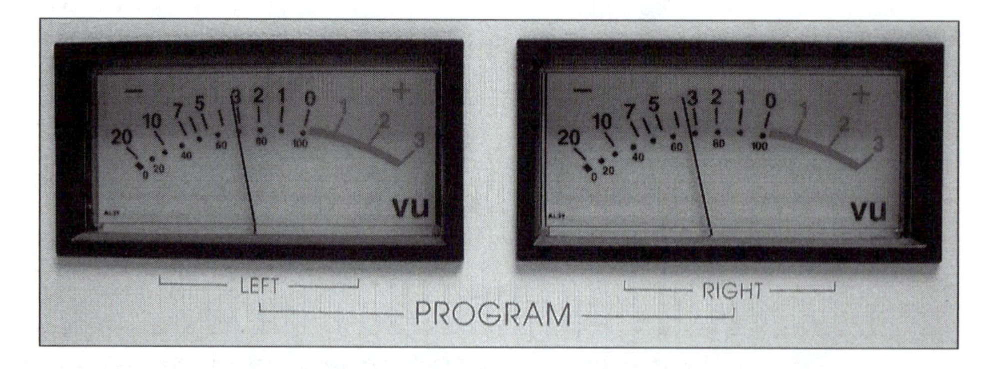

Figure 4.9 A VU meter has two scales—positive and negative. The negative side has two sets of numbers. The larger numbers show volume units from –20 VU to 0. The positive side is from 0 to +3 VU. The smaller numbers on the negative side represent the percent of modulation. At levels above 0 VU, distortion and clipping are likely.

sional peaks into the red area of the VU, since there is some headroom before actual distortion occurs.

A production person has to keep in mind that pushing the limit in a digital recording with a VU meter will certainly get you into trouble. This is because the VU meter only shows *average* levels and cannot physically move fast enough to show instantaneous audio peaks, which can cause the audio to break up. A good guide to follow when recording digital audio using a VU meter is to keep average levels at about –10 VU, to allow enough headroom for transients and instantaneous audio peaks that might occur. Digital recording techniques are discussed in detail in chapter 5.

Peak Meters. VU meters are preferred in the United States. In Europe, the peak program meter, or ppm, is the preferred method of metering audio. Although the VU meter displays only average levels, the peak meter displays the signal peaks, or the absolute maximum signal level present. This is the kind of meter best suited for digital recording, because by indicating the peak signal levels it allows you to avoid digital overloads, or splatter. Unlike the VU meter, the peak meter is calibrated only in decibels (see figure 4.10).

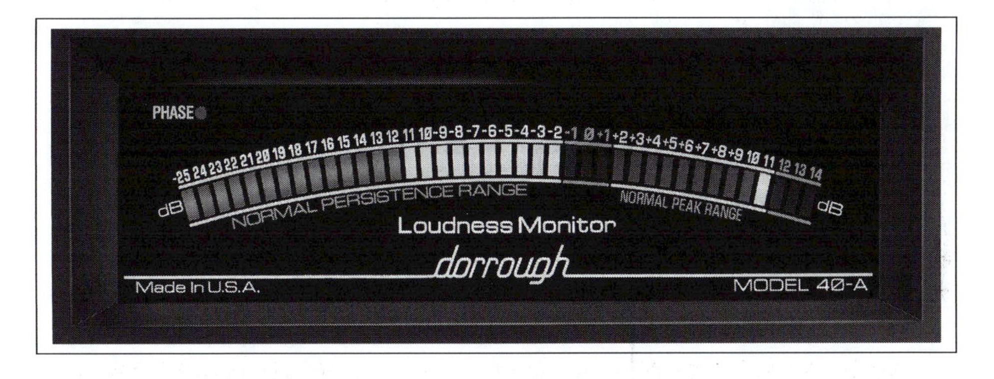

Figure 4.10 A VU meter only displays average audio levels, whereas the peak meter displays signal peaks, or the absolute maximum signal level present. A peak meter is well-suited for digital recording because it indicates the maximum signal level so that digital overloads, or splatter, can be avoided. Unlike the VU meter, the peak meter is calibrated only in decibels.

Electronic Bar-Graph Metering. Any VU or peak program meter that depends on a moving needle to indicate levels is at a distinct physical disadvantage from the very start. It is limited by the laws of physics as to how fast the needle can go in one direction, stop, and instantly reverse. Liquid crystal displays (LCD) and light-emitting diode (LED) displays do away with the bouncing needle; they rely on an electronic display that responds instantly to the changes in the audio level (see figure 4.11).

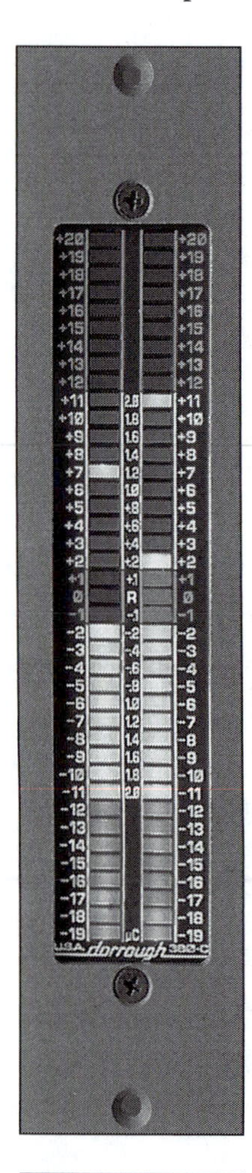

Figure 4.11
A bar-graph LED meter has a row of LEDs arranged in a vertical or horizontal pattern. Unlike the moving needle of a standard VU meter, an LED bar-graph meter can react instantaneously to changes in the audio signal and can be extremely accurate.

An LED meter has a row of LEDs arranged in a vertical or horizontal pattern. As the signal passes through the meter, the LEDs range from green to red. The green LEDs indicate an acceptable audio level and the red LEDs indicate an overload. Each LED is marked on either a VU or peak scale. An LED meter can be quite accurate.

An LCD meter is basically the same as a laptop computer screen. The meter is a graphic display and can indicate VU or peak program levels. Many larger consoles with LCD displays offer a programmable selection of VU, peak program, or both types of metering simultaneously (see figure 4.12).

Hearing Your Work: The Monitor Section

It is critical that you be able to hear your work accurately. The monitor section of a control board provides you the controls to do just that. The one control that can get you into the most trouble is the monitor volume control for either the headphones or monitor speakers.

Many beginners like to work with the monitors cranked (nearly loud enough to tear your clothes off). Big mistake. Learn to monitor your work at lower levels. Your listeners are not listening on studio monitors and certainly not at the levels you can listen at. And although it may seem easier to hear everything in the mix at louder levels, a good rule of thumb is to mix at a low level to make sure that everything in the mix can be heard clearly at that level. Most importantly, the voice needs to stand out above the mix.

As with an on-air console, assignment switches on the monitor section determine what audio source is sent to the monitor speakers and headphones. Using the example production console (again, see figure 4.6), the operator can choose to listen to audio from submaster pairs 1–2 and 3–4, or the main mix. The operator can also choose to monitor the return feed from a digital recorder; in other words, to listen to the audio *after* it has been recorded in real time to ensure it has been properly recorded.

All of these controls are overridden any time the solo switch is depressed, and the selected solo channel, or channels, appears on the monitors exclusive of any other sound.

The Master Section: Controlling the Final Console Output

Mastering is the final stage of audio production. It involves tweaking and making adjustments to the mix so that it sounds

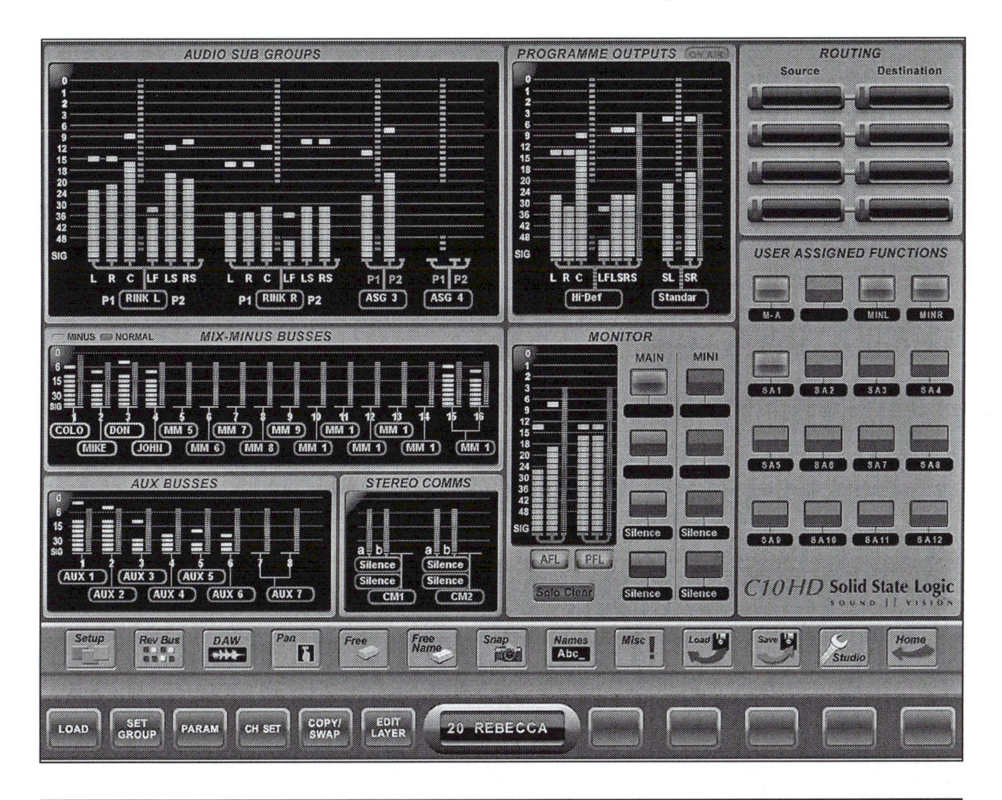

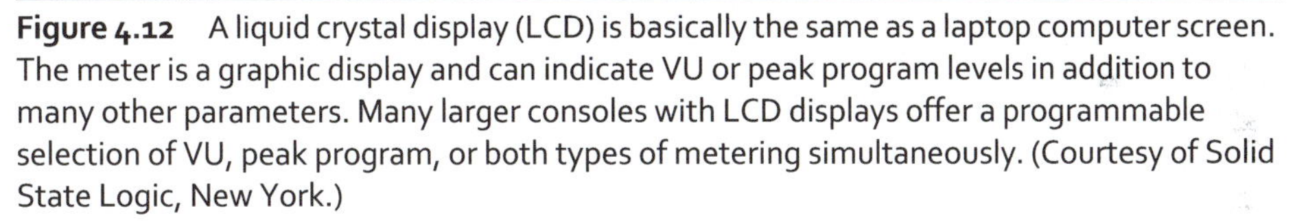
Figure 4.12 A liquid crystal display (LCD) is basically the same as a laptop computer screen. The meter is a graphic display and can indicate VU or peak program levels in addition to many other parameters. Many larger consoles with LCD displays offer a programmable selection of VU, peak program, or both types of metering simultaneously. (Courtesy of Solid State Logic, New York.)

exactly as you, or the client, wants it to. During a mastering session the project is listened to critically. Final editing adjustments are made, the mix levels are fine-tuned, and the left/right stereo balance is adjusted to deliver the overall sound picture you are shooting for.

The master section of the console consists of the submaster and master faders. Submasters are used to group together channels that the operator wants to control with one fader or a stereo pair of faders. As an example, in a recording studio five microphones on a drum set are mixed using five different console channels, one channel for each microphone. All five channels are then assigned to the same submaster stereo pair so that one stereo fader can raise and lower the overall level of the entire drum set. In turn, each submaster is assigned to the right- or left-channel master in a stereo mix.

In the case of the example production console, each of the 16 channels can be assigned to one of the four submasters and each of the four submasters can be assigned to either the left or right master output faders. The master faders control which audio leaves the console and enters the recorder or some other device.

Console Automation

By now, the question may have occurred to you: If there are more than just a few faders on a console, how in the world can one person handle all those faders simultaneously during a mix?

Console automation is an advanced feature that records the physical movements of each channel strip's controls, including the movement of the fader, mute, pan, EQ, filters, solo, and so forth. In the automation write mode, as you adjust each channel's level, a computer records the control board movements you are making and translates them into stored data. If you don't get a perfect mix the first time, you can go back and do it again, rewriting or rerecording the automation for that particular channel over and over until you are happy. In read mode, the automation plays back the audio and all of your recorded control movements that go along with the music.

This "one channel at a time" method of automation is how producers are able to create complex mixes involving numerous tracks of music, sound effects, dialogue, and other audio. There is nothing more incredible than watching a console literally come to life with all of its controls and faders moving on their own during the automated playback of a complicated mix. Automation is also a feature of digital audio workstations that is covered in detail later in chapter 5.

Virtual Control Consoles

Digital audio is nothing more than computer data. As such, it is quite easy for a digital audio workstation (computer) to record, mix, and edit digital audio. Digital audio workstations have virtual audio consoles as a part of their operating software (see figure 4.13). Virtual consoles look and function similarly to their physical counterparts. The primary difference is that a virtual console is controlled by a keyboard, a mouse, touch screen, or an outboard control surface.

An outboard control surface is a physical representation of what is on the computer screen (see figure 4.14). Move a control surface fader up and down and the corresponding fader on the computer screen moves up and down. Some operators prefer to use keyboard shortcuts, others a mouse or trackball, and some prefer the control surface. It is a matter of personal choice. Virtual consoles within digital audio workstations are commonplace in recording and production studios.

A logical extension of a virtual console is to use one to replace most or all of the physical knobs, faders, and switches of a production console. It is not difficult for a computer to emulate all of the functions of a production console (see figure 4.15 on p. 124). The example console shown in figure 4.15 is based upon a tablet that serves as the touch screen control surface of the console and also provides recording, playback, and storage capability for audio. In essence, it is a "console/studio in a box." An added feature is that the tablet can be removed from the console and wirelessly control the console from anywhere in the station!

Figure 4.13 Digital audio workstations use virtual audio consoles as a part of their operating software. Virtual consoles look and function similarly to their physical counterparts. The primary difference is that a virtual console is controlled by a keyboard, a mouse, touch screen, or an outboard control surface. (Adobe product screen shot reprinted with permission from Adobe Systems Incorporated.)

Figure 4.14 An outboard control surface is a physical representation of software controls on the computer screen. Some operators prefer to use keyboard shortcuts, others a mouse or trackball, and some prefer the control surface to run the software. It is a matter of personal choice.

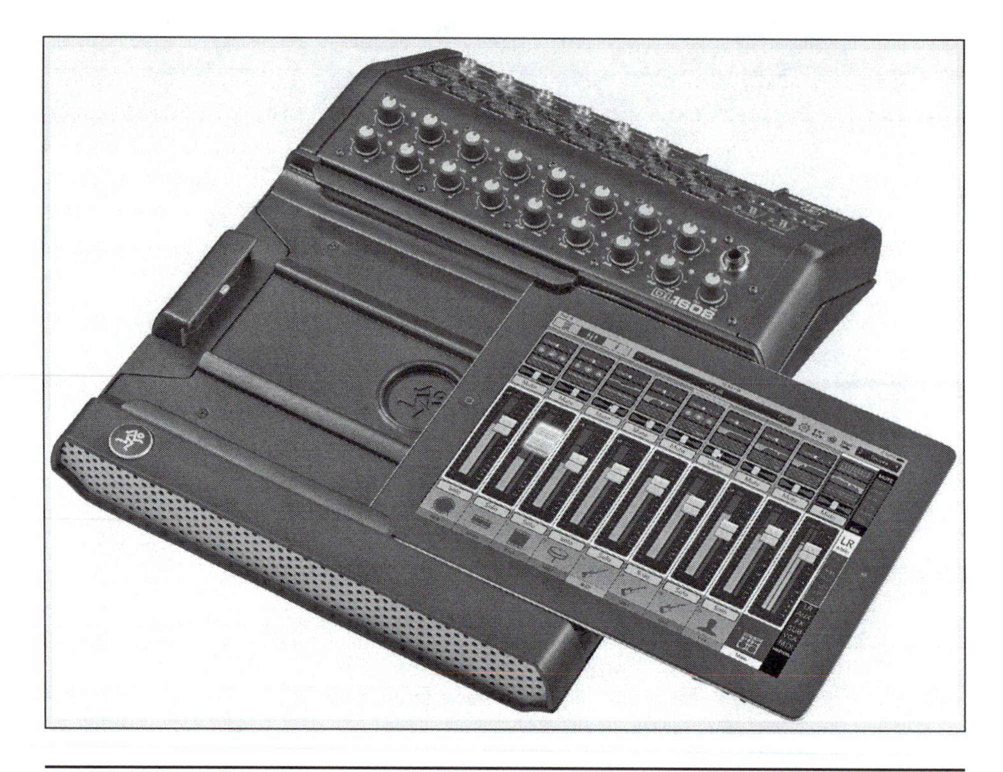

Figure 4.15 This production console is based upon a tablet that serves as the touch screen control surface of the console and also provides recording, playback, and storage capability for audio. In essence, it is a "console/studio in a box." An added feature is that the tablet can be removed from the console and wirelessly control the console from anywhere in the station!

Virtual consoles offer a number of functional, operational, and cost-saving advantages. The example console with tablet can be set up for less than $2,000. The biggest obstacle to the use of virtual consoles is personal preference. Those who are accustomed to a physical control board can be set in their ways and not easily swayed to change. People new to the industry tend to accept virtual consoles and find them easy to use.

Portable Mixers

Overview

Size does not matter. Portable field mixers are the smallest of all of the audio control consoles and, despite their very compact size, offer many of the same features and audio quality as do much larger control consoles. Although often referred to simply as mic mixers, most field mixers offer a choice between mic or line-level inputs. Portable mixers can be either mono or stereo.

Two of the several major advantages portable professional mixers offer are that they are small and that they are built to withstand a lot of hard use. Most are housed in steel or machined aircraft-aluminum enclosures. Portable mixers are able to run for several hours off internal battery power or can be plugged in with a **wall-wart** power supply. On long remote broadcasts, where AC power is not readily available, techs often power portable mixers with deep-cycle marine batteries that are similar to car batteries. Additionally, most professional mixers can provide 48-volt phantom power for condenser microphones.

Portable mixers are used extensively in radio for remote broadcasts from sponsors' businesses, for sports broadcasts, and major breaking news stories. Because the mixers have so much built-in flexibility, about the only limitation to their use is the producer's imagination.

Portable mixers tend to be expensive because of all of their advanced features and their small size. Prices for a quality portable mixer, such as the stereo example shown in figure 4.16, start at about $1,900 and go to over $4,000, depending on the number of inputs and features.

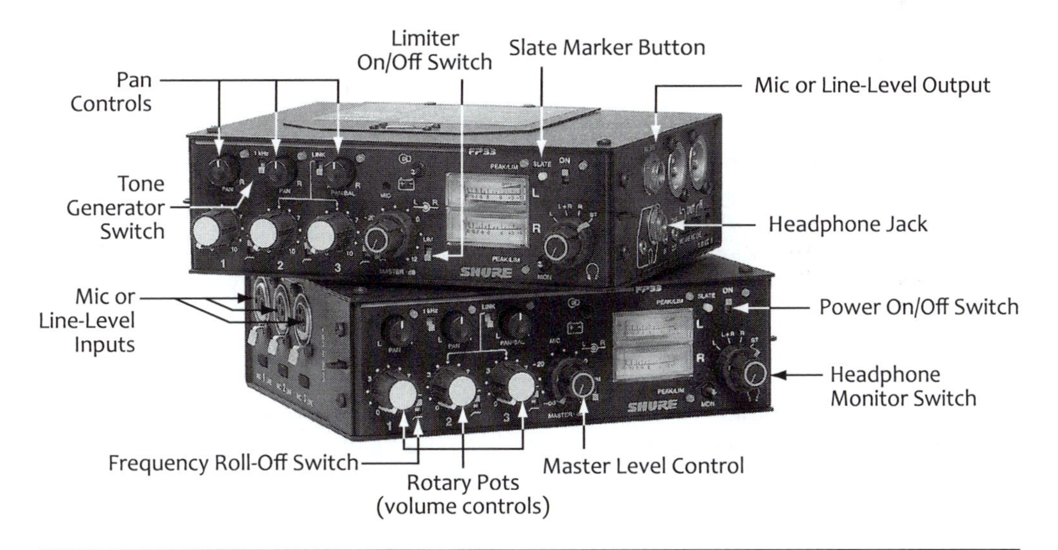

Figure 4.16 Portable mixers are very compact and offer many of the same features found on larger consoles. Mixers are used for station remotes, sportscasts, and for location news coverage. (Copyright Shure Incorporated, used with permission.)

The Portable Mixer Input Stage

Using our example portable mixer (see figure 4.16), audio enters through the switchable mic or line-level-balanced XLR inputs on the left side and is routed to the preamplifier, whose output is controlled by a pot. Although many portable mixers have sliding faders, the most compact portable field mixers use rotary pots to save space.

At the pot you have two sets of choices. The first concerns frequency: either the frequency response of the signal is left flat, or a selector switch rolls off (turns down) the frequencies below 150 Hz. The roll-off switch is located to the lower right of each pot and is marked with a flat line for flat-frequency response and a line that drops off to indicate the roll-off position. The roll-off feature can be particularly useful in a location where there is a lot of low-frequency background noise and rumble you don't want to muddy your show or recording.

The second choice you can make at the pot concerns the sonic placement of the microphone within a stereo image. The pot's pan control, located directly above each pot, allows the pot's input to be assigned to the left channel, to the right channel, or to remain in the center channel for a mono signal. If the three channels offered are not enough mixing capability, two of the mixers can be linked together by a bridging cable that connects the two, providing six channels to mix with.

The Portable Mixer Output Stage

When audio leaves the pot, it is routed to the master output-level controls. To save space and keep the mixer as compact as possible, a knob within a knob, located on either side of the meters, is used to control the left- and right-channel master volumes so the stereo pair can be evenly balanced. The master output levels are routed through the meters and finally to the mixer outputs. The left and right outputs on the right side of the mixer can be switched between balanced mic or line level, depending on what equipment the mixer is feeding.

Additional Features on Portable Mixers

Tone Generator. Other features of portable mixers include a 1 kHz tone generator to send a reference tone to recorders or any other devices connected to the mixer output, so that all of the equipment can be adjusted to operate at the same audio level. On our example (again, see figure 4.16), pulling out the pot 1 control knob activates the tone generator, and the level is adjusted using the pot.

Slate Marker. A **slate** button at the upper right corner of the meter on the example mixer generates a one-second long 400 Hz tone and activates an electret-condenser microphone, called the slate mic, built into the front of the mixer, which feeds directly to the mixer output. The operator can slate the take—in other words, record important identifying information about the take on the recorder for later reference. Another use of a slate mic is to feed an intercom system so that the field producer can talk with the main control room operator during a live remote broadcast.

Limiter. Located to the lower left of the meters on the example mixer is the limiter switch, labeled LIM (again, see figure 4.16). A limiter acts as an automatic volume control to prevent overload. If the audio level suddenly spikes too high, the limiter turns it down to a safe level. This feature is intended to help an operator avoid hard clipping, or distorted audio, due to transients and sudden level changes. The operator has the choice of switching the limiter on or off. Above and below the meters on the right side are PEAK/LIM indicators.

The Portable Mixer Monitoring Stage

Location monitoring is usually done with a pair of headphones plugged into the headphone output (on the right side of the example mixer). The headphones can be switched to monitor the left channel only, right channel only, left plus right channel (mono), or stereo. Normally, a radio station sets up a public address system at live remotes and plays the station's air signal directly from a radio. However, in some unusual circumstances, such as when you are outside your station's signal coverage area, it is customary to feed the mixer output to the public address system so that listeners can hear playback of the broadcast.

Monitors

While you are on air, and during the production process (as we already mentioned in talking about the monitor section of the control board), you have to be able to hear accurately what you are working on. Monitor speakers are the preferred method for studio work, and headphones are the choice for studio talent and location work. For remotes, however, radio stations often use portable monitor/public address speakers.

Monitor Speakers

It has already been established that there is no such thing as the best microphone; likewise, there is no such thing as the best monitor speaker. Just as you try to select the best microphone for a given job, an engineer tries to select the best monitor speaker for the job. The audio chain, including the microphone, preamplifier, control console amplifier, audio processor, digital recorder, and speaker amplifier—even the speaker cable—all depend on the monitor speaker to change electrical energy back into acoustic energy that you can hear as sound. Although you probably will not get to choose the monitor speakers used in your radio station, or where to place them in your studio, you should be familiar with how speakers function so that you are more able to judge your work accurately.

One of the very first pieces of equipment budgeted for when building a studio, and one that often gets a good portion of the budget, is the monitor-speaker system. The speakers are the direct link to your ears. Just as the microphone was critical in capturing the sound, the speaker is even more critical in playing back the sound; it is imperative that you accurately hear the audio you are working with.

Most radio production studios work with a single set of **near-field monitors** or speakers. Near-field speakers are bookshelf-sized and are extremely accurate in their sound reproduction (see figure 4.17 on the next page). They are designed for smaller rooms, where the producer sits closer to the speaker. Some stations also use a set of small, full-range speakers to emulate the "at work" or "in car" lis-

Figure 4.17 This production studio has a single set of near-field monitors or speakers. Near-field speakers are bookshelf-sized and are extremely accurate in their sound reproduction. Near-field monitors are designed for smaller rooms, where the producer sits closer to the speaker.

tener's environment. Such small speakers are often referred to as cubes because the speaker cabinet is shaped like a small cube measuring just a few inches on each side.

In addition, extremely critical production people often take test mixes on a USB flash drive or smart phone to the parking lot and listen to them on several different car stereos. By using different monitor speakers, a production person is able to judge how the end product is going to sound to the consumer.

How Monitor Speakers Work

Chances are you are never going to build a set of speakers, but you should be familiar with how they work and understand the terms associated with them so that you can better judge the accuracy of a speaker. A microphone is the start of the audio chain, and the monitor speaker is at the end of the audio chain. Interestingly, both microphones and monitor speakers are transducers that change one form of energy into another. In the case of a monitor speaker, it converts electrical energy into acoustic energy (see figure 4.18). In simplest terms, a monitor speaker is a giant dynamic microphone working in reverse. Just like microphones, monitor speakers are judged by a set of specifications and how they sound to your ears.

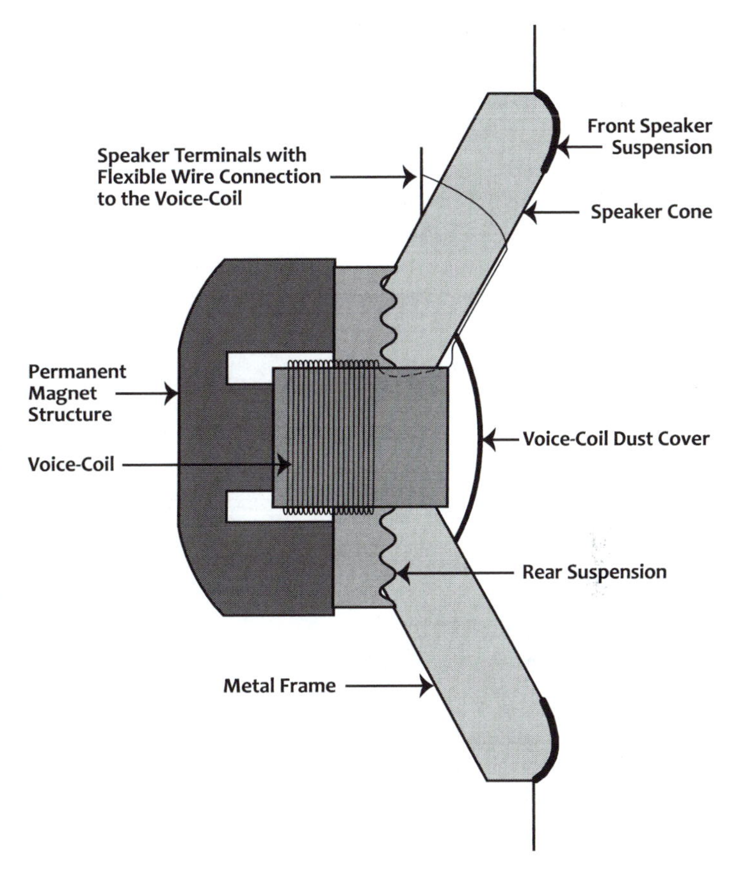

Figure 4.18 Microphones and monitor speakers are both transducers that change one form of energy into another. In the case of a monitor speaker, it converts electrical energy into acoustic energy. In simplest terms, a monitor speaker is a giant dynamic microphone working in reverse.

A speaker typically features a bowl- or dish-shaped speaker cone that comes in various sizes, depending on the application. Attached to the back of the speaker cone is a coil of wire suspended within a large, permanent magnet structure mounted on the back of the speaker. When audio is sent to the speaker, the voltage enters and travels through the speaker coil. When a coil of wire is charged with electricity, it becomes an electromagnet, and when two magnetic fields are placed near one another, one of two things happens—the two either attract one another or they repel one another. As the magnetic field of the speaker coil pushes away from the speaker's permanent magnet, the speaker cone moves forward, setting air molecules in motion and creating compaction. When the speaker cone moves backward, it reduces air pressure, creating rarefaction. These cycles of back-and-forth movement of the speaker cone correspond to analog electrical waveforms and cause the speaker to create sound waves.

Woofers and Tweeters

As you learned in chapter 2, different frequencies have different properties and sound wavelengths. It is very difficult, if not impossible, to design one speaker assembly that can reproduce all of the frequencies equally, or with a flat-frequency response. By dividing the audio frequency spectrum into different bands and directing the audio to different speaker designs within the same cabinet, the frequency response of a monitor speaker can be made much more accurate. The individual speakers within a speaker cabinet are referred to as **drivers**.

The simplest division that can be made in a speaker is between bass and treble, or the low end and the high end of the frequency spectrum. The low end of the frequency spectrum, or bass, requires more power and a larger physical structure to create the long, low-frequency wavelengths. These larger, more powerful speakers are called *woofers*. Practical woofers range in size from 6 to 15 inches in diameter, with permanent-magnet structures that can weigh more than 50 pounds (see figure 4.19).

The high end of the frequency spectrum, or treble, requires much less power and a smaller physical structure to create the high-frequency wavelengths. These smaller, less-powerful speakers are called *tweeters*. Practical tweeters range in size from three-quarters of an inch to 2 inches in diameter, with much smaller magnet structures than woofers. Tweeters can be dish- or bowl-shaped, like the woofer, or they can be just the opposite, and be shaped like a dome; it's the designer's choice (again, see figure 4.19).

Both the woofer and the tweeter are typically housed in the same speaker cabinet, with the tweeter located above the woofer on the speaker face. The size of the speakers varies greatly, based upon the decibels of sound pressure that the designer wants them to be able to reproduce. Bigger is not always better, though. As an example, using exotic materials such as titanium, carbon fiber, or other synthetic materials for the speaker cone, designers can produce an incredibly strong and physically small monitor speaker. The smaller, stronger drivers can handle large amounts of power and generate over 105 decibels of sound pressure.

Figure 4.19 The larger, more powerful speaker is called a woofer. Practical woofers range in size from 6 to 15 inches in diameter, with permanent magnet structures that can weigh more than 50 pounds. The smaller, less powerful speaker is called a tweeter. Practical tweeters range in size from three-quarters of an inch to 2 inches in diameter, with much smaller magnet structures than woofers.

The Crossover Network. The number of drivers within a single speaker cabinet identifies the type of speaker system. A two-way system has a woofer and a tweeter. A three-way system has a woofer, tweeter, and a midrange driver to handle the middle frequencies (see figure 4.20).

Dividing the audio from the amplifier to send it to the appropriate driver within a speaker is the job of the crossover network. A crossover is an electronic circuit that splits the audio frequencies and routes them to the proper speaker, so that the bass goes to the woofer and the higher frequencies go to the tweeter. The actual frequency that is used as the dividing point is called the crossover frequency.

Figure 4.20 The number of drivers within a single speaker cabinet identifies the type of speaker system. A two-way system (left) has a woofer and a tweeter. A three-way system (right) has a woofer, tweeter, and a midrange driver to handle the middle frequencies.

The crossover that divides the frequencies can be passive or active. A **passive crossover** network is comprised of fixed-value electronic components. The crossover frequencies are fixed and cannot be changed. A passive crossover is located within the speaker cabinet.

An **active crossover** network allows the user to change the crossover frequency to suit his or her tastes or particular acoustical needs by adjusting a knob or electronic control. An active crossover requires electrical power and can be mounted within the speaker or as an outboard piece of rack-mounted gear.

Self-Powered Monitor Speakers

A traditional monitor-speaker setup feeds a signal from the control board to an amplifier and through the speaker cable to the monitor speaker. **Self-powered speakers** are self-contained monitors that include the power amplifiers mounted on or within the speaker cabinet.

Powered speakers have specially designed amplifiers matched to each of the drivers in the speaker. For example, in a two-way speaker system, there is an amplifier for the woofer and a separate one for the tweeter, or bi-amplification. In a three-way system it is tri-amplification.

An advantage of the powered monitor speaker is that the amplifier power ratings are matched to each specific driver. Most professional studio monitor speaker systems use bi- or tri-amplification.

Monitor Speaker Placement and Stereo Imaging

Ideally, monitor speakers should be at the mixer's ear level and should be centered in front of the mixing position at the console. Where should you sit in the studio to get the most accurate stereo image from your monitors? Stereo uses two channels of audio and two matched speakers to create a lifelike sound field comprised of width and depth. As you learned earlier in this chapter, panning sound sources from left to right and placing them at a given point within the sound field creates width. Reverbs and delays create a feeling of depth to sound, positioning it either at the forefront, close to the listener, or in the distance, depending on the timing of the effect. Equalization can give height to a sound by placing it above or below other sounds.

For a tight space, use a simple equilateral triangle as a guide. If the speakers are 4 feet apart, then the ideal listening position is at a point equally distant from both speakers, or 4 feet from each, forming the triangle (see figure 4.21). Ideally, the

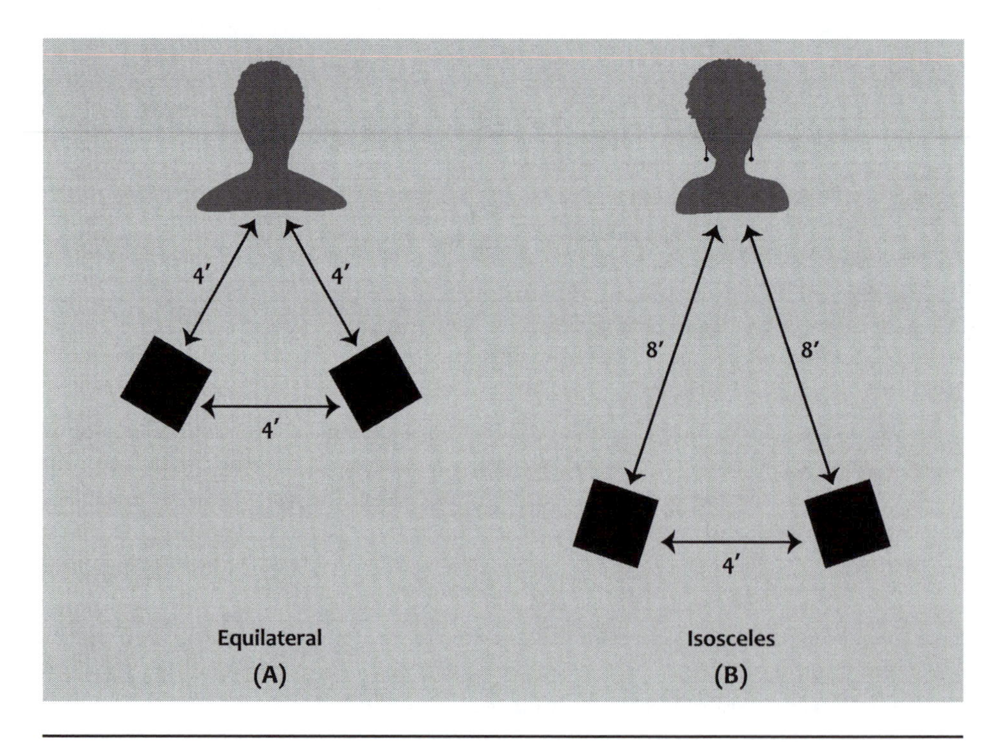

Figure 4.21 (A) For a tight space, use a simple equilateral triangle as a guide. If the speakers are 4 feet apart, then the ideal listening position is centered between the two speakers, 4 feet from each, forming the triangle. (B) For a wider overall sound field, if the speakers are 4 feet apart, double the distance (8 feet) between the two speakers to the mixing position. This forms a stretched, or isosceles, triangle.

speakers are angled inward and focused directly on the listener. The only drawback to this positioning is that the proper stereo image appears only in a very narrow area and the mixer can easily move out of the sweet spot just by turning his or her head. If more than one person needs to listen to this setup, the speakers should be angled outward to enlarge the sweet spot.

For a wider overall sound field, double the distance between the two speakers to get the appropriate distance from the speakers to the mixing position. Using the preceding example, if the speakers are 4 feet apart, then a wider sound field is 8 feet from each, forming a stretched, or isosceles, triangle (again, see figure 4.21). This position is not as sensitive to the movement of the mixer as is the simple equilateral triangle.

Suggested Activities

1. Seek out the different types of portable mixers and control boards you have available to you as a part of your class work. Spend time with each. While speaking into a microphone, route it from the input to the output using the different audio buses available. On production control boards, route the microphone through the channel assignments to the submasters and masters. Likewise, experiment with the different monitoring options that you have.
2. Find an example of as many different kinds of audio meters as you can and do a comparison of how they display your voice. Which do you prefer, mechanical meters, LCD, or LED meters?
3. With a pair of stereo speakers, experiment with listener placement using an equilateral triangle and an isosceles triangle placement. Which gives you the best stereo image in your listening environment?
4. Arrange for tours of a radio station, television station, and recording studio to see firsthand what the different types of control boards and mixing desks look like and how they are used in the different facilities. Ask ahead of time if they will let you bring a reference recording that you are familiar with so that you can play a track to hear it on their monitors.

Websites for More Information

For information about:

- *monitor speakers and placement of monitor speakers*, visit Genelec at www.genelec.com
- *on-air consoles for radio and television*, visit the Wheatstone Corporation at www.wheatstone.com
- *portable mixers*, visit Shure, Inc. at www.shure.com and Professional Sound Corporation at www.professionalsound.com
- *production consoles for radio and television*, visit Mackie at www.mackie.com
- *recording studio consoles*, visit Solid State Logic at www.solid-state-logic.com

■ Pro Speak

You should be able to use these terms in your everyday conversations with other professionals.

active crossover—A powered speaker crossover network with adjustable crossover frequencies.

bus—A common point in an audio device into which several inputs feed and are joined together.

combo mode—When the on-air radio talent serves as the show's engineer as well, running the control board and other elements of the show in addition to announcing.

driver—An individual speaker within a speaker cabinet; each driver consists of a moving-coil transducer that converts analog electrical energy into acoustic energy.

dry signal—An original, unprocessed audio signal sent from an audio console to an outboard audio processor, such as a reverb or delay unit.

feedback—A condition that occurs when a microphone is left open near a monitor speaker, creating a loud howling or whistling sound. The microphone picks up the amplified sound from the speaker and sends it back through the system, regenerating the signal over and over.

hard clipping—An audio condition in which the audio is driven beyond distortion to the point that the edges of the sine waves start breaking up.

headroom—The additional capacity of an audio device to handle level increases above average working levels to protect from transients, overload, and distortion.

high-pass filter—A filter that blocks audio frequencies below a specific frequency, allowing those frequencies above the specified frequency to pass; sometimes referred to as a low-cut filter, a term that is more descriptive of what the filter actually does.

interruptible foldback (IFB)—A method for sending a monitor signal to the talent while also being able to talk to him or her through an earpiece or headphone; a sophisticated intercom system.

low-pass filter—A filter that blocks audio frequencies above a specific frequency, allowing those frequencies below the specified frequency to pass; sometimes referred to as a high-cut filter, a term that is more descriptive of what the filter actually does.

near-field monitors—Smaller monitor speakers placed very near the mixing position, often at the front of the console.

pan (panoramic) control—A sonic positioning control that allows the input to a mixer channel to be assigned to the left channel or the right channel, or to remain in the center channel for a mono signal.

passive crossover—A crossover network comprised of electronic components whose values are fixed, with the audio requiring no electrical power to pass through the network.

pot—Pot is the abbreviation for potentiometer, which is the technical term for a volume control.

self-powered speakers—Speakers that have amplifiers built into them so that an audio feed from a mixer or control board can be fed directly into the speaker.

slate—The radio equivalent of the beginning of a scene in a television or motion picture shoot, when someone holds a slate, or clapboard, in front of a camera with the production name, scene number, take number, roll number, director, producer, and date to identify the material later. In radio, this is done audibly with a slate mic so that a production person can identify the audio cuts.

unity gain—A gain of one; the position on a fader at which the ideal operating level of 0 VU occurs. A device with unity gain does not raise or lower the volume, or gain, of a signal.

wall-wart—A self-contained power supply for running portable equipment. It plugs directly into the wall and down-converts 110 volts AC to a DC voltage.

wet signal—A processed audio signal that is returned to a control board after audio processing, such as reverb or delay.

5

Basic Concepts in Digital Recording

HIGHLIGHTS

You live in an analog world, a natural world in which all of the sounds that you perceive are made up of sound waves. But since Harry Nyquist's early work at Bell Labs in the late 1920s, the theories behind digital audio have been based not on nature's sound waves but on binary mathematical principles developed in the 1700s. In fact, Nyquist was so far ahead of his time in the 1920s that his theory on digital audio could not even be tested until the invention of the transistor in 1947!

Although analog audio is fairly simple and remained relatively unchanged for more than 50 years, digital audio is a moving target and changes as fast as hardware and software developers come up with new gear. Because of its incredible scalability and flexibility, digital audio is more technically involved. After all, we're talking computers and data. However, the basic principles of digital audio apply across a broad spectrum of equipment and production techniques ranging from a desktop computer to your smart phone. In this chapter you are going to get to work with the raw audio from a multitrack studio recording session. The music was especially written and recorded for this chapter so that you could learn about recording, storing, manipulating, transporting, and delivering digital audio.

Digital Recording

In earlier chapters, you learned the basics of how to select a microphone, place it, cable it, and route it through a preamplifier into a control board. The next stage in the production process is to learn the basics of digital recording and playback.

Recording

The first step in recording involves routing the signal into a recorder and laying down, or recording, the various tracks. The primary goal of recording any sound is to find the sweet spot (critical distance) with the microphone that most accurately captures the essence of the live sound; technically, you're after the best signal-to-noise ratio. Likewise, if you are recording music, sports broadcasts, or a news or public affairs program, your goal is to capture the sound at a good usable level. As was pointed out in chapter 4, it is critical in digital recording that you rely on your VU or peak meters to adjust the recording levels.

During the initial recording session, your only goal is to capture the sound accurately; editing, mixing, and mastering are done later, in postproduction. As a radio production person, most of the time you are going to be a one-person band serving as the recording engineer, mixer, and mastering engineer. Occasionally, you might receive assistance from the production director with fine-tuning or tweaking a mix to ensure that a commercial or program captures the essence of what the client wants.

A word of caution is in order about digital recording. When audio is fed into a digital recorder, it's important to make the best use of the recorder's available dynamic range. A recorder's dynamic range is measured in decibels and extends from the softest sound to the loudest sound the recorder is capable of recording. Set the recording level too low, and background noise and hiss become noticeable. Set the recording level too hot (loud), and the signal overdrives the inputs, resulting in hard clipping and seriously distorted audio. In a worst-case situation, the audio splatters and drops out.

The input audio level should register as high as possible on the meters without causing clipping or driving the meters into the red. A good rule of thumb for recording digital audio is that the average audio levels should be at about –10 on a VU meter, so that the audio peaks present in the signal have the necessary headroom to record without driving the recorder into hard clipping (see figure 5.1).

Digital audio software typically has two recording modes, nondestructive and destructive. In the nondestructive mode, each time a track is placed into record, the software automatically labels and saves the previous recording prior to the start of the new recording. In other words, nothing is erased; you can keep recording takes over and over until you are satisfied, and should you desire to, you can return to any one of the earlier saved takes. Destructive recording is just what the name

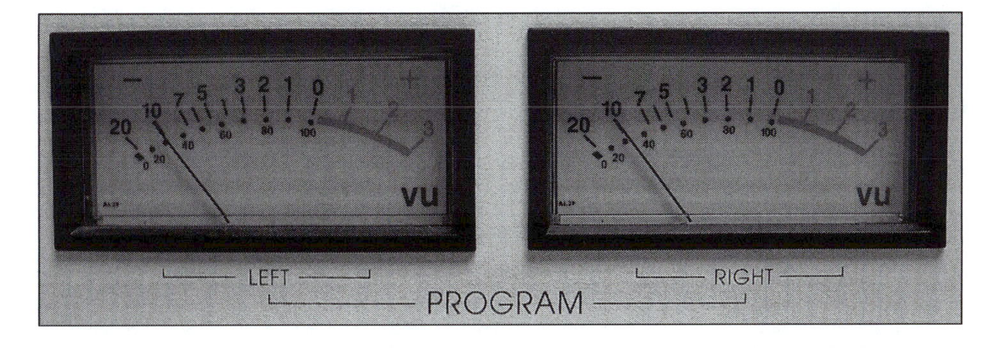

Figure 5.1 It does not look like much signal level, but a good rule of thumb for recording *digital* audio is that the average audio levels should be at about −10 on a VU meter, so that the instantaneous audio peaks present in the signal have the necessary headroom to record without driving the recorder into hard clipping.

implies; each time a track is placed into record, any previously recorded material on that track is recorded over and is lost or destroyed.

When you have completed either type of recording (nondestructive or destructive), always "safe" the recorder, locking it out of the record mode, so there is no chance of the recorder being accidentally enabled and recording over your work. Multitrack recorders let you do this on a track-by-track basis as you work on a project.

Editing

Editing is the ability to cut, paste, and copy audio, just as you would do with a computer for a paper or any other written work you might create.

Digital audio editing not only lets you listen to the audio to pick an edit point, you can also see the audio waveform displayed (see figure 5.2 on the following page). Although we depend on our ears to hear, most people (even musicians) rely on sight as their primary sense. Using the visual display, it is possible to zoom in or out on the waveform and select the particular point to make the edit, right down to a point within a single sine wave (see figure 5.3, also on the following page).

Digital editing, as opposed to recording, is known as nondestructive editing, because although you may be cutting, pasting, and copying on the visual display, absolutely nothing is physically happening to the original audio on the hard drive. The audio is never physically cut or moved about on the hard drive. Manipulating the visual display is how the computer commands are written that tell the hard drive what audio segments to play back and in what sequence. Of course, just as with every other edit function on a computer, if you make an error while editing prior to saving the session, there is always the Undo command on the software's menu bar.

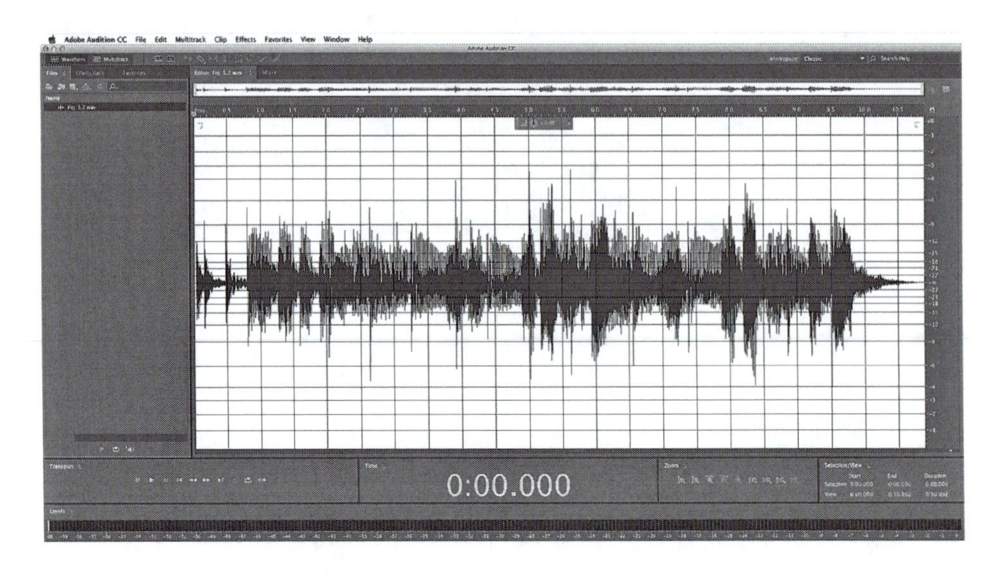

Figure 5.2 Although we depend on our ears to hear, most people (even musicians) rely on sight as their primary sense. The visual display is used to examine the waveform and select a particular point to make an edit or some other adjustment to the audio. (Adobe product screen shot reprinted with permission from Adobe Systems Incorporated.)

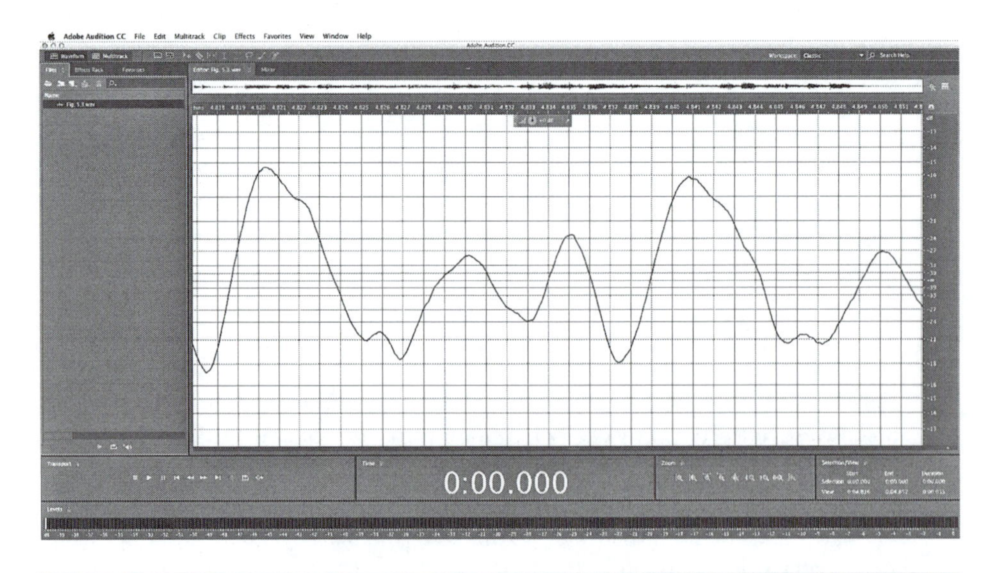

Figure 5.3 Using the visual display, it is possible to zoom in or out on the waveform and select a particular point to make the edit, right down to a point within a single sine wave. (Adobe product screen shot reprinted with permission from Adobe Systems Incorporated.)

Mixing

Mixing is the art of adjusting the audio levels of each of the recorded tracks and blending them to create the sound picture you or the client is seeking. Although it may seem a bit unusual, most production people already have some vision in their heads of the final mix or sound they want before they sit down at the console or computer to mix a project.

In its most basic form, mixing involves five elements, with the most primary being balancing the playback-volume levels of the various tracks in the recording. This is referred to as *balance.* If you are doing a voice-over-music commercial and adding a few sound effects (SFX), you have to balance the levels of the music, sound effects, and voice so that the voice is prominent. Sound effects are mixed to be either in the foreground or in the background, depending on how critical they are to the commercial.

A second aspect of the mix is the *sonic panorama*, or placement of the tracks. In a stereo mix, some tracks are left channel and some right, and a few may end up at various positions between the left and right channels.

As you mix and blend the sound from the various tracks, you may notice that some sounds seem to get lost and others dominate the mix, but neither can be corrected with simple level changes. *Equalization* (EQ) is used to constructively correct and enhance the frequency range of the mix to make it brighter, bigger, clearer, and so forth. And whereas anyone can adjust the controls on an equalizer, it takes training to EQ a mix properly. EQ is addressed in detail in the next chapter.

Dimension, or space, is something that should ideally be captured in the original recording, using natural room ambience or reverb. However, when that quality is not present in the original recording, there are a number of tools, such as electronic reverb, echo, and delay, that can be used to give the recording life and add depth to the sound field. See chapter 6 for more details on dimension.

The *dynamics* of the mix refers to the difference between the softest sound and the loudest sound. Typically, though, as a part of the mix, dynamics processing techniques such as **compression**, **limiting**, and **gating** are used to control the dynamics and to increase the average audio level, making the mix sound louder. Just like EQ, dynamics processing can enhance a recording but, if improperly applied, can also seriously degrade a recording. Dynamics processing is discussed in-depth in chapter 6.

An additional component of mixing is automating the mix, using the audio software's controls to record your individual track-by-track adjustments and to control all of the functions previously described in this section simultaneously during the playback.

As a general rule, it is an excellent idea to save your mix prior to mastering. Some people save the original tracks, or session file, as one file and the mix of those tracks as another. By saving these as separate files, you make it easy to come back to the project at a later date and modify it.

Mastering

Mastering is the last stage of audio production and involves the final tweaking and adjustments to the mix. Generally, the focus of a mastering session is critically listening to the project and making any final editing adjustments, fine-tuning the mix, confirming the left/right stereo balance, ensuring that the frequency response is relatively flat, and checking the overall sound picture.

For the average commercial produced in a radio station the mastering process is pretty simple. Most clients tend to listen more for creative content than for audio quality. In fact, most local clients are satisfied to approve their radio commercials by listening to an MP3 of the commercial sent via email or by listening to the commercials over the phone. Many stations have secure client websites where production projects can be posted for clients to listen to.

A good rule of thumb prior to mastering a large project or a commercial for a major client is to go ahead and move the audio file to a flash drive or to a smart phone. With the project in hand you are prepared to go to the parking lot and listen critically to the project on a number of car stereos. In particular, the types of cars your station's target demographic might own. Take detailed notes on what you hear and return to the production room to make adjustments in the audio tracks. After doing this a few times, and keeping notes, you will see a trend in your mixes and you will be making fewer trips to the parking lot! The goal of this process is to make sure your station's production sounds better than your competitors.

When completed, the final project is likely to end up in a number of locations. Using a local radio commercial as an example, one copy is transferred directly into the radio station's main server for on-air playback and a second copy of the commercial is converted to an alternate format and sent via email to the client or to another radio station running the ad. Some clients will specify a preferred format, such as an MP3 or WAV file. Finally, the production person should back up the project files in accordance with station policy. Clients often ask to reuse a commercial they liked from several months ago, with just one minor change. Having an archive copy of the session files makes it a simple matter of reopening the master and making the requested changes, saving a lot of valuable time and money rather than re-creating the commercial from scratch. Likewise, newspeople often want to extract audio clips from earlier programs that have long since aired.

Digital Audio Software: Using a Computer as a Recorder

Digital audio is based on converting an analogous sound wave into binary numbers. Once the audio has been transformed into data by the analog-to-digital (A/D) converter, the possibilities for recording, storing, manipulating, transporting, and delivery are limitless. As you will learn in this chapter, virtually any device that can process computer data can be creatively used to record, store, and play back digital audio. A question that arises when considering a computer to be used for a **digital audio workstation** (**DAW**) is whether the computer should have a Windows or Mac platform. (Which is better, chocolate or vanilla ice cream?) The correct answer is: the platform that the software you've chosen was optimized for and that you feel most comfortable working with. Because of the variety of equipment in the marketplace, many production people are dual-platform literate, understanding and operating both major computer operating systems.

Current-model computers, regardless of whether they are desktop or laptop models, can either serve as a digital audio workstation right off the shelf or be quickly and inexpensively modified to do so. It is simply a matter of matching the basic hardware to the requirements of the software.

Production studios are centered on a computer with audio software or some type of system based on a central server (see figure 5.4). Increasingly, the software may reside on a "cloud" server. A full-featured audio software program is a virtual studio. In addition to the basic recording and playback, the software usually features **digital signal processors** (**DSP**) or plug-ins. Sometimes manufacturers will refer to plug-ins as apps. A plug-in is a software-based audio processor such as a reverb, echo, delay, or dynamics controller. A popular plug-in used in radio is time compression and expansion. This plug-in digitally rearranges the audio to stretch

Figure 5.4 Production studios are centered on digital audio workstations (DAW). A full-featured DAW is a virtual studio within a computer. This studio features a DAW on the left. The monitor in the center is for the radio station's central audio server where all on-air production is stored. Note the laptop computer on the table to the far left, a portable digital audio workstation.

or compress the length of an audio cut. If a commercial is produced and ends up being 62 seconds, it is simply a matter of processing the commercial and compressing the time line to 60 seconds. This is done with no change in the pitch of the voices or other artifacts that the listener can hear. Pitch shift is another popular plug-in with both men and women announcers, allowing them to slightly lower the pitch of their voice for a rounder, fuller sound.

A feature of audio software that some people find disconcerting is that the computer's keyboard, mouse, touch pad, or trackball controls everything. Whereas some operators are quite comfortable using these controls, others feel more comfortable touching real faders and controls. The solution is a control surface that interfaces with the software (see figure 5.5). Move a fader on the control surface, and the software's corresponding fader moves; turn the knobs and the software's knobs turn. This kind of interface is often referred to as a knob-per-function user interface and makes the software much easier to record and mix with for some people.

An additional advantage of computer or server-based audio software is networking with other computers/workstations through a **local area network** (**LAN**). For example, a radio station might have a control room, two production rooms, and a newsroom, each with a computer and audio software. All four computers are networked together to share files; they are also networked to the station's main server. When a commercial or news story is produced, it is instantly accessible in the main control room or anywhere else in the radio station.

Figure 5.5 A digital audio workstation control surface provides a physical interface with the software. Move a fader on the control surface, and the software's corresponding fader moves; turn the knobs and the software's knobs turn. This kind of interface is often referred to as a knob-per-function user interface and makes the software easier to record and mix with.

Radio stations within regional or national groups can share files using **wide area networks** (**WAN**) or through **file transfer protocol** (**FTP**) on the Internet.

Computer-Based Hybrid Devices

In addition to computers, computer-based hybrid devices offer production people a steady stream of new creative tools and recording media. A hybrid device is one that combines the technology and functions of two or more distinctly different devices into a related third device.

An excellent example of a hybrid device is a studio in a box. A studio in a box is a digital, multichannel audio workstation. A complete mixing console and digital signal processors, such as reverbs, compression, and limiting, are included. Recording and play back are on an internal hard drive or flash memory card(s) and most units can be connected to a computer via a USB cable or Bluetooth technology. Audio never has to leave the box during the entire production process (see figure 5.6). Hybrid devices are very compact, portable, and powerful tools.

Figure 5.6 A studio in a box is a digital, multichannel workstation. A complete mixing console and digital signal processors, such as reverbs and echo, are included. Recording and play back are on flash memory cards or the unit can be connected to a computer via a USB cable. Audio never has to leave the box during the entire production process. In this example, at least four separate devices are combined to form an incredibly compact and powerful tool.

Digital Audio Software

Something that you should keep in mind as you review the array of digital audio software on the market is that creativity does not come out of a box from a software or computer company. Creativity comes from you and from those you surround yourself with. Sure, having good tools helps, but don't think for a moment that just because you have all the latest bells and whistles on your computer you are suddenly more creative. Nothing could be farther from the truth. Your focus must first and foremost be on communicating the message: on creating a great commercial, news, sports, podcast, or public affairs program for your radio station. As your creativity requires, use the tools at your disposal, but don't use them just because they are there and you can.

There are a number of software products available for turning a computer into an audio workstation. For the most part, each company has tried to seek out a particular price point and niche market for itself. The commercial music industry has favorite hardware/software combinations; broadcasters, in turn, have a number of broadcast-specific software products, designed for everything from live studio work to full station automation. Many broadcast systems interface with traffic and billing systems to schedule and play back commercials, promotional announcements, and station identification announcements.

In addition to software, some products require a physical interface to get the audio in and out of the system. These range from the sound card that came with the computer to outboard digital audio converters costing thousands of dollars (see figure 5.7). If necessary, the software provider specifies a particular type of input/output converter.

With the large number of digital audio software products on the market, it is well beyond the scope of this text to attempt to cover all of them. As is usually the case, two major products have found significant acceptance in their respective industries: Adobe Audition Creative Cloud (CC) and Pro Tools.

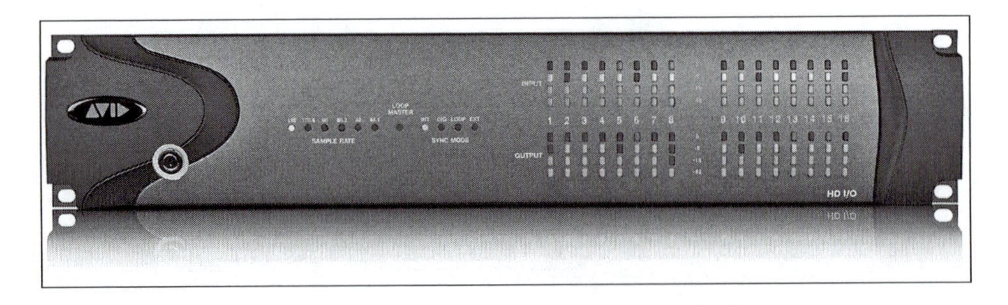

Figure 5.7 Digital audio software requires some type of physical interface to get the audio in and out of the system. These range from the sound card that came with the computer to a high end outboard digital audio converter like this one that costs thousands of dollars.

Adobe Audition CC

Adobe Audition CC is a powerful cloud-based program available by annual subscription with a version for both the Mac and Windows platforms—which brings up an interesting point. Radio broadcasters, as a general rule, operate their stations based on the Windows platform. This includes everything from the traffic and billing software to sophisticated automation systems. Why this preference developed over the years may have to do with radio embracing computers in the early 1960s for the tedious tasks of scheduling each day's commercials and preparing program logs. The early punch-card systems came from IBM—and thus the tie to Windows, the personal computer platform made popular by IBM. (If you have never heard of a computer that used paper punch cards to store data, look the subject up on the web.)

Adobe Audition CC is a typical, intuitive, point-and-click program with a fairly short learning curve (see figure 5.8). If you have any experience at all with computer recording, chances are you will skip the instructions and tutorials until you hit a roadblock. Depending on how your computer is configured, the software enables your computer to record an unlimited number of audio tracks and supports the highest digital audio sampling rates. Additionally, Adobe Audition CC features over 50 DSP effects such as echo, reverb, and a variety of other audio processors.

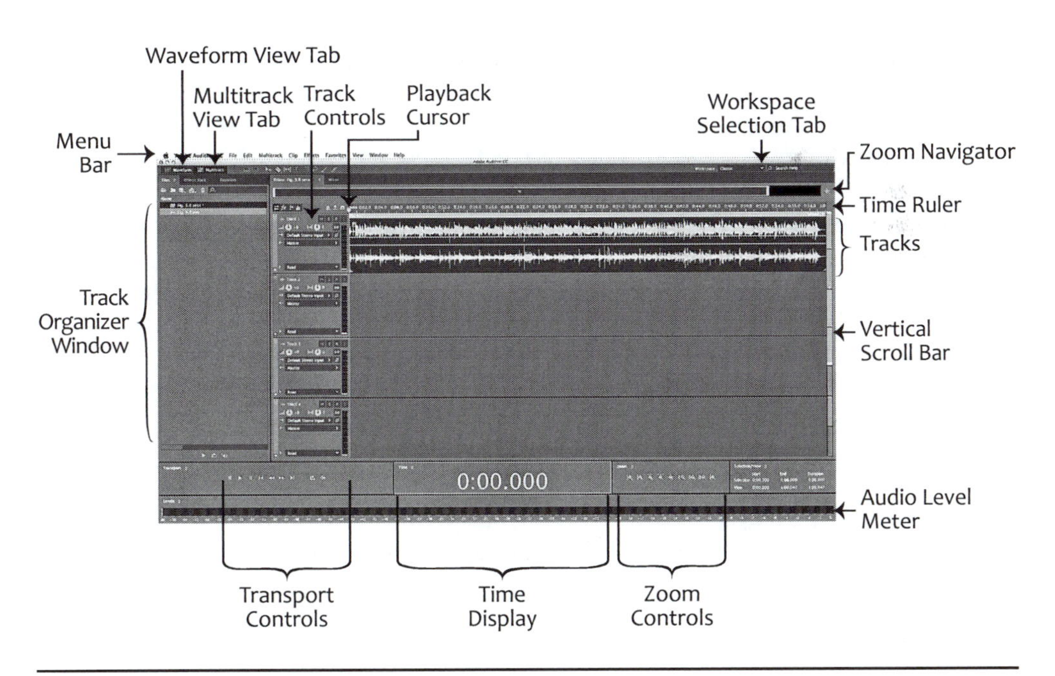

Figure 5.8 In the multitrack view, Adobe Audition CC represents a virtual recorder that can be used to record, play, edit, and add audio processing to an unlimited number of tracks (depending on how your computer is configured). (Adobe product screen shot reprinted with permission from Adobe Systems Incorporated.)

The screen captures and instructions below are based on Adobe Audition CC for the Mac platform.

With Adobe Audition CC's ability to support unlimited numbers of audio tracks and high sampling rates, your first thoughts might be about running out of hard disc storage space. From a practical standpoint, though, storage space is rarely a problem. It is the hard drive speed (measured in rpm, or revolutions per minute) and RAM (random access memory) that can potentially limit the sampling rates and the total number of tracks that can be recorded and played back simultaneously. However, most current computer models can easily handle digital audio.

Adobe Audition CC does not require the purchase of any specific analog-to-digital audio hardware for the software to function. It works with the sound card that came with your computer. However, to interface with professional audio equipment and to take full advantage of the software, it is advisable to add a professional sound card or interface with analog and digital inputs and outputs (more on that later in the chapter), depending on your specific needs. Such devices, depending on the number of inputs and outputs, range in price from $300 to $1,200.

Due to the intuitive nature of the software and the fact that it is platform neutral, Adobe Audition CC is popular with broadcasters. Some of the largest broadcast groups in the United States, and internationally, use Adobe Audition CC.

Producing a Project with Adobe Audition CC. Before you continue on in this chapter, if you have not done so already, visit the Adobe Audition CC website and download a free trial version of the software on your personal computer. Become familiar with the software. There is a lot of excellent information in the Help menu. The production music available for download on the book's website (http://www.waveland.com/Connelly/) will also be beneficial for this section of the chapter if you wish to follow along. One of the music selections is a traditional blues progression (see music demo, cut 1) and features not only the final mix of the music but the five raw individual studio tracks that were used to create the selection (see music demo, cuts 2 to 6). What is unique about these five tracks is that any one of them could be used by itself behind a project. Additionally, any two or more tracks can be mixed together to form a new musical selection. In all, just by mixing and matching the five tracks the way they are, more than 30 new versions of this blues progression can be produced; with variations in level, audio processing, and editing to each of the tracks, the possibilities are limitless!

Adobe Audition CC typically opens to a default workspace. There are several workspaces to choose from and you can even create your own custom workspace.

For the purposes of this text, the "classic" workspace has been chosen for simplicity. You can select the desired workspace in the upper right-hand corner of the display on the drop-down menu labeled Workspace. Within the workspace, Adobe Audition CC has two main views or screens—the multitrack view (again, see figure 5.8) and the waveform or edit view (see figure 5.9). After you have selected the workspace, you can select either the waveform or multitrack view by clicking on the Waveform or Multitrack tab in the upper left-hand corner of the workspace. You can toggle back and forth between the two workspaces.

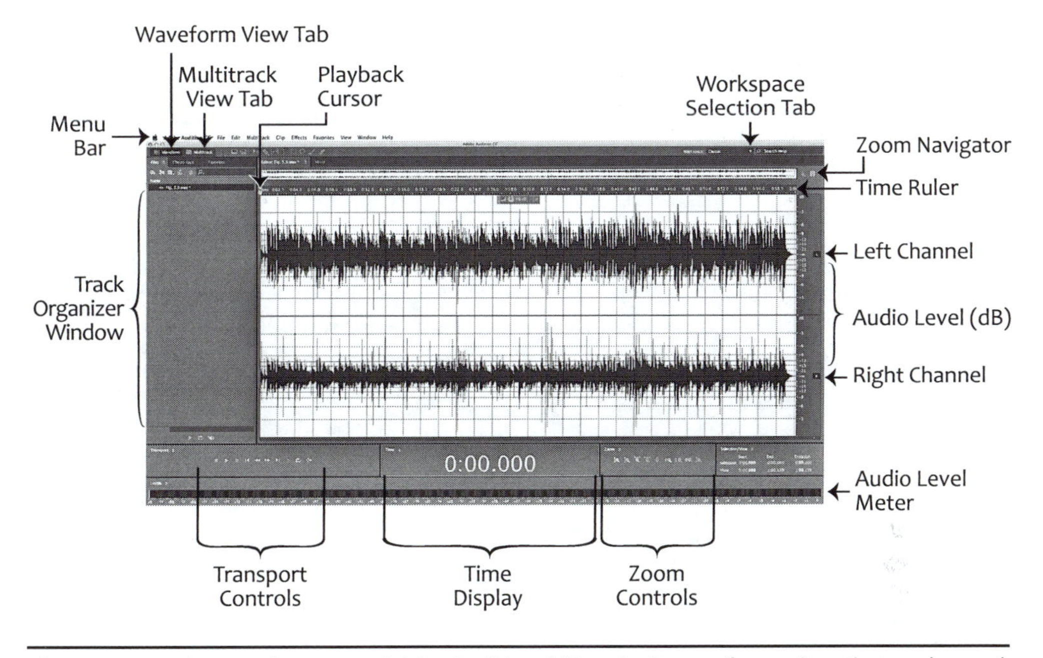

Figure 5.9 In the waveform view, Adobe Audition CC is a single waveform editor that can be used to record, play, edit, and add audio processing to mono or stereo tracks. From the waveform view, tracks can be transferred to the multitrack view. (Adobe product screen shot reprinted with permission from Adobe Systems Incorporated.)

In the multitrack view, Adobe Audition CC permits you to record, edit and mix, and sonically position (pan) each track between the left and right channels. Depending on your computer's configuration, Adobe Audition CC allows you to use an unlimited number of tracks of audio on a project. In the waveform or edit view, Adobe Audition CC is a single waveform editor that can be used to record, play, edit, and add audio processing to mono or stereo tracks.

Recording with Adobe Audition CC. Open the program and start in the waveform view, classic workspace. There are any number of ways to record material in Adobe Audition CC. For live recording, you can record audio from a microphone or any other device (e.g., a control board) you can route into your computer from either an analog or digital source. To use the audio and music cuts provided from the book's website, click on File and on the drop-down menu select Open. You can then locate the audio folders and files on your computer and click on the ones you want to open. When you click on the audio file to open it, the file appears on the left side of the Adobe Audition CC screen in the Files window and as a waveform in the Edit window. The same procedure can be used for downloaded material such as production music from a studio.

The only thing left to do is to move the audio cut you just opened into the multitrack view. Click on the Multitrack tab and then click on the file in the Files window and drag it to the track where you want to place it (see figure 5.10). Once in the track, you can position the file by placing the cursor on the audio waveform title bar and clicking and dragging, usually to the left or start of the track. You also can click and drag the file to any track you wish. To play the file, use the Play button or use the space bar on your keyboard to start and stop playback.

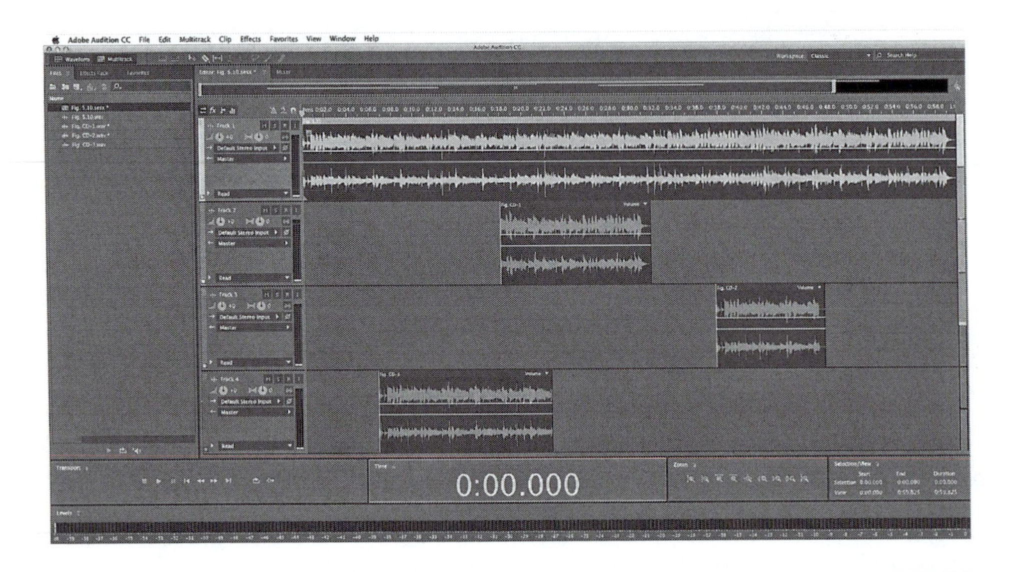

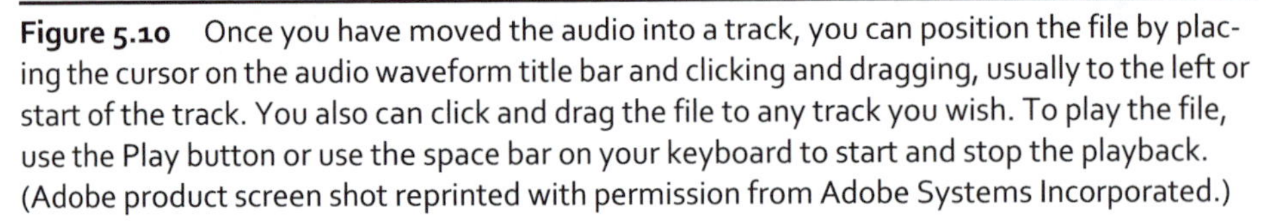

Figure 5.10 Once you have moved the audio into a track, you can position the file by placing the cursor on the audio waveform title bar and clicking and dragging, usually to the left or start of the track. You also can click and drag the file to any track you wish. To play the file, use the Play button or use the space bar on your keyboard to start and stop the playback. (Adobe product screen shot reprinted with permission from Adobe Systems Incorporated.)

To rip the music from a CD, place a CD in your CD drive and in the waveform view, click on File and on the drop-down menu select Extract Audio from CD. The Extract Audio from CD window appears (see figure 5.11). In the Drive window select the CD drive you placed the music in; often this is done automatically. A list of the cuts appears in a window below. Place a check mark next to the track you wish to rip into Adobe Audition CC and click OK at the bottom of the window. The Extract Audio from CD window opens and shows you the ripping progress. When finished, the audio file appears on the left side of the screen in the Files window and as a waveform in the Waveform window. The only thing left to do is to move the audio cut you just ripped into the multitrack view. Click on the Multitrack tab and then click on the file in the Files window and drag it to the track where you want to place it (again, see figure 5.10). Once in the track, you can position the file by placing the cursor on the audio waveform title bar and clicking and dragging, usually to the left or start of the track. To play the file, use the Play button or use the space bar on your keyboard to start and stop playback.

Figure 5.11 The Extract Audio from CD window permits you to select the CD drive you wish to rip the audio from. Place a check mark next to the audio cut you want to import and press OK to rip the audio into the waveform view. From the waveform view, the audio cut can be moved into the multitrack view for mixing into a project. (Adobe product screen shot reprinted with permission from Adobe Systems Incorporated.)

In the multitrack view you may want to change the track name by clicking on the track number and highlighting it. Type in any name you wish to use to identify that track; you can change it as many times as you wish.

Now is a good time to save your work, before you start editing. On the menu bar, click on File, and on the drop-down menu, select Save As. You will be asked to give your session a name and where you would like to save it. When you are working on a project it is a good idea to save your work often, as a power outage or computer problem could otherwise wipe out all of your unsaved work.

Editing with Adobe Audition CC. Once you have a track recorded, the next step is to trim or edit each end of the track so that the music starts and ends cleanly. To demonstrate, we are going to edit the very beginning of the track to remove any dead space or noise, so that the music starts instantly on playback. In the waveform view place your cursor anywhere in the audio waveform and click and drag to highlight an area you would like to edit (see figure 5.12 on the next page). Looking at the top of the Waveform Display window you will see the zoom bar. This is a small version of the waveform display and shows you the relative position of the audio you just highlighted within the audio cut. Now you are ready to zoom in on the area using the zooming controls located just to the right of the time display below the Waveform window (see figure 5.13, also on the next page). Adobe Audition CC offers you nine control buttons to zoom in or out on an area. The far left button in the group is used to zoom in vertically (amplitude). Each time

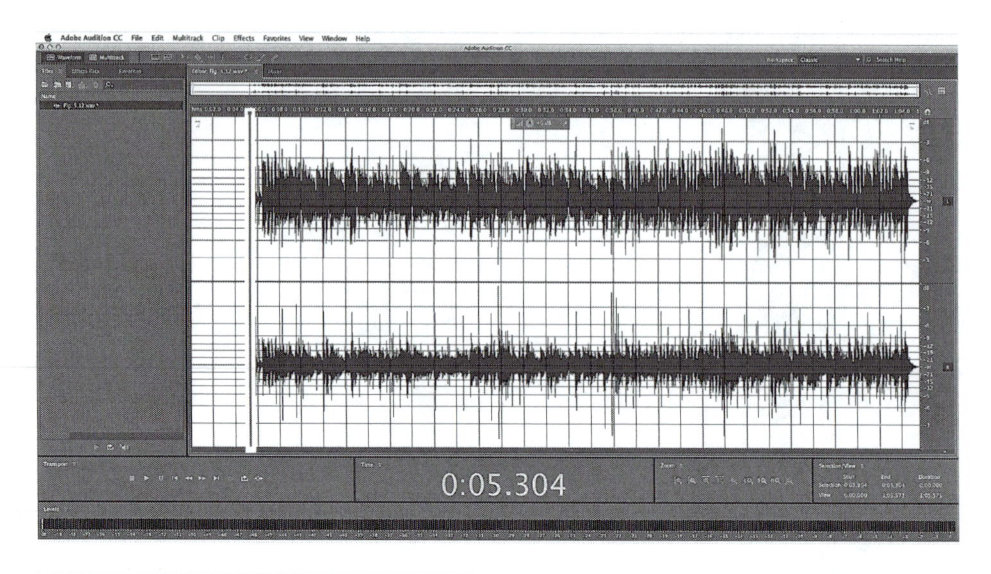

Figure 5.12 This audio cut has some blank space before the music begins that can be edited out. There are a couple of ways to place the cursor to start the edit. The cursor can be positioned by clicking on the small triangle at the top of the playback cursor and dragging it wherever you want, or, click on the line extending down from the playback cursor and drag it over the area you want to edit. (Adobe product screen shot reprinted with permission from Adobe Systems Incorporated.)

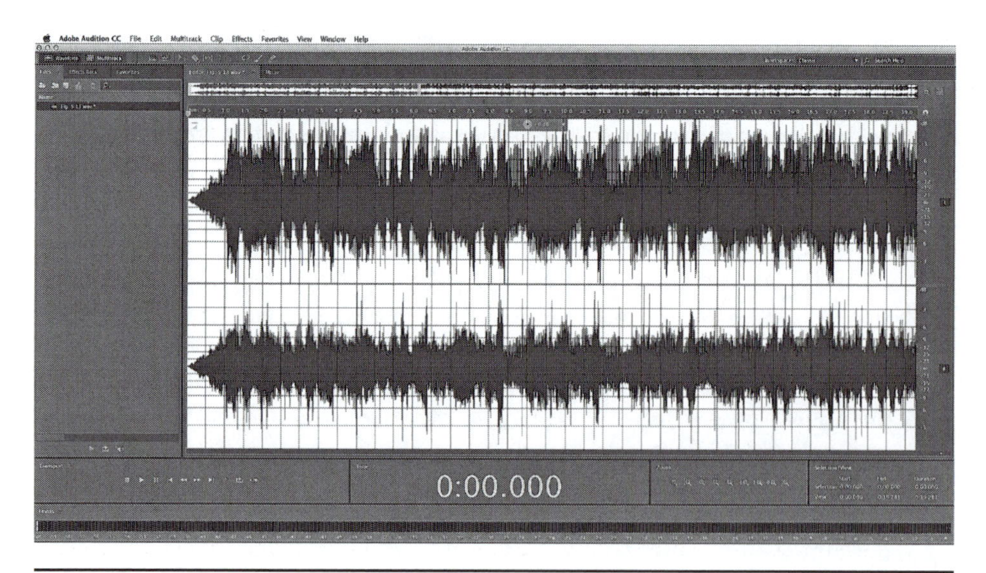

Figure 5.13 This audio track shows little detail and is hard to see. The zooming controls are located just to the right of the time display window. They permit you to change the wave-form view so as to display more detail with greater accuracy. (Adobe product screen shot reprinted with permission from Adobe Systems Incorporated.)

you click the button you zoom the view in closer. The next button in the row is used to zoom out vertically (amplitude) (again, see figure 5.13). Notice that as you zoom in and out vertically, you change the size of the track (see figure 5.14). The third and fourth buttons zoom in and out horizontally (time line) and operate just like the amplitude zoom buttons. At the extremes of zooming in you will be able to see the individual sound waves that make up the audio.

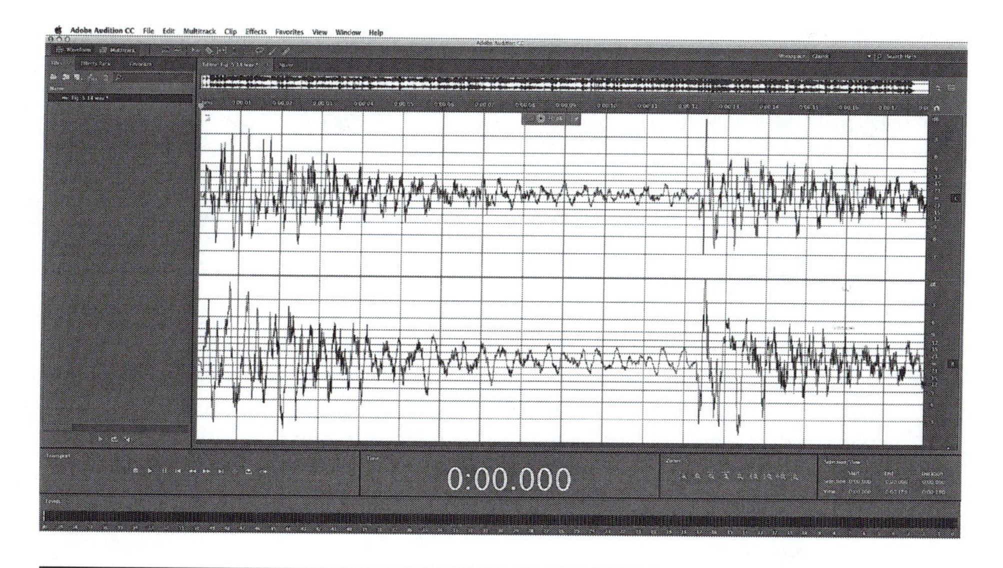

Figure 5.14 Adobe Audition CC offers you nine control buttons to zoom in on and out of an area of your audio track. The far left button in the group is used to zoom in vertically (amplitude). Each time you click the button you zoom in closer, displaying more detail. The third button from the left is used to zoom in horizontally (time). This example is from the same audio cut shown in figure 5.13 by zooming in horizontally. (Adobe product screen shot reprinted with permission from Adobe Systems Incorporated.)

To edit out the silence preceding the beginning of the music, zoom in so that you can easily see the very start of the music waveform. Place your mouse's cursor at the starting point of the music, then click and drag the cursor to the left, highlighting the area you want to remove (see figure 5.15 on the following page). If you don't manage to highlight exactly the area you want the first time, just place the cursor anywhere on the track, click once, and the highlighting will disappear. Keep trying, and once you have highlighted the area you want to edit out, press the delete key on your keyboard to remove the area. (Any time you have an "oops" and have removed too much, or made any other kind of error, all you have to do is go to the Edit drop-down menu and use the Undo command.)

With the beginning trimmed, you are ready to edit the end of the musical cut using the same technique. Once that is done, record at least two more tracks of the music so that you are ready to move on to mixing.

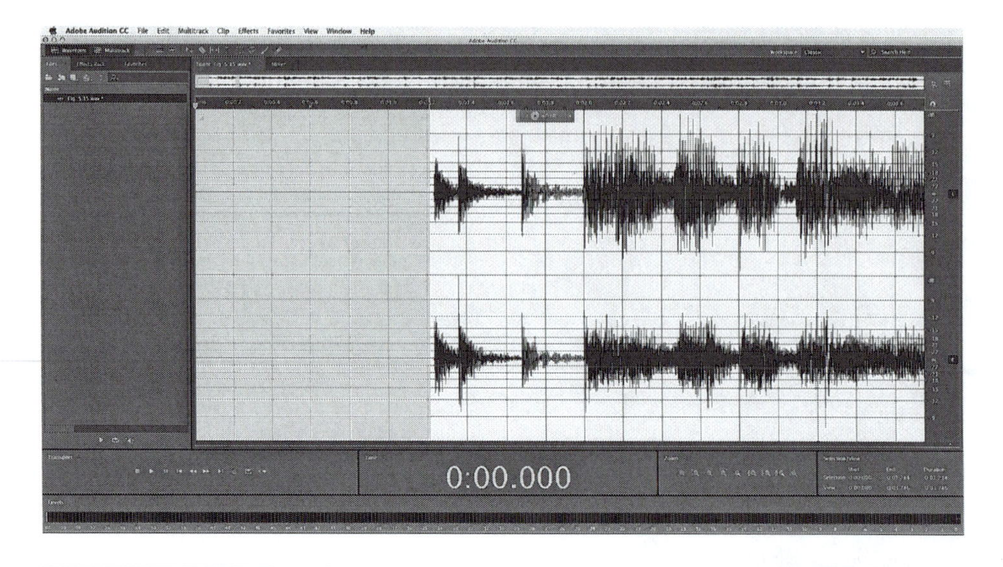

Figure 5.15 The silence at the start of this audio cut was highlighted by clicking and dragging the cursor over the area. Once you have highlighted the area you want to edit out, press the delete key on your keyboard and the area is removed. (Adobe product screen shot reprinted with permission from Adobe Systems Incorporated.)

Mixing with Adobe Audition CC. If you laid down more than a couple of tracks and tried to play them back, you probably noticed that they were all playing at the same level and it was hard to tell what in the world was going on. Mixing is the art of adjusting the volume and sonic position of each track, or blending the sounds from the tracks together. Adobe Audition CC makes this easy with controls on each track.

In the multitrack view if you look just to the left of each track you will see a Track Controls window (see figure 5.16). On the left side of the window, right under the track name, there is a volume control that is at the center position. The reading beside it is +0, indicating unity gain. By placing the cursor over the control and clicking you can adjust the track's audio level from negative infinity (nothing) through unity gain (+0) to +15 dB. You can set it wherever you like, keeping in mind that the 0 position is unity gain and going past that can cause distortion. Should you wish to mute all of the other tracks and listen only to this track, press the S or Solo button in the upper right corner of the Track Controls window.

You can sonically position the track with the pan control located just to the right of the volume control. Moving it all the way to the left or right position will cause the track to appear in the left or right channel of a stereo pair of speakers. Leaving it in the center position will cause it to come out of the stereo speakers equally. This control allows you to position a sound effect or music however you like between the left and right channels of a stereo pair.

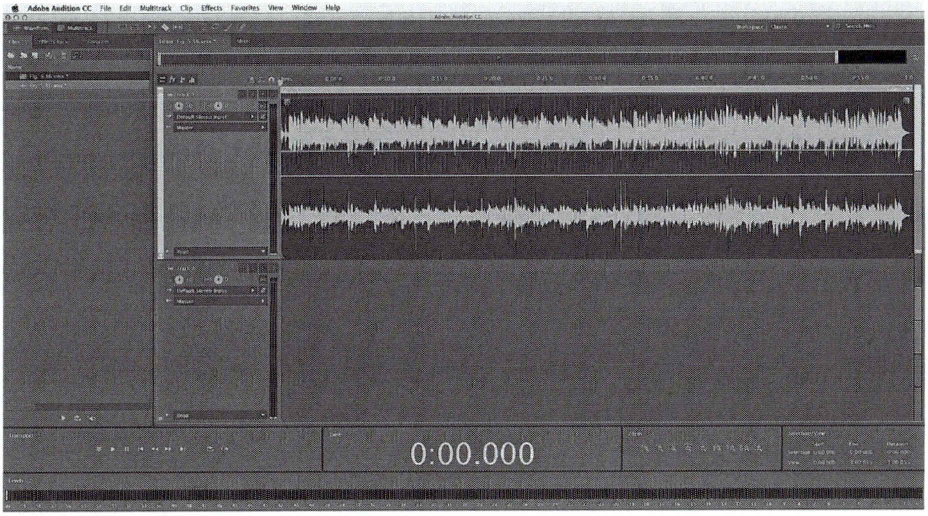

Figure 5.16 Just to the left of each track is a Track Controls window. On the left side under the track name is the volume control. Zero indicates unity gain. You can adjust a track's audio level from negative infinity (no sound) through unity gain (0 dB) to +15 dB. The pan control is located just to the right of the volume control. Moving it all the way to the left or right position will cause the track to appear in the left or right channel of a stereo pair of speakers. (Adobe product screen shot reprinted with permission from Adobe Systems Incorporated.)

Adjust all of the properties of each track individually until you have blended the tracks the way you want them. When you are satisfied and ready to save your multitrack project you will need to save your work as a two-track stereo mix. To do this use the Mix Down Session to New File function located under Multitrack on the menu bar. Mix Down Session to New File creates a two-track stereo file that includes all of the unmuted tracks and all of the waveform properties you used such as volume, pan, EQ, and so forth. Once mixed down, the stereo file can be saved and distributed electronically to other users or the station's server as backup. The mix-down process is sometimes called bouncing to disc.

Naturally, there are many more controls, shortcuts, and adjustments that can be made in Adobe Audition CC. After this basic introduction, you are ready to explore the software using the excellent Help menu and Adobe Audition CC tutorials available online. And, of course, there is always your own trial and discovery process.

Pro Tools

Pro Tools is the de facto professional recording standard for the music recording industry and television/motion picture production/postproduction industry. It is also used in some larger-market radio stations. The software is available in both Mac and Windows versions. Due to the sophisticated versatility and scalability of the software and hardware, Pro Tools has a steeper learning curve than many other products on the market.

There are several versions of Pro Tools, ranging from an entry-level consumer version to a full professional package. Pro Tools is scalable from the standpoint that the on-screen interface remains essentially the same from the entry-level product through the top-of-the-line software. Once the entry-level product is mastered, you should be able to adapt quickly to the extensive features in the professional versions (see figure 5.17).

Whereas Adobe Audition CC opens with a virtual transport and is in a near ready-to-record mode, Pro Tools requires you to create your own custom work surface. Each operator selects the transport and software features to suit his or her particular project and personal preferences. The software settings are easily scaled

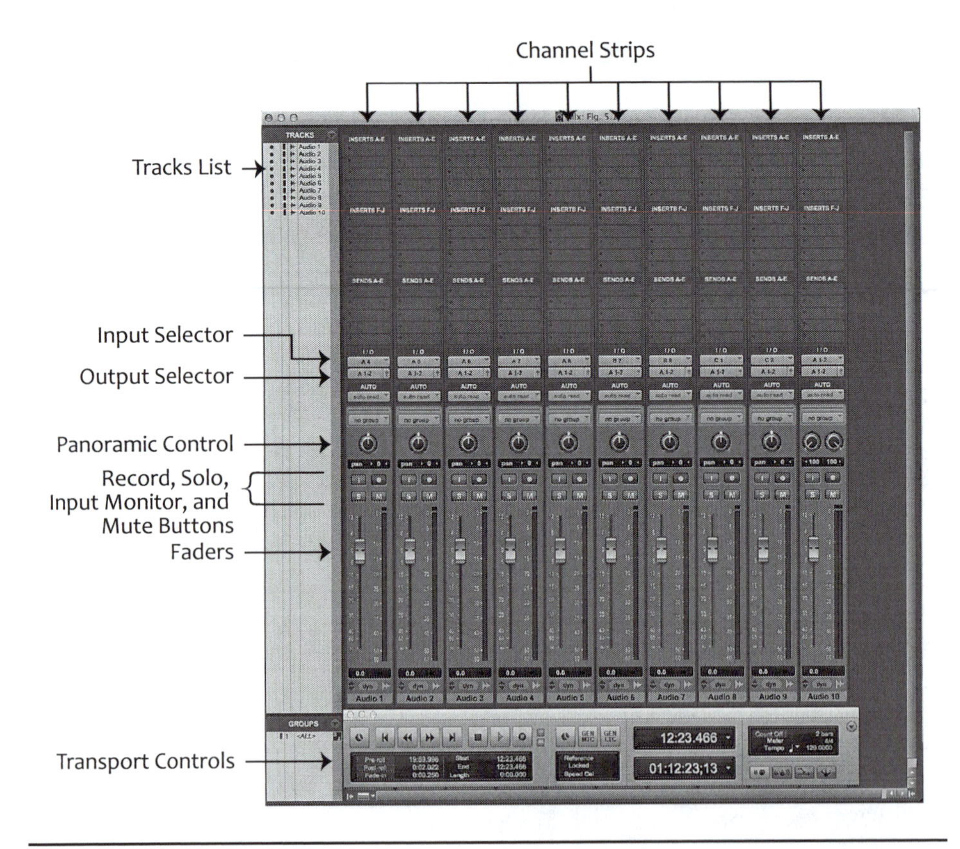

Figure 5.17 Pro Tools is scalable from the standpoint that the on-screen interface remains essentially the same from the entry-level product through the top-of-the-line software. Each operator selects the transport and software features to suit his or her particular project and personal preferences. The software settings are easily scaled to the project at hand, allowing you to work with only the features you need. The Mix window above is configured to record and play back 10 tracks of audio. (© 2016 Avid Technology, Inc. All rights reserved.)

to the project at hand, allowing you to work with only the features you need. Pro Tools also manufactures a number of control-surface interfaces, allowing the operator physically to work with as many as 24 faders at a time (see figure 5.18).

Figure 5.18 A control surface allows an operator to physically handle the controls illustrated within the digital audio software. Pro Tools has a number of control-surface interfaces, allowing the operator to work with as many as 24 faders at a time.

Producing a Project with Pro Tools. Although not as intuitive as Adobe Audition CC, Pro Tools comes with a well-documented reference guide. Beginners should spend some time studying the guide to better understand the capabilities of Pro Tools and the numerous options it offers producers. Regardless of the system, it is necessary to install and configure properly the analog-to-digital hardware and software before use.

Pro Tools has three work areas, the Mix window, the Edit window, and the Transport window or controls. The Mix window displays the Pro Tools virtual console. The console features individual channel strips with a fader for playback level control and Mute, Solo, Record, and Input monitoring buttons just below the Pan control (again, see figure 5.17). Automation controls are located just above this group of buttons. At the top of each strip are the insert buttons for activating audio processing plug-ins. Below them are the audio sends for routing audio to external devices or to any one of the internal audio bus paths. You can create as many channel strips as you need for your project, up to the limit of the particular system you are working on.

The Edit window displays the recorded audio waveforms on each track (see figure 5.19). Additionally, there are a large number of edit controls for functions such as zooming in/out and cutting and pasting the waveform. Across the top of the window are the available rulers for the various timebases, including minutes and seconds, bars/beats, tempo, and meter. To the left of the Edit window is the Tracks list that includes a listing of every track that has been created. To the right of the Edit window is the Regions list that includes a listing of every audio segment (region) that has been recorded, imported, or created by editing. Items in the list can be dragged onto any available track in the Edit window to work with.

The Transport window features the controls to operate the system and can be configured to suit your particular needs (see figure 5.20). The primary control buttons, from left to right, are the Online Indicator, Return to Start, Rewind, Fast Forward, Go to End, Stop, Play, and Record. When activated, the Online Indicator to the far left allows the playback of an audio cut to be started by an external device if your machine is so configured.

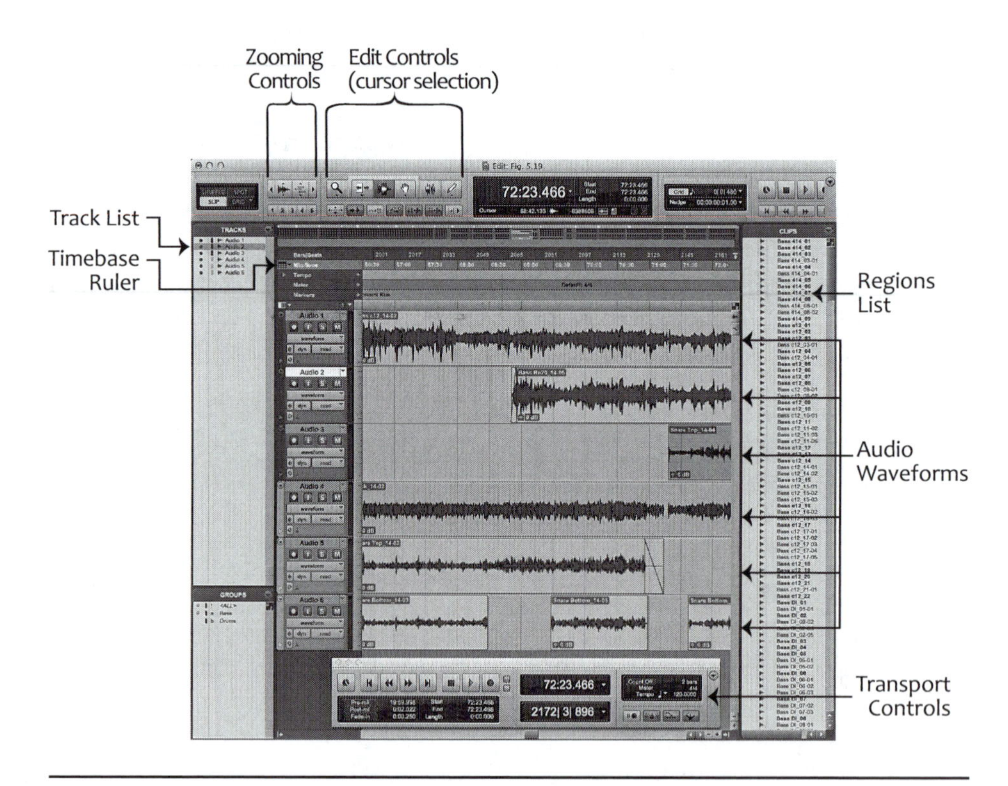

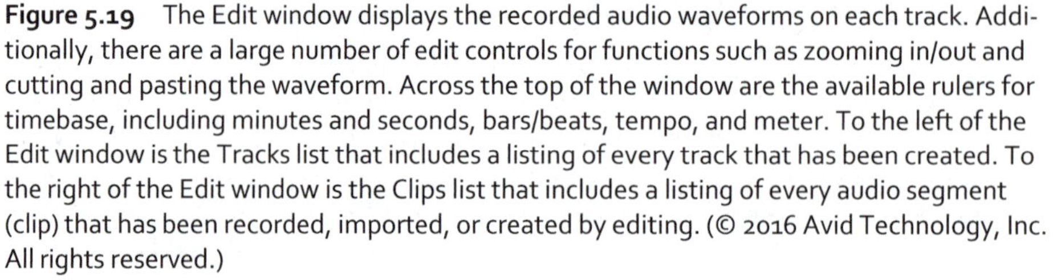

Figure 5.19 The Edit window displays the recorded audio waveforms on each track. Additionally, there are a large number of edit controls for functions such as zooming in/out and cutting and pasting the waveform. Across the top of the window are the available rulers for timebase, including minutes and seconds, bars/beats, tempo, and meter. To the left of the Edit window is the Tracks list that includes a listing of every track that has been created. To the right of the Edit window is the Clips list that includes a listing of every audio segment (clip) that has been recorded, imported, or created by editing. (© 2016 Avid Technology, Inc. All rights reserved.)

Figure 5.20 The Transport window features the controls to operate the system and can be configured to suit your particular needs. The primary control buttons, from left to right, are the Online Indicator, Return to Start, Rewind, Fast Forward, Go to End, Stop, Play, and Record. When activated, the Online Indicator to the far left allows the playback of an audio cut to be started by an external device if your machine is so configured.

In addition to the primary controls, MIDI (musical instrument digital interface) controls can be added to the Transport window when composing and recording music. Likewise, there are a number of time displays and clock options that can be configured as a part of the Transport window (again, see figure 5.20).

Recording with Pro Tools. Just as with Adobe Audition CC, there are two ways to record audio using Pro Tools. Downloaded audio can be opened in the system, or you can record material from any other source by playing the material back in real time and using the analog or digital inputs of the system. Since you are working with the text's production music and we don't know which Pro Tools system you have access to or what operating platform, we can't walk you through the ripping process. If you elect to rip the music in, refer to your specific Pro Tools reference guide for the correct procedure.

After starting Pro Tools, you should be in the Edit window view. If not, on the menu bar click on Windows and on the drop-down menu select Edit. You should see the Edit window and the transport controls (see figure 5.21 on the following page). To create the tracks, or channel strips, that you want to work with, click Track on the menu bar and on the drop-down menu select New. In the window that appears, enter the number of tracks you wish to work with and click Create. Since the blues progression sample we are working with has five tracks, you can start with five and add more if you need to.

Switching to the mix view, the bottom of each channel strip is labeled Audio 1, 2, and so on. You can change the label by double-clicking on the label and entering a new name in the window that appears. The names can be changed as many times as you wish.

To prepare a channel strip to record, click the channel's Input selector and a drop-down menu will appear, listing the system inputs (see figure 5.22 on the next page). Select the appropriate audio input for your system. Likewise, you need to select the channel's output, or where you want the audio to go during playback (see figure 5.23 on p. 163). Clicking the Output selector causes a drop-down menu to appear listing the system outputs. Select the appropriate output for your system. You have now routed audio into and out of each channel strip.

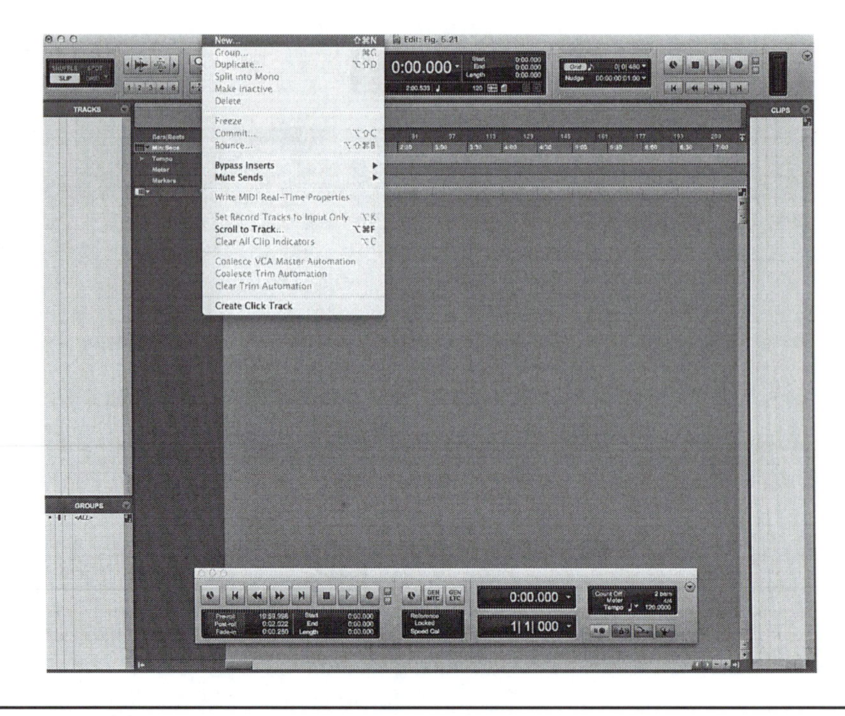

Figure 5.21 After starting Pro Tools, you should be in the Edit window view. At this point all you have to work with are the transport controls. To create the tracks, or channel strips, that you want to work with, click Track on the menu bar and on the drop-down menu select New. In the window that appears, enter the number of tracks you wish to work with and click Create.

Figure 5.22 To prepare a channel strip to record, click the channel's Input selector and a drop-down menu will appear listing the system inputs. Select the appropriate audio input for your system.

Figure 5.23 After selecting a channel's input you need to select the channel's output, or where you want the audio to go during playback. Clicking the Output selector causes a drop-down menu to appear listing the system outputs. Select the appropriate output for your system.

To record the first track, click on the channel's Record button and place the channel into the record mode. The button turns from gray to red, indicating that the channel is ready to record. To place the transport into the record mode, click on the Record button to the far right of the transport (see figure 5.24 on the following page).

Adjust the input level either at the source or on the control board feeding Pro Tools. Click the Play button, and the transport clocks begin rolling, indicating that you are recording. After recording track 1, click the Stop button on the transport to stop the recording. Before you play back, click the Record button again to take the channel out of the record mode so you don't accidentally record over the track later. If you wish to lock out the Record button so that it takes two steps to put the track back into record, consult your guide, as the procedure is slightly different for the Windows and Mac versions of the software.

To play the track back, click on the Return to Start button, turn up Pro Tools on the control board, and press play. You can also use the space bar on the keyboard to start and stop playback. If you are happy with the track, now is a good time to save your work, *before* you start editing. On the menu bar click on File, and on the drop-down menu select Save. When you are working on a project it is a good idea to save your work often, as a power outage or computer problem could wipe out all of your unsaved work.

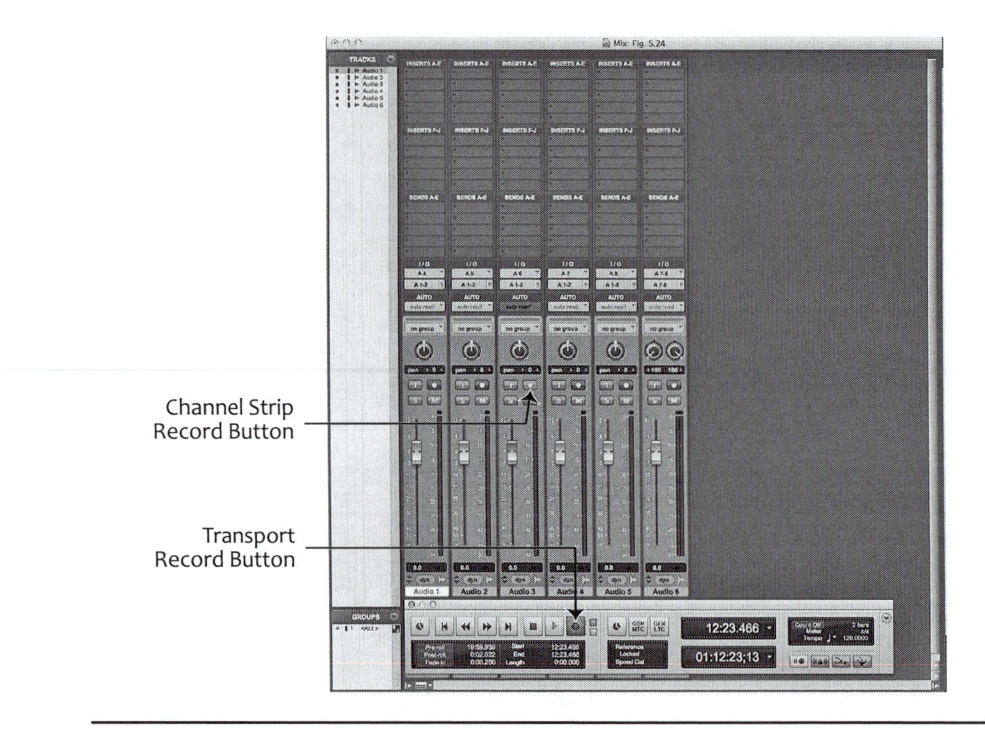

Figure 5.24 To place the transport into the record mode, click on the Record button to the far right of the transport, changing it from gray to flashing red, indicating that the transport and the selected channel, or channels, are ready to record. ()

Editing with Pro Tools. When the first track is recorded, the next step is to trim the ends of the track so that the music starts and ends cleanly. For demonstration purposes, we are going to trim the beginning of the track to remove the dead air or noise so that the music starts instantly upon playback. To get to the Edit window, go to the menu bar and click on Windows; on the drop-down menu select Edit. The audio waveform you recorded is displayed under the timebase ruler (see figure 5.25). A waveform is referred to as a region in Pro Tools.

The toolbar across the top of the window has a number of controls for viewing and editing the waveform. Starting in the upper left corner are the Edit mode controls. With the Shuffle mode selected, when an edit is made, the edited ends of the audio snap together automatically. In the Slip mode, the two ends of an edit remain where they are and you can click and drag the audio regions wherever you like.

To the right are the next group of buttons, zooming controls used to magnify or resize the audio waveform (again, see figure 5.25). To zoom in horizontally, click the right Zoom button until the view you want is displayed. To zoom out, use the left Zoom button and click until the view you want is displayed. To zoom in vertically, click on the small arrow button with the large waveform indicator. Likewise, to

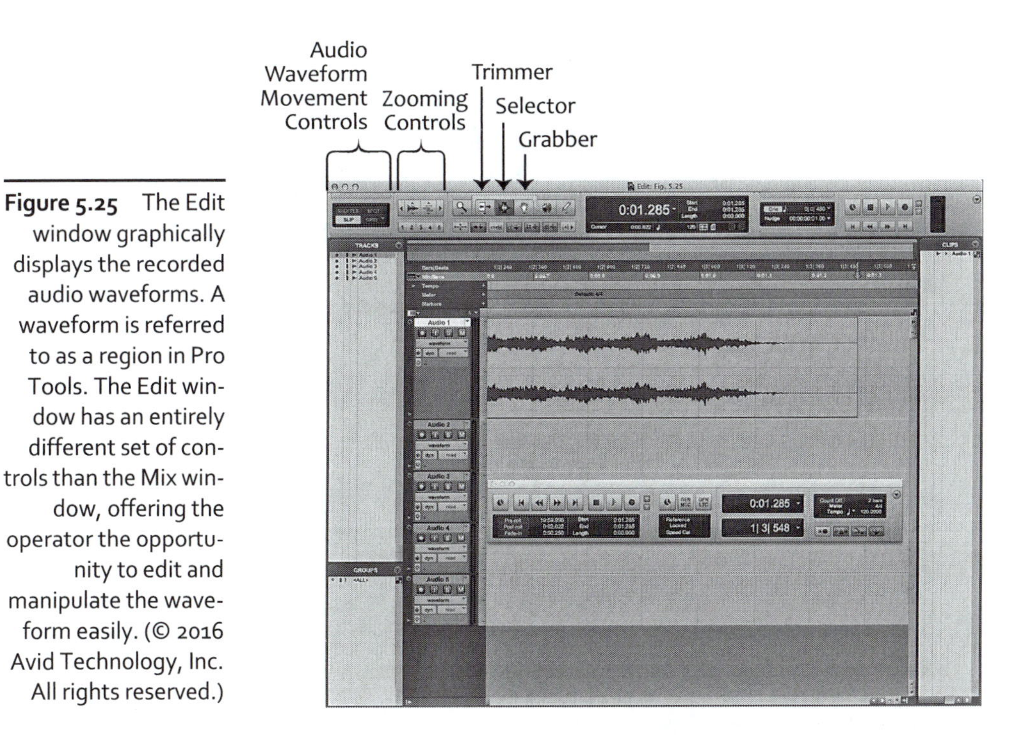

Figure 5.25 The Edit window graphically displays the recorded audio waveforms. A waveform is referred to as a region in Pro Tools. The Edit window has an entirely different set of controls than the Mix window, offering the operator the opportunity to edit and manipulate the waveform easily. (© 2016 Avid Technology, Inc. All rights reserved.)

zoom out vertically, click on the small arrow button with the small waveform indicator. Below the Zoom buttons are five preset zooms numbered 1 to 5. Selecting one of these snaps the view to one of the preset zoom views. To change the size of the track, click on the Track Height selector and a drop-down menu appears with the available sizes ranging from Micro to Fit to Window (again, see figure 5.25).

Pro Tools uses two cursors in the Edit window. The first is a solid black vertical line that moves horizontally across the waveform as you play back the audio. This is the playback cursor and indicates the exact point of playback as it moves. The second cursor is the edit cursor and is a blinking vertical line that only appears in the track in which you place it using the Selector.

To edit the beginning and end of the musical cut, there are two tools to choose from, the Trimmer and the Selector (again, see figure 5.25). To trim the beginning of the waveform with the Trimmer, first select the Slip edit mode. Click on the Trimmer button, and move your cursor into the start of the waveform. Notice that the cursor has turned into a [cursor. Click and drag the Trimmer to the right; as you do, notice that you are trimming the unwanted noise and blank space (see figure 5.26 on the next page). When you release the Trimmer, the audio is trimmed to that point. The same procedure can be used for the end of the waveform, with the Trimmer automatically turning into a] cursor at the end of the cut. Should you make an error and trim too much, click Edit on the menu bar and on the drop-down menu select Undo.

The second tool you can use to edit a waveform is the Selector (again, see figure 5.25). With the Edit window still in the Slip mode, click on the Selector button

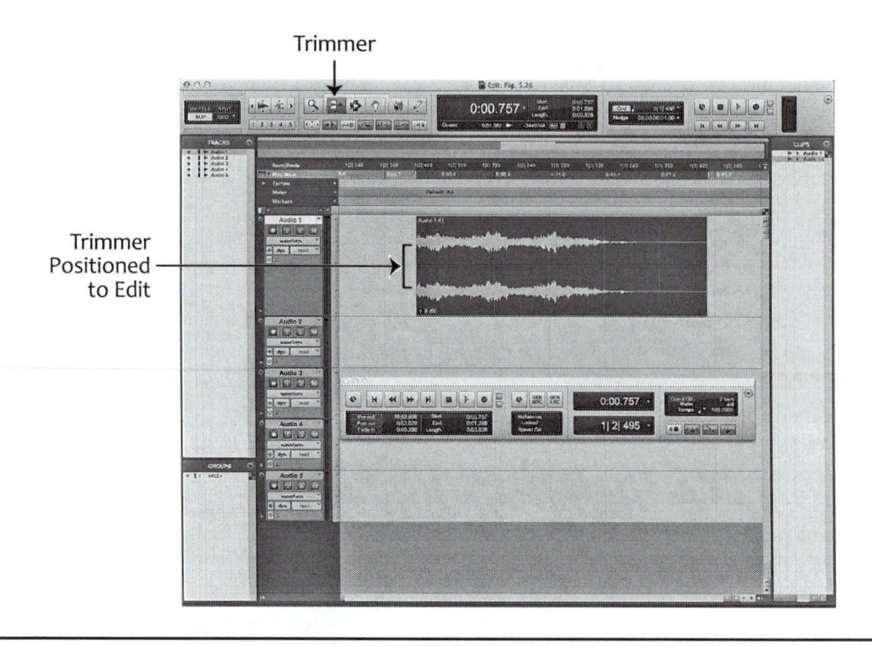

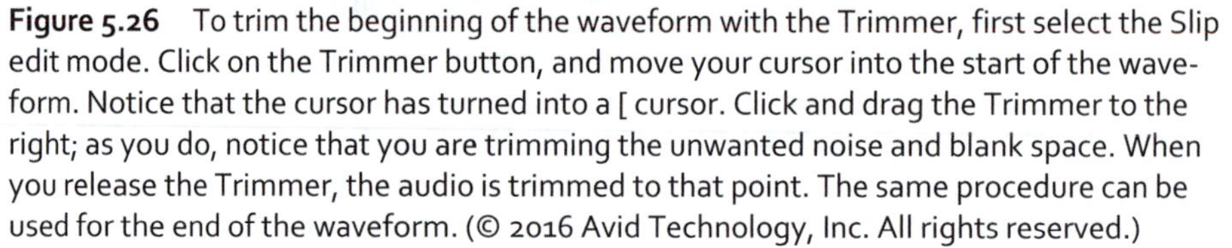

Figure 5.26 To trim the beginning of the waveform with the Trimmer, first select the Slip edit mode. Click on the Trimmer button, and move your cursor into the start of the waveform. Notice that the cursor has turned into a [cursor. Click and drag the Trimmer to the right; as you do, notice that you are trimming the unwanted noise and blank space. When you release the Trimmer, the audio is trimmed to that point. The same procedure can be used for the end of the waveform. (© 2016 Avid Technology, Inc. All rights reserved.)

to highlight it. Moving the cursor into the waveform changes the cursor into the shape of an I (see figure 5.27). Place the I at the start of the audio waveform, and click and drag it to the left, highlighting the area to be removed. If you don't highlight the exact area you want the first time, place the cursor anywhere on the screen outside the waveform and click once, and the highlighting will disappear. Try again.

With the area you want to remove highlighted, press the delete key on the keyboard. The area is removed, leaving a space between the start of the audio and the start of the track. Click on the Grabber button located on the tool bar to activate the Grabber function (again, see figure 5.25); as you move the cursor into the waveform, notice that the cursor is now in the shape of a hand. Click on the track, and you can now slide it to the left or right, positioning it exactly where you want it. (Remember that had the Edit window been in the Shuffle mode when you hit the delete key, the waveform would have automatically snapped to the start of the track.) Anytime you are editing and have removed too much or made an error, all you have to do is go to the Edit drop-down menu and use the Undo command.

After trimming the beginning, you are ready to edit the end of the musical cut using the same technique. Once that is done, record at least two more tracks of music so that you are ready to move on to mixing.

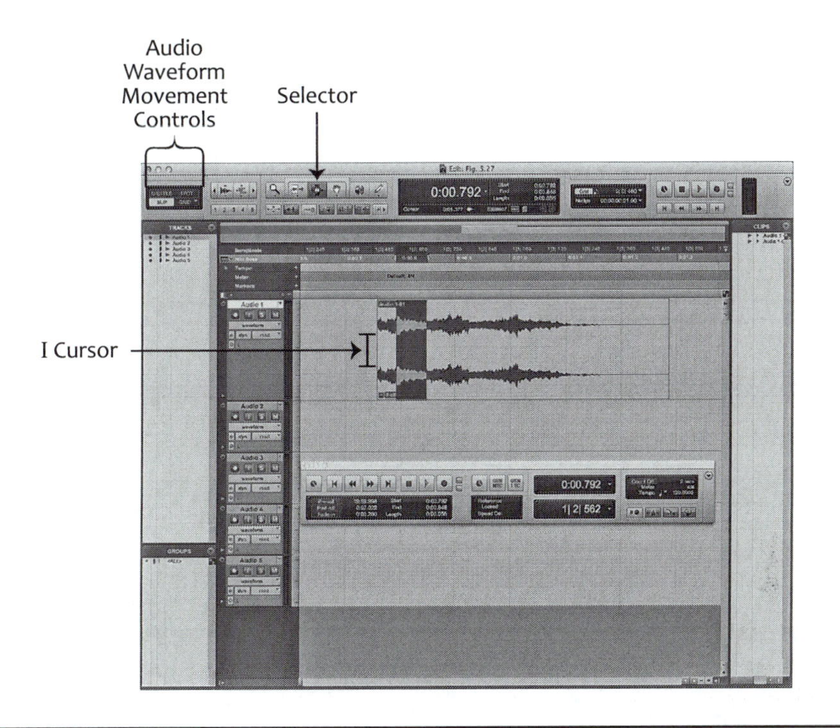

Figure 5.27 Another tool you can use to edit a waveform is the Selector. With the Edit window still in the Slip mode, click on the Selector button to highlight it. Moving the cursor into the waveform changes it into the shape of an I. Place the I at the start of the audio waveform, and click and drag it to the left or right, highlighting the area to be removed. (© 2016 Avid Technology, Inc. All rights reserved.)

Mixing with Pro Tools. To get back to the Mix window, go to the menu bar and click on Windows. On the drop-down menu select Mix (again, see figure 5.17). Before you mix anything, make sure that all of the individual channel strips are out of the record mode. Just as you did with Adobe Audition CC, use the channel strip faders to adjust the audio levels from each channel to achieve the sound you desire. Just above the faders are the pan controls to sonically position the audio from each channel into either the right or left channel. When satisfied with your project, you can save it to a flash drive or email it for electronic distribution.

As you have no doubt noticed, Pro Tools has many more controls, shortcuts, and adjustments with which to work. Generally, Pro Tools offers the user a number of different options to perform each task. Although this is handy, it does increase the learning curve necessary to take full advantage of all of the available options. However, with these basics, you are ready to understand the additional Pro Tools features by working with the software and reference guide.

Portable Digital Audio Recording

Portable audio recording lends itself to some very creative solutions. One of the most popular is a flash drive recorder (see figure 5.28). Most flash drive recorders have a variety of input and output options, including a built-in microphone(s), jacks for external microphones, jacks for line inputs, USB outputs, and analog and digital outputs. Depending on the size of the memory card, some can record nonstop for days.

Both Adobe Audition CC and Pro Tools can easily be set up on a laptop computer for field use, which has many practical applications. For example, a laptop with news-writing software and Adobe Audition CC makes a professional, mobile news facility. The newsperson splits the computer's display, with the audio soft-

Figure 5.28 Portable audio recorders include products such as high-quality flash card recorders. Digital location recording is very affordable for even small radio stations and producers. In addition to the built-in stereo microphones, professional microphones can be plugged in to the recorder. Above the recorder is the windscreen for the built-in microphones.

ware on one side and the news software on the other. While recording the city council meeting, the person can be writing his or her story at the same time. Editing audio cuts is easy and fast. When it's time to feed the story, the reporter reads the copy and plays back the audio cuts, all from the same laptop. By using a WiFi hotspot or by adding 4G LTE wireless Internet service to the laptop, the reporter becomes a virtual newsroom.

Project Backup and Storage

Backup

It will happen, and it is nothing new. No matter how hard you try or how careful you are, you are going to lose a project at some point in your career. A hard drive can crash and it is not unheard of to forget or lose a flash drive. It's Murphy's Law: if something can go wrong, it will. Having a backup plan and being prepared for the inevitable is the key to saving your neck. Always plan for the worst and hope for the best!

If you are working with a free-lance announcer, the last thing you want to tell him or her after a perfect take is that your software glitched and you lost the track. In a radio station you can usually move to another studio, but on location there is no other studio to move to. This applies particularly to live location events such as a concert, political speech, or important station event. The solution is simple: run two recorders simultaneously—your best recorder as the primary and another one as backup. You can even use the audio recording feature that most smart phones are equipped with as a backup. Should the primary fail for some reason, you still will have captured the event, or take, on the backup. For major events it is not uncommon to have multiple recorders in use running different audio sampling rates ranging from an MP3 to a studio-quality 24 bit, 96 kHz sample rate.

A true backup, though, is a copy of a project on another medium. Archival project backup can be done with a flash drive, or by uploading to a cloud server. Another effective and sensible way to backup projects is with a portable hard drive that is easily removed and stored in a secure location. For maximum safety, you should backup to two different media, such as a hard drive and a flash drive, keeping each in a separate, secure location.

Thanks to the reliability of the present generation of hard drives, many computer users are lulled into a false sense of security. Keep in mind, though, that a hard drive is an electromechanical device and it *will* fail at some point. Flash drives are another good way to archive material. Most flash drives, while holding huge amounts of data, are physically small. Your author has found that attaching a station lanyard (the kind given away as swag) works well in helping to identify your drive and making it a little easier to see and not lose. You might also want to label a file on your flash drive with your name and contact information.

A personal backup also serves another valuable purpose. When archived by the production person, it is a backup in case anything happens to the original, including getting accidentally erased from a server or destroyed by someone else in your department. The other use for a personal backup is an archive of your work for possible inclusion in an audition.

Project Storage

When planning for project storage and backup keep in mind that each minute of uncompressed stereo audio (44.1 kHz, 16 bit) consumes about 10 megabytes of storage. Bump the sampling rate up to 96 kHz at 24 bits, and the same minute of stereo audio gobbles up nearly 34.5 megabytes per minute. Stereo audio sampled at 192 kHz, 24 bits takes up a monstrous 66 megabytes per minute. Many broadcast and production facilities monitor server space and require staff to archive on a regular basis, moving outdated projects from the main server. This is particularly true with long-form news and public affairs programs.

The Digital Recording Process

Now that you know what some of the possibilities are in digital recording, the question is: how? What are the processes, components, and media that make it possible? From this point on, it may sound as though we are talking about computers rather than radio equipment.

The process of digital recording begins with signal acquisition, either from a microphone or a digital signal source, such as a control board or playback device like a flash drive recorder. Analog audio uses mic cables and wires to transfer audio from one piece of equipment to another. Digital audio, though, requires unique cabling systems designed to handle computer data rather than analog audio signals. Getting the audio from point A to point B is where we start.

Digital Inputs and Outputs

Digital audio is transferred from one piece of equipment to another using standardized data-transfer protocols (methods) over twisted-pair wire, category 6 computer cable, coaxial cable, or fiber-optic cable. The data is transferred at very high frequencies, similar to a video signal.

Digital audio cable has the same appearance as analog mic cable, a twisted pair of conductors surrounded by a shield, but that is where the similarity ends. Because of the very high frequency at which the data is sent, the cable's impedance (resistance to the flow of electricity) has to be higher than that of an analog cable, and cable quality is critical, especially for long runs. Typically, digital audio cable can be run a maximum distance of anywhere from 350 to just over 1,000 feet, depending on the quality of the cable. Cable manufacturers specify the maximum working distances of their particular brand of digital cable.

What makes things a little confusing is that common XLR microphone connectors are used for both analog and digital audio cables. In a facility where analog and digital equipment is mixed, engineers typically use a different color cable or clearly mark cables as digital. Digital audio quality is seriously affected if routed over an analog cable, and in some cases it can result in a complete loss of the digital signal.

To simplify the transmission of digital audio signals, several industry standards have been adopted to help eliminate confusion and to make the production person's job easier. However, keeping track of all the standards requires attention to detail.

AES/EBU Digital Audio Protocol. The Audio Engineering Society and the European Broadcast Union jointly established the AES/EBU professional protocol for transferring digital audio. The standard uses a balanced cable and sends two encoded audio channels (stereo) within a serial data stream on the cable. A serial data stream breaks down each audio sample's digital words into the individual bits (binary digits), sending them one bit after the other down the line. The receiving equipment reassembles the bits into the original digital words at the proper sampling frequency, and finally into audio using a digital-to-analog converter (D/A).

The standard AES/EBU audio cable connector is the familiar XLR connector (see figure 5.29). For long cable runs exceeding 1,000 feet, coaxial cable (similar to that used for the cable TV hookup in your home) is used. The standard connector for AES/EBU digital audio coaxial cable is a locking **BNC** connector (again, see figure 5.29).

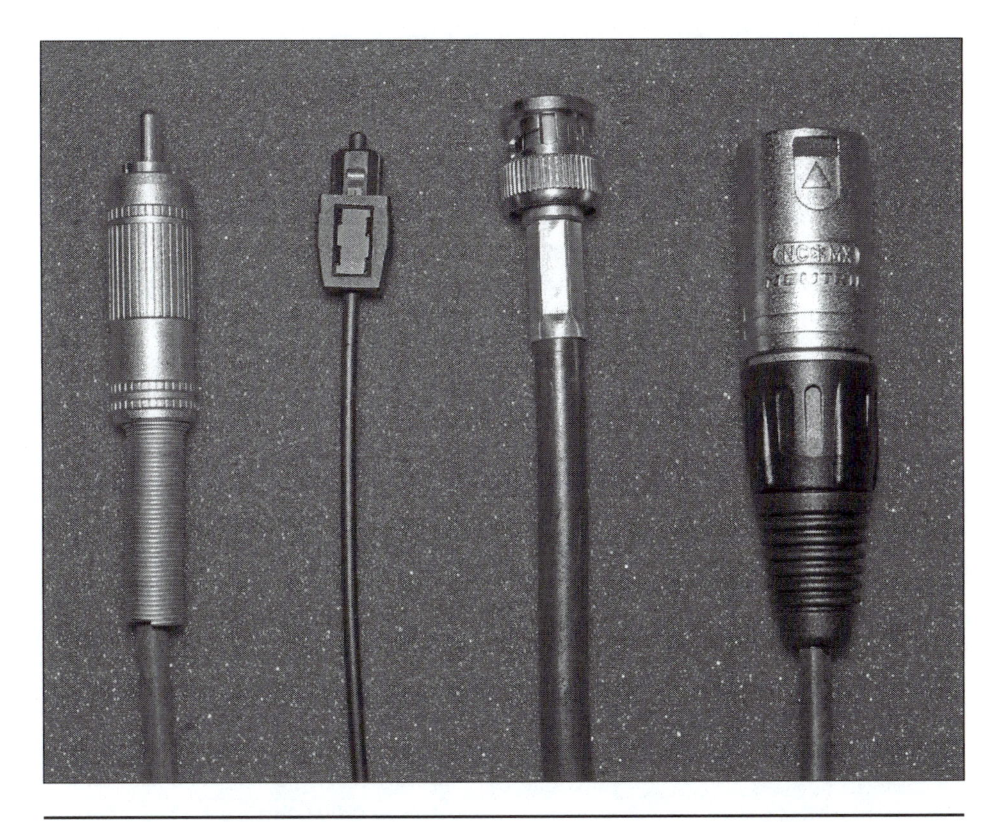

Figure 5.29 Some digital audio connectors look identical to their analog counterparts. From left to right, a SPDIF connector, a fiber-optic TOS link connector, a locking BNC connector, and an AES/EBU XLR connector.

Sony/Philips Digital Interface Audio Protocol. The **Sony/Philips Digital InterFace**, or **SPDIF**, is the consumer version of the AES/EBU standard. Commonly found on digital equipment, it uses small-diameter, lightweight, coaxial cable with an RCA phono connector on each end for connecting equipment together (again, see figure 5.29).

A **TOS link** fiber-optic cable can also be used to send SPDIF audio over short distances (again, see figure 5.29). Fiber-optic cables have fewer inherent data errors than coaxial cables, and they have the potential to deliver higher audio quality. Typical consumer TOS link cables are generally no more than 15 feet in length. Most consumer/prosumer electronic devices have both RCA phono and fiber-optic SPDIF inputs and outputs.

Firewire. Firewire was developed by the Institute of Electrical and Electronics Engineers (eye triple e) to transfer data between computers and external components using a special cable and data transfer protocol. Firewire was a popular method of connecting digital audio control surfaces, external hard drives, and external analog-to-digital audio converters to computers. However, USB 3.0 has replaced Firewire. And while Firewire is no longer being supported, you may see some legacy equipment in operation that still uses Firewire.

USB. Universal Serial Bus (USB) is a popular means to connect digital audio components to a computer. Each end of a USB cable has a different connector to facilitate ease of connection between equipment (see figure 5.30). Many digital audio devices such as MIDI controllers, portable hard drives, and outboard audio effects processors have USB connections. Virtually all computers have USB connections. A USB cable can transmit data at speeds up to 10 gigabits per second and provide electrical power to external components. USB cables are limited to 10 to 15 feet in length.

Figure 5.30 USB 3 and Thunderbolt 2 are two very popular protocols to transfer data between computers and external components. The cables are easy to tell apart. USB 3.0 can transfer data at up to 10 gigabits per second (left). Thunderbolt 2 is capable of transferring data at speeds of up to 20 gigabits per second (right).

Thunderbolt. **Thunderbolt** is another popular digital connection found on Mac products. And while the Mini DisplayPort (MDP) connectors are tiny, the cable's ability to carry data is powerful (see figure 5.30). A Thunderbolt 2 cable provides two digital channels on the same connector/cable at up to 20 gigabits per second and provides electrical power to external components. Most Thunderbolt copper cables are 6 to 9 feet long. Thunderbolt 3 is able to deliver 40 gigabits per second and using fiber-optic cables can reach as far as 300 feet.

Category 6 Cable. Category 6 cable (Cat 6) is used to interconnect computers, control boards, and any number of digital audio and control devices. A radio station making maximum use of digital audio will likely be wired with Cat 6 cable (see figure 5.31). The cable has four twisted pairs of wire within it and uses modular Ethernet connectors. Cat 6 is the standard cable for gigabit Ethernet and computer connections. Cat 6 cables are limited to 100 meters or about 330 feet in length.

Figure 5.31 This cable terminal room has all of the digital connections for five radio stations. Note that Cat 6 cable is used exclusively in the equipment racks and in the cable trays above the racks. Literally miles of audio cables have been eliminated by the use of Cat 6 network cable to distribute digital audio in the station group.

Digital Audio Interconnection

Professional equipment using AES/EBU digital audio inputs and outputs has either XLR or locking BNC connectors; some equipment features both types of connectors. Consumer equipment using SPDIF digital audio inputs and outputs has either RCA phono jacks or TOS link fiber-optic connectors, or both, to interconnect with other equipment.

Some equipment is designed to interface with just about any digital audio source. Such equipment has a variety of connectors including AES/EBU XLR connectors, AES/EBU BNC coaxial connectors, SPDIF RCA phono connectors, SPDIF TOS link fiber-optic connectors, and a USB connector. A number of prosumer and professional control boards and individual pieces of audio equipment are designed in this manner, allowing them to interface with virtually any digital audio source.

Digital Sample Rates and Clock Synchronization

Sample Rates. As you learned in chapter 2, the first step in digital recording is the analog-to-digital (A/D) conversion. The analog waveform is sampled, and each sample is converted into a binary number. The number of samples taken per second (sample rate) and the length of the digital word (bits) directly affect the audio quality. By their very nature, sampling and bit rates are rather technical; however, unless you are exporting audio to some other facility or media, the chances of your needing to know the many different sampling rates are pretty slim. It is when you want to take a project outside your studio that the sampling and bit rates become important.

So that professional equipment is compatible, the AES/EBU established preferred sample rates, depending on the application. A sampling frequency of 48 kHz is recommended for the recording, processing, and exchange of digital audio. However, there are some exceptions. A sample rate of 44.1 kHz is recognized for consumer audio applications such as the CD. For high-quality audio applications, such as in a recording studio, 96 kHz is the recognized sampling frequency.

Professional broadcast equipment often is sample-frequency agile, meaning the sampling frequency is selectable depending on the intended use. As an example of how sample rates are used, a CD has a fixed sample rate of 44.1 kHz at 16 bits; Adobe Audition CC offers recording sample rates from 44.1 kHz through 192 kHz at 16, 24, and 32 bits.

As a final note about sample rates, once hardware and software is established and operating properly in a studio, it is rare that you would ever need to adjust sample rates. In fact, doing so will most likely result in either recording or playback problems within the studio.

Clock Synchronization. You have seen the spy movies where all the spies stand around and synchronize their watches so they can all be on time to catch the bad guy. The same thing has to happen when multiple pieces of digital audio equipment are interconnected. The digital clock in each piece of equipment has to be synchronized with the ones in other pieces of equipment.

When a serial digital audio signal is sent to another piece of equipment, each sample's digital words are broken down into their individual bits and the bits are sent one bit after another down the line. The receiving equipment reassembles the bits into the original digital words at the proper sampling frequency and into audio using a digital-to-analog (D/A) converter.

The data streams also contain additional information that tells the receiving equipment the signal's sampling frequency and where each sample begins and ends. This is called a self-clocking signal, since the signal includes information that keeps the sampling frequency and bit stream synchronized during the audio transfer.

Just as with the spies, someone has to keep the master clock for everyone to synchronize to. When two digital devices are connected together, one of them keeps the time or maintains the master word clock (a digital word is comprised of data bits) so that all of the digital words and bits arrive in synchronized order. One piece of equipment serves as the master and one, or more, as the slave.

Since the word clock, or synchronization signal, is already built into the AES/EBU and SPDIF signals, the process happens automatically most of the time. The recording equipment looks for the word clock at its input and synchronizes itself to the digital source feeding it. The source serves as the master, and the recorder becomes the slave. A master clock is also referred to as an external clock, and a slave clock is regarded as an internal clock. If two digital devices are connected and neither has an external, or master, clock the equipment receiving the digital audio does not know where the samples begin and end or how to put the bits back together into digital words. In other words, the two pieces of equipment can't communicate with one another, and the audio just becomes random computer data and is lost.

Normally, when a number of digital audio devices are connected together, one is set as the master and the others are slaves. Another way to synchronize the equipment is with a separate master clock (house clock) that feeds all of the equipment in the radio station a single, synchronized, master word-clock signal.

Digital Recording Media

Once a signal has been converted into binary numbers, the recording process involves storing the data just as you would store any other computer data. The stored data represents audio that can be played back, edited, mixed, processed, and eventually mastered to the final product.

Magnetic Tape

High-quality magnetic tape *was* a suitable medium for digital audio recording. It is a thing of the past. All of the major manufacturers of magnetic digital audiotape no longer make or service the machines, including Sony, who invented the Digital Audio Tape (DAT) format.

Hard Drives and Solid-State Memory

Computer hard drives are an ideal storage medium for digital audio due to their ability to store and retrieve quickly large quantities of data. Stereo (44.1 kHz, 16 bit) digital audio data requires just over 10 megabytes of hard disc space for each minute of stereo audio. At the opposite end of the spectrum, recording-studio quality (96 kHz, 24 bit) audio requires nearly 34.5 megabytes for each minute of stereo storage.

Internal Hard Drives. An internal hard drive is mounted within the computer or server being used as a digital audio workstation. A hard disc used for digital audio recording needs to access large amounts of data very fast and often records and plays back continuously for several hours on end, a much more demanding duty cycle than what is asked of a typical office computer hard drive. It is important that the appropriate hard drive be selected and that it be configured for the specific tasks it must perform.

Hard drives have increasingly gotten larger as the cost per gigabyte of storage has gone down. Drives of two to five terabytes are common (one terabyte equals 1,000,000,000,000 bytes) and offer more than ample storage capacity for audio projects. However, bigger is not always better. Since a large hard drive has such a wide area to search, it takes more time to retrieve your project and can slow down search and playback times.

Large hard drives are often partitioned, or divided, into smaller sections, which are then given separate drive numbers even though they are all physically located on the same magnetic disc. For example, a one-terabyte drive might be divided into two 500-gigabyte segments. The hard drive then only has to work in, and search, one smaller section of the disc at a time. Drive speed, or how fast the disc spins, also affects how quickly the data can be accessed and retrieved, and it also determines the maximum sample rate and the number of tracks that can be recorded simultaneously. Typically, the minimum hard drive speed for professional audio applications is 7,200 rpm.

Modular or Portable Hard Drives. Modular or portable hard drives are referred to as stand-alone drives, since they are either rack-mounted or set on the console and connected to the computer (see figure 5.32). Several drives can be stacked or added to a computer to greatly enhance its storage capacity. In addition to normal use for storage and playback, a stand-alone drive is sometimes used as a mirror drive running in synchronization with the primary drive and recording everything the primary drive records. Should the primary drive fail, the mirror drive seamlessly takes over.

Another type of modular drive is a small portable hard drive. Capacity can range up to 2 to 4 terabytes and many will fit in a man's shirt pocket. Such drives are self-contained and get their power to operate from the device they are connected to. Most are powered by a Thunderbolt or USB cable that connects them to the computer.

Figure 5.32 Modular/portable hard drives are referred to as standalone drives, since they can be set on the console or be rack-mounted. Several drives can be stacked or added to a computer to greatly enhance its storage capacity.

Memory Cards and Solid-State Media. Portable digital recorders are popular for fieldwork in radio. Such devices use removable flash memory cards available under a number of commercial names, including memory sticks or flash cards. Referred to as flash recorders, they are capable of storing days of high-quality audio on a single card. A 32-gigabyte memory card can store over 50 hours of uncompressed stereo audio (again, see figure 5.28). An added benefit of a flash recorder is the ability to remove the flash card, insert it into a computer port, and transfer the audio instantly to the computer for editing or playback. Many such devices also have a USB connector. Audio can be transferred to a computer by simply dragging and dropping the file from the recorder to the computer.

Another handy device for storing audio is a USB flash drive. This drive is smaller than a pack of gum and contains a memory chip having up to 1 terabyte of data storage space. The drive is plugged into a USB port on the computer and can be loaded with audio, carried on a keychain, and plugged into another computer and downloaded. A more popular place to carry a USB flash drive is around your neck, on a lanyard or media badge holder.

Depending on the audio storage format, a USB flash drive can easily hold hundreds or thousands of commercials or many hours of audio and can be used either as a temporary backup or to transfer commercials from one studio to another in your station. A 32-gigabyte flash drive can hold over 3,000 60-second commercials or about 40 to 45 hours of uncompressed audio. The same flash drive could hold about 8,000 MP3 songs at an average of 4 minutes each.

Optical Discs

With the development of the CD in the late 1970s and its commercial release in the early 1980s, the world adopted a standard in music playback.

There are several reasons why the CD was so popular. It has excellent audio quality, the machines for recording and playback are inexpensive and easy to operate, and the medium itself costs only pennies per disc. At one time most desk or laptop computers were equipped with a CD recorder/player of some type, giving individuals as well as professionals the ability to capture and create quality recordings. With the use of the CD waning, most computer manufacturers now only offer the CD recorder/player as an option. Most music is now purchased via direct digital transfer from the web. Many reading this book may have never even purchased music on a CD!

A CD is a clear polycarbonate (tougher than regular plastic) disc that holds up to 700 megabytes of data, or up to 80 minutes of 44.1 kHz, 16-bit stereo audio. (The history of the original size and audio capacity of the CD is a quirky story. Mr. Norio Ohga, president of the Sony Corporation and a former classical vocalist, decreed that the new technology had to be able to play Beethoven's Ninth Symphony in its entirety, without interruption. So the original length of a CD was set at 74 minutes.) The laser that records and plays the disc looks up through the disc from below.

There are three layers of material on top of the disc that make it possible to record and play back audio. The first layer is a heat and photosensitive organic dye; this is where the information is stored. When a tightly focused laser beam is aimed at the disc during the recording process, it burns or blackens a tiny oblong area called a pit. On top of the dye is a layer of gold or silver that, wherever there is no pit, reflects the laser beam back off the disc to be read by the recorder/player.

During playback, a less-powerful laser is aimed at the disc, and it reads the reflections of the shiny flat areas and the darkened pits, respectively, as ones and zeros. If the laser's reflected signal changes (it sees a blackened pit), the binary value is read as a zero. If the reflection is constant, it is read as a one.

The third layer, on top of the gold or silver, is a coating of clear lacquer extending all the way over the inner and outer edges of the CD to seal the materials onto the disc. Once sealed, the disc's label is printed on top of the lacquer.

With regard to the gold or silver reflective layer, the gold is projected to last about three times longer than silver. One major manufacturer estimates that a silver-based CD can last 100 years, versus 300 years for the gold disc. Although the gold obviously has long-term advantages over the silver, for short-term use the less-expensive silver is more than adequate. Discs that have a blue or yellow cast to them are silver discs in which the dye color is apparent.

Suggested Activities

1. With the help of your facility engineer, trace the audio signal path through your production facility from the microphone to the final mastering medium. Is the path digital or analog? If digital, what kind of signal path is being used in the facility, AES/EBU, SPDIF, Cat 6 Ethernet, or a combination of the three?
2. If you have not done it already, visit the Adobe Audition CC website (www.adobe.com) and take a look at the software and other products to see what is available to turn your computer into a digital recorder.
3. Visit the Pro Tools website (www.avid.com/pro-tools) and explore the many different versions of Pro Tools available.

Websites for More Information

For more information about:

- *Adobe Audition CC,* visit Adobe at www.adobe.com
- *Audio Engineering Society's digital audio standards,* visit the Audio Engineering Society at www.aes.org
- *digital audio cable,* visit Canare Corporation at www.canare.com
- *Pro Tools,* visit Avid Technology, Inc. at www.avid.com/pro-tools

Pro Speak

You should be able to use these terms in your everyday conversations with other professionals.

BNC—An unbalanced coaxial cable connector with a bayonet-type locking system invented by Paul Neill and Carl Concelman, thus the name Bayonet Neill-Concelman. Typically found on video and digital audio cables. The center conductor of the cable is connected to a pin, and the shield is connected to a barrel surrounding the pin. A rotating ring outside the barrel locks the cable to any female connector.

compression—An audio processing technique in which the highest audio peaks are automatically turned down, reducing the dynamic range of the audio signal in a direct ratio to the audio input level; a form of automatic volume control.

digital audio workstation (DAW)—The term generally applied to a computer-based digital recording system that incorporates the ability to record, store, manipulate, transport, and deliver digital audio products.

digital signal processor (DSP)—A device for the manipulation and modification of audio signals, such as reverb, delay, or dynamics processing.

file transfer protocol (FTP)—A system for exchanging computer files with another computer using the Internet as opposed to the World Wide Web.

gating—An audio processing technique that acts just as its name implies, as a gate. When audio is above a preset level, the gate allows it to pass. When the audio level drops below the preset level the gate closes, blocking the background noise.

limiting—An audio processing technique that automatically controls audio levels that approach a preset maximum audio level. When the preset level is reached, the audio is turned down so that it does not exceed the preset level.

local area network (**LAN**)—A computer interconnection system for sharing data between or among computers within a company or media facility at one location.

Sony/Philips Digital InterFace (**SPDIF**)—A standard digital audio file transfer format using 75-ohm coaxial cable.

Thunderbolt—A Thunderbolt 2 copper wire cable provides two digital channels on the same connector/cable at up to 20 gigabits per second and provides electrical power to external components. Thunderbolt 3 is able to deliver 40 gigabits per second over a fiber-optic cable.

TOS link—The name given a system of fiber-optic cables and connectors developed by Toshiba and used for the transmission of SPDIF digital audio.

wide area network (**WAN**)—A computer interconnection system for sharing data among computers within a company or media facility at multiple geographic locations.

Audio Processing

Highlights

An Introduction to Audio Processing

A chocolate cake is pretty good by itself, but with a thick, rich layer of dark chocolate fudge icing, the cake is irresistible. It excites our taste buds and creates an insatiable desire for more. It's the same thing with a good piece of production. Add audio processing that complements the original work, and suddenly the production comes to life, demanding the listener's attention. Audio processing is the icing on the cake to our ears.

Audio processing is used either to help correct a problem or to add a unique artistic signature that sets a recording apart from the competition. Why do those announcers on national commercials all have the little-extra-something special about the bottom end of their voice (see audio demo, cut 6-1)? If you play a musical instrument, have you ever listened to a song and wondered how in the world the musician on the recording got that sound out of that instrument? The answer to these questions is most often some form of audio processing.

To this point, most of what you have learned with regard to audio production is basic science. In this chapter you will begin to merge the science with your talents and abilities. Audio processing is not an exact science; it is mostly based upon artistic taste. There are no charts or tables that list exactly how much reverb to add to a voice track to make it blend with the production music that it's being mixed with. Your skills in this area develop from listening, observing, and reading about how others work, along with your own experimentation. There is one finite aspect of audio processing that you need be clear about. A bad recording is a bad recording: garbage in equals garbage out. No amount of audio processing can turn a bad recording into a good recording.

Throughout this chapter audio processors are referred to as if they are individual pieces of equipment. While many audio processors still exist as individual pieces of equipment, many have been replaced with software and apps found in programs like Adobe Audition CC and Pro Tools.

There are three primary types of audio processors: dynamic processors, frequency equalizers, and dimensional processors.

Dynamic Processing

The traditional recording technique is a simple process. Select the studio, or performance venue, with the best acoustics. Select the best pair of stereo microphones for the job; one for left channel, one for right. Audition the musical group, taking care to place the mics in the **sweet spot** of the studio or performance venue to achieve the desired direct-to-ambient-sound ratio. Have the musicians play their loudest musical passage, and set the recorder input levels at about –10 VU, leaving adequate headroom. Start the recorder and do not touch the input-level controls until the group is finished. If you have done everything correctly, the recording likely sounds very natural, with a wide dynamic range from the softest to the loudest sound. This is a timeless recording technique for jazz and classical music that highlights the subtle nuances of the music, just as they occurred in the live performance. The very same basic recording technique can be applied to recording an announcer in your production studio.

Unfortunately, we live in a very noisy world, and listening to such traditionally recorded material is difficult in a car, office, or when exercising, because the quieter passages get lost in the ambient noise that surrounds us. This is one of the reasons producers turn to audio processing to tweak, or tune up, a recording or radio signal and make it more commercially viable.

As a general rule, dynamic processors are the first ones used to improve or alter a recording. A dynamic processor changes the dynamic range of the program material, either reducing or increasing its dynamic range. Dynamic processing includes compression, limiting, expansion, and gating.

Compression

Successful recording and mixing techniques require producers to develop a unique sound image that stirs emotions and draws attention to a project. The project must stand out from all of the other material in the marketplace if it is to be successful.

A compressor is an automatic level control that reduces an audio signal's dynamic range by turning down, or compressing, the loudest audio peaks. By reducing the high-level audio peaks, the producer can then raise the overall average level without the peaks causing hard clipping. For example, if a compressor is used to lower the level of the loudest audio peaks by 6 dB, then the average level can be raised 6 dB, producing a louder, more uniform audio signal. The compressed audio peaks still sound loud, and because you have turned up the average level of the audio by 6 dB, the softest sounds are now also louder (see figure 6.1). A compressor enables you to control sounds, such as vocals, that have a wide dynamic range. A compressor is very effective at making an announcer sound louder and making his or her voice dominate anything it is mixed with (see audio demo, cut 6-2). Even a whisper, when properly compressed, can dominate a commercial or promo and although it is loud, it can still sound like a whisper (see audio demo, cut 6-3). A compressor is the most basic tool used for processing.

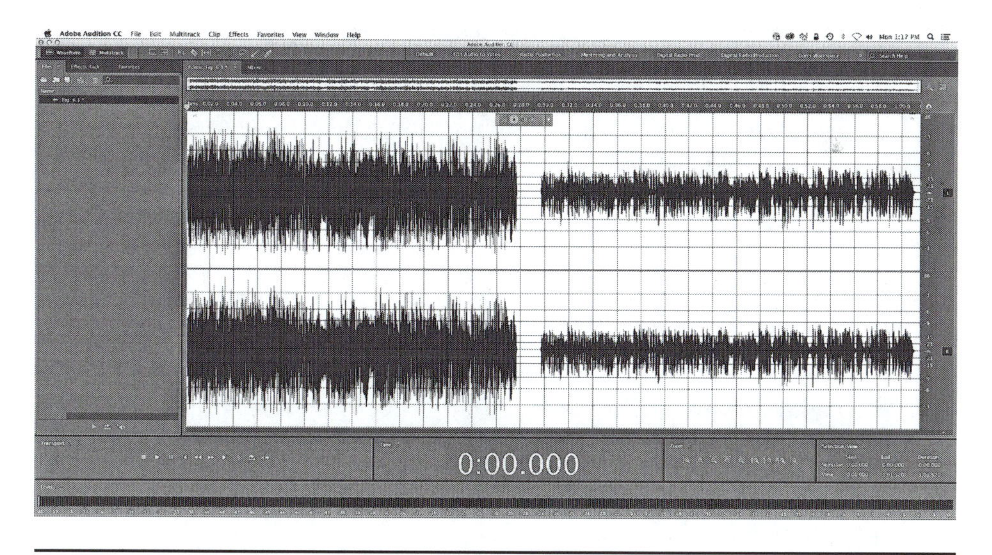

Figure 6.1 A compressor enables you to control sounds, such as vocals, that have a wide dynamic range. By reducing the dynamic range, the overall level can be increased. A compressor is the most basic tool used for processing. If a studio has only one audio processor, count on it being a compressor. In this example, the audio on the left has been compressed and the gain increased to maximized levels while the audio on the right remains uncompressed. (Adobe product screen shot reprinted with permission from Adobe Systems Incorporated.)

How Compressors Work. When audio is fed to the compressor input, the compressor decides whether the audio is above or below an adjustable **threshold,** set in decibels below 0 VU. If the audio is below the threshold, the compressor does nothing. If the audio rises above the threshold, the compressor reacts, turning down the level. How quickly the compressor reacts to the incoming audio is the **attack time,** measured in milliseconds (thousandths of a second).

The compressor reduces the audio level of the signal based on an adjustable gain ratio that compares the input level of the compressor to the output level. For example, with the ratio set at 3:1, when the audio rises above the threshold, every 3 dB of audio input produces only 1 dB of audio output. When the audio drops below the threshold, the compressor releases control of the audio and it returns to normal. How slowly the compressor releases control of the audio is the **release time,** also measured in milliseconds. Once the audio peaks are compressed, the overall level can be raised using the compressor's gain control.

The procedure for adjusting a compressor is pretty straightforward. The first adjustment is the **compression ratio,** or how much the audio will be turned down. Ratios of between 3:1 and 6:1 produce natural-sounding level control (compression) that is not generally noticeable. Excessive compression ratios produce dull-sounding recordings because the dynamic range is reduced to the point that there are no loud or soft passages—it's all one fixed level, producing lifeless-sounding production and causing **listener fatigue** (see audio demo, cut 6-4).

The second set of adjustments on a compressor is the attack and release times (see figure 6.2). Start with a fast attack time of 1 to 1.5 milliseconds and a release time of 350 to 500 milliseconds. You want the compressor to take action quickly and to release slowly, so that it is not noticeable. However, depending on the sound you want to achieve, you can start with the slowest attack time and the fastest release time, delivering a very punchy sound, one that jumps out at you.

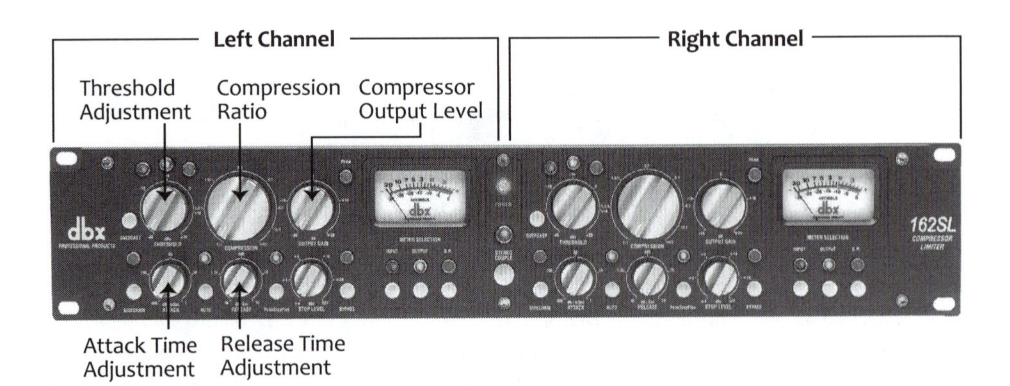

Figure 6.2 If a radio station has only one audio processor in production, it is likely a compressor/limiter. The versatility of such a device allows you to create a number of unique effects and improve the sound of audio tracks. The controls labeled above are used for compression.

Announcers love this kind of effect as it makes their voice sound bigger and gives it more impact.

An indicator of excessive compression is pumping or breathing. This is caused by improperly adjusting the attack and release times so that the millisecond the audio is sensed the compressor kicks in and dramatically lowers the level. The moment there is a pause in the audio the compressor releases and the background noise rushes up, producing breathing or a pumping sound as the audio fluctuates up and down. This is a very annoying effect (see audio demo, cut 6-5).

The third adjustment is the threshold: slowly turn down the threshold control of the compressor until the compressor meter reads 3 to 6 dB of compression on the loudest audio peaks (again, see figure 6.2).

Since the compressor is turning down the audio, the compressor's meter reads the opposite of a VU meter. The needle, or indicator, sits at the right side of the meter at 0 and pulls back to the left, showing the decibels of audio reduction. An LED meter will react in the same way, appearing to function opposite of what it normally does. After compression, to compensate for the reduced audio level, use the compressor's gain control to boost the final output to the desired level.

These basic compressor settings are just that, a place to start. With each different project and brand of compressor, the settings are going to vary depending on the effect you want. Some compressors have a broader range of adjustment than others. Software-based compressors typically have a number of preset configurations for you to choose from in drop-down menus. As you will find with most audio processors, the best way to learn how a compressor works is to experiment with it. Feed it an audio signal and see how many different ways you can make the audio sound.

Limiting

A limiter is basically a compressor with a gain ratio in excess of 10:1; some go as high as 100:1. A compressor can be adjusted to act as a limiter, and many compressors are sold as a compressor/limiter. Where the compressor's output is always a ratio in relation to its input and varies, the output of a limiter is fixed at a maximum amplitude or level (see figure 6.3 on the following page). Limiters are like a brick wall through which the audio peaks cannot pass, regardless of the audio input level (see figure 6.4, also on the following page). Once you hit the limiter's predetermined level, it really doesn't matter how much you turn up the input, the audio output level stays the same.

Limiters are used to prevent hard clipping and to increase average audio levels. Compressors and limiters are often used in production to make sure that each piece of production that leaves your studio is at the same maximum level. This makes the commercial sound louder and makes it easier for the jocks to do their job, since they don't have to be on guard constantly for commercials that are too loud or too soft. Limiters have similar controls to a compressor and are set up and used in a comparable manner (see audio demo, cut 6-6).

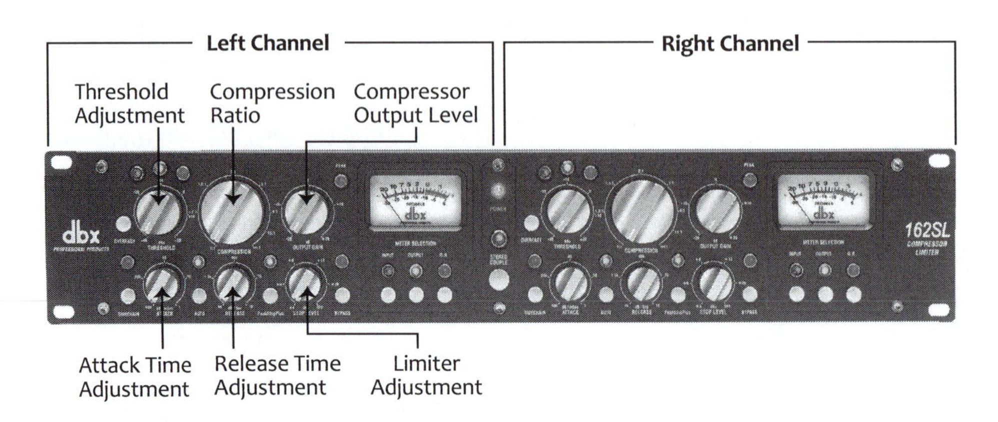

Figure 6.3 A limiter is basically a compressor with a very high compression ratio; some go as high as 100:1. Limiters are used to prevent hard clipping and to increase average audio levels. On this model, the limiter control is called the stop level. In other words, you set the maximum audio level and no matter how much you turn up the input, the output will not exceed your preset maximum level.

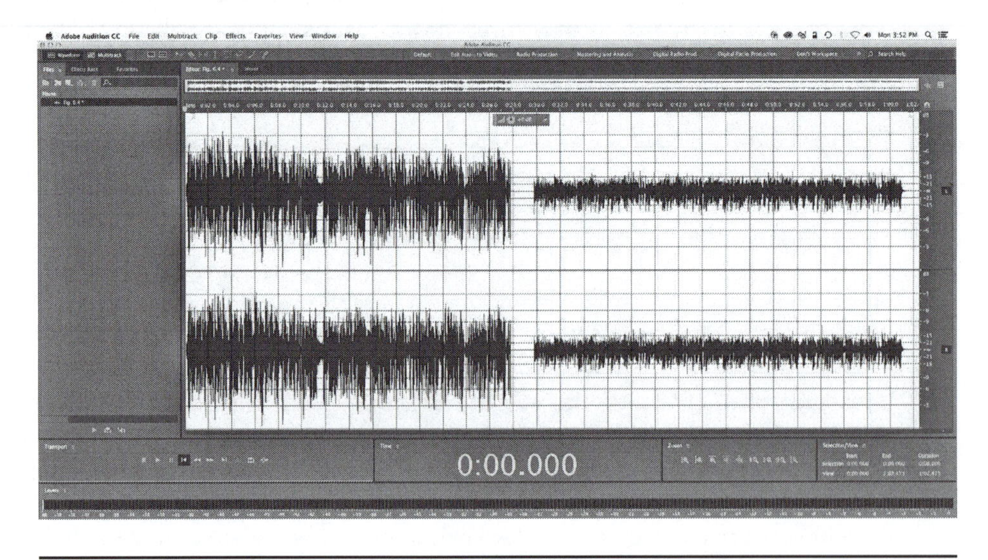

Figure 6.4 Limiters are like a brick wall through which the audio peaks cannot pass, regardless of the audio input level. The audio on the left has not been processed. The audio on the right is passing through a limiter adjusted to prevent audio peaks. Once the audio hits the limiter's predetermined maximum level (in this case, −17 dB), it really doesn't matter how much you turn up the input, the audio output level stays the same. (Adobe product screen shot reprinted with permission from Adobe Systems Incorporated.)

Expansion

An expander is the opposite of a compressor. A compressor reduces a signal's dynamic range, and an expander increases, or restores, it after compression. An upward expander makes loud audio segments even louder, and a downward expander turns softer audio segments down. Upward expansion is unusual since there are so many other ways available to make audio louder. A downward expander, though, is used to lower background noise.

A compressed audio signal has a reduced dynamic range, and the average audio level has been increased or made louder by several dB. When the average level is increased, the softer passages and background noise level are also made louder. Loud average levels normally mask any background noise until a soft passage comes along and the noise suddenly becomes apparent.

A downward expander functions like a compressor in reverse and turns down the background noise during the softer passages, making the audio appear to have a wider dynamic range from the softest sound to the loudest sound. An expander works well to turn down the studio or background noise present when an announcer pauses between words. An expander has a threshold and a variable ratio setting similar to those of a compressor, except that the ratios are reversed (see figure 6.5). The expander reacts to the audio when it drops below the threshold, and instead of a 3:1 compression ratio (for example) the expander might have a 1:3 expansion ratio (see audio demo, cut 6-7).

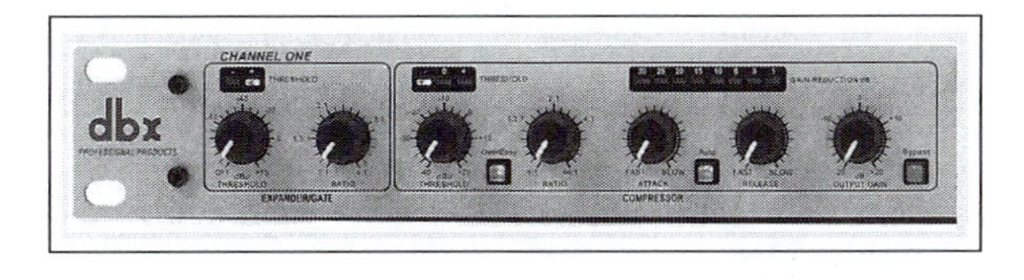

Figure 6.5 A downward expander functions like a compressor in reverse and turns down the background noise during the softer passages. An expander is an excellent tool to turn down the studio or background noise present when an announcer pauses between words. The expander shown at the far left is a component in a compressor/gate.

Gating

A gate is a dynamic processor used to reduce background noise. Just as its name suggests, a gate normally remains closed until the audio reaches a predetermined threshold, at which time it opens, allowing the audio to pass. When the audio level drops back below the threshold, the gate can either lower the audio to a preset level or close the channel completely, blocking any background noise.

Sometimes referred to as noise gates, gates are used to open and close audio channels automatically to reduce unwanted background noise (see figure 6.6). Noise gates can also be used to create special effects, such as shortening the reverb or delay time on a recording. As the audio fades out and drops below the preset level, the gate reduces the audio level or closes, muting the channel (see audio demo, cut 6-8). Noise gates have thresholds, attack and release times, and ratio set-

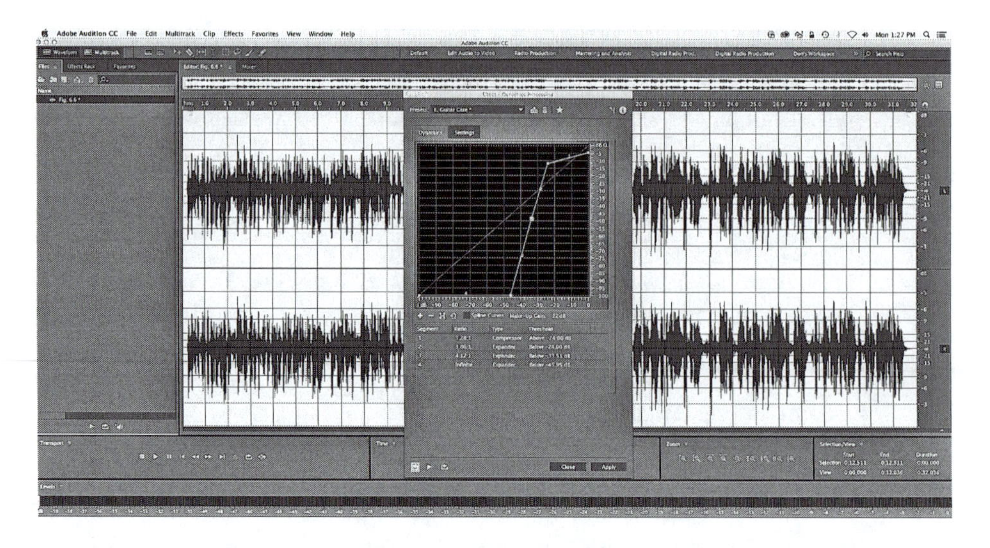

Figure 6.6 Sometimes referred to as noise gates, gates are used to open and close audio channels automatically to reduce unwanted background noise. As the audio fades out and drops below the preset level, the gate reduces the audio level, or closes, muting the channel. In the example above, when the audio falls below −12 dB the gate closes. Likewise, the audio has to rise above −20 dB before the gate will reopen. (Adobe product screen shot reprinted with permission from Adobe Systems Incorporated.)

tings. Just as a compressor can be adjusted to become a limiter, an expander can be adjusted to become a noise gate.

Although still widely accepted, gates do not see as much use as they once did because automated consoles and digital audio software are capable of easily opening and closing faders, eliminating unwanted background noise just as effectively as a gate.

Frequency Equalization

The telephone company was the first to use frequency equalizers to compensate for signal losses over early telephone lines. Audio production equalizers are really frequency unequalizers, since they are generally used not to make all frequencies equal in strength but to boost or cut specific frequencies to make sounds stand out or blend into the background. Equalizers are primarily used to shape the frequency response and tone of a recording. The industry term for equalization is simply EQ.

Of all the audio processors, EQ is the one that requires the most skill and training to use. It is also the audio processor that is most often abused, usually due to a lack of knowledge about the purpose and power of EQ. The first rule of EQ is that if you are not sure what to do, *don't do anything*, and leave it alone.

The second rule of EQ is that if you have been working on the project for any length of time, walk away and go have lunch, let your ears rest and flush the project out of your head. When you come back to the project you are now starting with rested ears and a fresh perspective. Pros sometimes give serious projects 24 hours before they come back to do the final EQ.

There are three basic reasons to EQ a track. The first reason is to make an instrument or voice sound clearer, or to achieve more clarity in the overall sound. The second reason is to add size and make an instrument or a vocal bigger or larger than life. The final reason is to make all of the elements in a mix fit together by adjusting the frequencies surrounding each voice or instrument so that it has its own space, so to speak, or its own distinct frequency range.

Regardless of how careful a production person is, there are some sounds that for one reason or another blend well during a project mix, some that may stick out, and still others that may seem to disappear. EQ is subjectively used to make things sound brighter, clearer, bigger, and sometimes **fatter**. Other terms used to describe signals before or after EQ include **muddy**, **crispy**, **tinny**, **hollow**, and **bright** or **dark**.

There are two popular types of equalizers—graphic and parametric.

Graphic EQ

A graphic equalizer gets its name from the physical layout of the controls on its faceplate (see figure 6.7). It consists of a series of slide faders vertically arranged side by side. Each fader represents a particular fixed frequency (center frequency) and a small band of frequencies just above and below the center frequency, called the bandwidth. Observed on an oscilloscope, the center frequency is at the top of a curve surrounded to the left and right by the rest of the bandwidth (see figure 6.8 on the next page).

Although the number of frequency adjustments varies from equalizer to equalizer, the frequencies are based on third-, half-, and full-octave intervals. A full-octave EQ usually begins near 50 Hz, with the succeeding filters at 100 Hz, 200

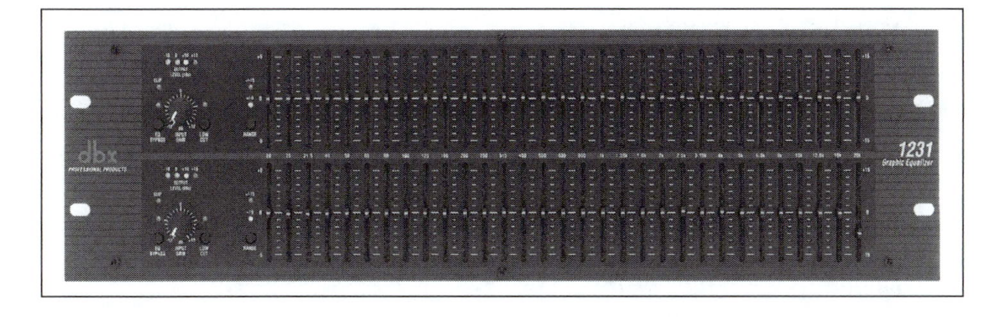

Figure 6.7 A graphic equalizer gets its name from the physical layout of the controls on its faceplate. Because each frequency band has its own fader, more than one can be used simultaneously. Raising the fader up boosts the frequency, and lowering it cuts the frequency. When adjusted, it is easy to see the pattern the faders have created, giving a visual, or graphic, reference of the frequency response created.

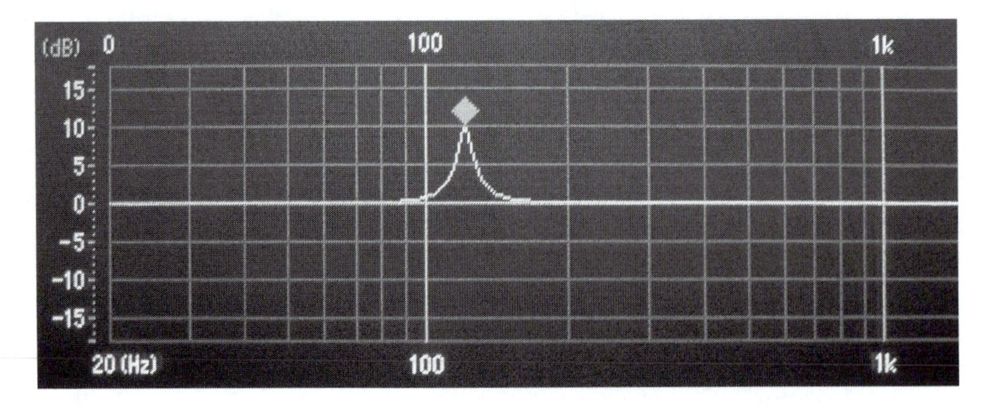

Figure 6.8 Each fader on a graphic equalizer represents a particular fixed frequency (center frequency) and a small band of frequencies just above and below the center frequency, called the bandwidth. Observed on an oscilloscope, the center frequency is at the top of the curve, surrounded to the left and right by the rest of the frequency bandwidth. This display indicates that the center frequency is 120 Hz and the bandwidth extends from 90 Hz to 150 Hz.

Hz, 400 Hz, 800 Hz, and 1.6 kHz, and going all the way up to 12.8 kHz. For finer adjustments, a half-octave or third-octave EQ is used. Most equalizers have between 10 to 15 frequency bands.

The mid position of each fader is 0. Raising the fader up boosts the frequency, and lowering it cuts a frequency. Because each frequency band has its own fader, more than one can be used simultaneously. When adjusted, it is easy to see the pattern the faders have created, giving a visual, or graphic, reference of the frequency response created (again, see figure 6.7). The only drawback to a graphic equalizer is that the center frequencies and bandwidths surrounding them are fixed, and you are limited to working with the equalizer's preset available bands.

Parametric EQ

A parametric EQ has a series of adjustable frequency controls with additional adjustable parameters; thus the name parametric. A parametric equalizer provides much more control and precision than does a graphic equalizer in selecting a specific center frequency and bandwidth to boost or cut. A parametric EQ has as many as eight controls on its face that let you select a center frequency for adjustment (see figure 6.9). With the frequency selected, the user then adjusts the amount of boost or cut for the selected frequency.

A third adjustment allows you to select the bandwidth surrounding the center frequency. This adjustment is called Q, or quality factor. The higher the Q value is set, the higher and sharper the peak is around the center frequency of the bandwidth. The lower the Q value is set, the wider and smoother the bandwidth peak is around the center frequency (see figure 6.10). The end result is an equalizer that has continuously variable frequencies and bandwidths.

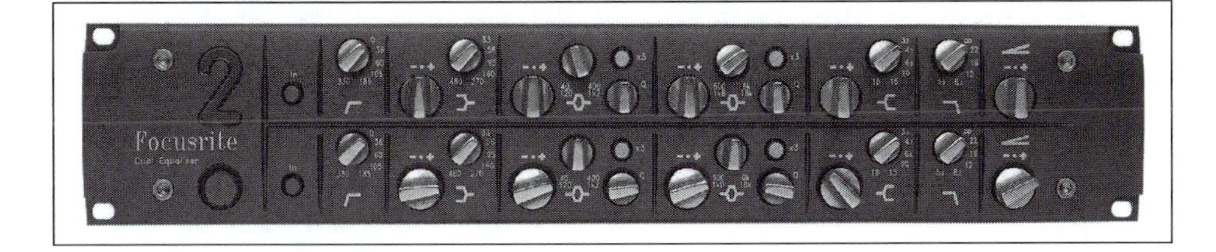

Figure 6.9 A parametric equalizer offers maximum flexibility. The user selects the frequency, the bandwidth surrounding the frequency, and the amount of cut or boost for the frequency. The end result is an equalizer that has continuously variable frequencies and bandwidths.

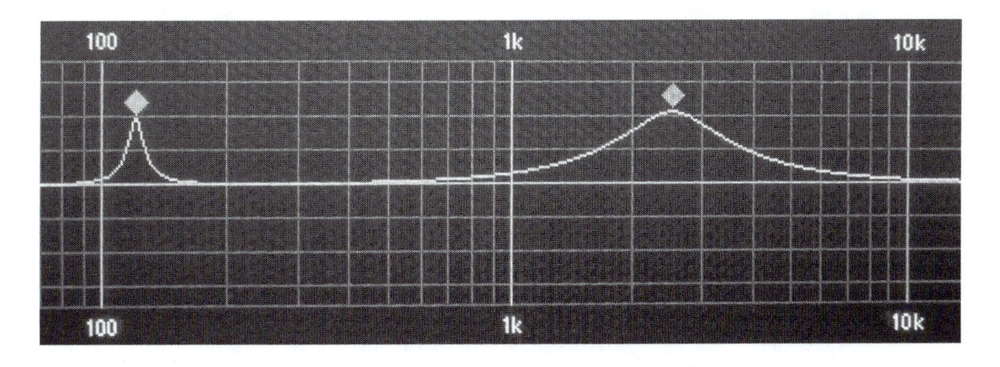

Figure 6.10 The Q factor determines the bandwidth on a parametric equalizer. The frequency curve on the left has a center frequency of 120 Hz boosted 10 dB. Note the sharp peak and that its narrow bandwidth extends only from 90 Hz to 150 Hz. The curve on the right has a center frequency of 2.5 kHz boosted 10 dB. Its wide, smooth bandwidth extends from 650 Hz to 10 kHz. The higher the Q value is set, the higher and sharper the bandwidth peak. The lower the Q value is set, the wider and smoother the bandwidth peak.

Filters

Filters are the components of an equalizer used to boost or cut the center frequency and the bandwidth on either side of the frequency. Filters, though, can serve another valuable purpose. A filter can be set to block certain frequencies while at the same time allowing others to pass. Filters are used to remove unwanted sounds, including everything from pops and clicks to building and traffic rumble. There are three types of filters.

A **high-pass filter** (often called a low-cut filter) blocks all of the frequencies *below* a preset frequency and allows those higher than the preset frequency to pass unaffected. As an example, high-pass filters are found on a lot of microphones to roll off or attenuate all of the frequencies below 75 Hz. This gets rid of most building rumble from air conditioners and traffic vibrations (see audio demo, cut 6-9).

A **low-pass filter** (often called a high-cut filter) blocks all of the frequencies *above* a preset frequency and allows those lower than the preset frequency to pass unaffected. As an example, a low-pass filter might be used on a vocal to eliminate room noise and hiss occurring in the studio above the vocal (see audio demo, cut 6-10). A human voice typically has no usable frequencies above 10 kHz. By cutting those frequencies above 10 kHz you are cleaning up the track and eliminating the possibility of picking up another sound source you don't want in the recording.

Another kind of filter is a **band-pass filter**. This filter establishes a low and a high cutoff frequency and allows the band of frequencies in between to pass unaffected. A band-pass filter is usually used to isolate a sound or correct a problem with a track.

Finally, there is the **notch filter**, which is just as its name implies. A very narrow frequency band is blocked, allowing the frequencies above or below the notch filter to pass unaffected. A notch filter is typically used to eliminate a specific sound such as a click or pop or the rumble of a building's air conditioner (see audio demo, cut 6-11).

De-essing

Between the frequencies of 6 kHz and 8 kHz is an annoying vocal quality known as sibilance. Sibilance refers to the hissing sound people make when they say words with the "ess" (*s*) sound. Some people have little or no sibilance when they talk; others, however, create a very annoying "ess" sound that is a cross between a hiss and a sizzle.

A **de-esser** is a specialty compressor that has an adjustable frequency range so that the frequency of the sibilant "ess" can be determined and attenuated with very narrow bandwidth compression. A threshold is set, and when the "ess" sound goes above it, compression kicks in and the "ess" is de-emphasized. De-essers have to be adjusted to each particular person to be effective (see audio demo, cut 6-12).

Guidelines for Using EQ

Generally you start the EQ process in the midrange, keeping in mind that in most cases it is better to *cut* frequencies than to boost them. This is particularly important between the frequencies of 250 Hz and 10 kHz. Humans are most sensitive to the frequencies around 5 kHz. Boosting the frequencies in this range too much (and it does not take a lot) can turn a mix into a harsh or hard-sounding mix. For example, if your mix sounds muddy, cut some of the lower frequencies, between 100 Hz and 300 Hz, by a few dB rather than cranking up the midrange or high end, which could cause harshness in the mix.

If your project has a nasal sound to it, try cutting in the 250 Hz to 1 kHz range. If a vocal or an instrument sounds too punchy, or it stands out above the rest of the mix, try cutting a few decibels between 2.5 kHz and 4 kHz. If you are trying to make things sound better, *cut*. Only if you are trying to create a different sound for effect should you boost.

When it comes to the bass in a project, listen to some commercial music mixes carefully. You will likely find that the bass has *not* been boosted but rather filtered and compressed. Using a high-pass filter, everything below 40 Hz has been rolled

EQ Chart		
Frequency Range	**Description**	**Acoustic Effect**
20–80 Hz (sub bass)	Felt more than heard; creates a sense of power.	Rumble zone; too much sub bass makes music unclear and sound muddy.
80–250 Hz (bass)	Fundamental notes of the rhythm section and drums; bottom end of guitars.	Too much makes things sound boomy or muddy.
250 Hz–2 kHz (low mids)	The bottom end of most instruments and their low-end harmonics.	Boosting 500–800 Hz makes things sound flat and hard; 1–2 kHz improves intelligibility, but be careful or it will sound like a tinny car horn.
2–4 kHz (mid highs)	Human ear is most sensitive to this range, which contains speech-recognition sounds.	Too much boost in this area causes listener fatigue. If a mix sounds harsh, cut here first.
4–6 kHz (presence)	Clarity of the mix and vocal definition, including articulation and breathiness.	Boosting makes music seem closer, but be careful—too much boost creates a biting sound.
6–16 kHz (brightness)	Brightness and clarity come from this range.	Boosting creates sizzle, vocal sibilance; a little boost in this region adds an open airiness.

off. There are two reasons for this. One is that, although there is sound energy below 40 Hz, it is only the frequencies from 50 Hz to 100 Hz that most people perceive as bass. The other reason is that most car and home speakers don't reproduce these low frequencies anyway, so by rolling off these low frequencies, the bass that remains appears cleaner and better defined. Rather than boosting the EQ in this bandwidth, use a multiband compressor to increase the average level and give the bottom end some punch.

The high frequencies have the ability to give your mix a clean, open sound. If you compare your mix to a commercial mix and it sounds a little flat in the upper end, this is not a serious problem. Use a parametric EQ with a center frequency between 12 kHz and 15 kHz with a low Q setting (wide, smooth curve). Don't overdo it; compare your project to similar commercial projects. Last, but not least, don't be afraid to roll off the upper end of your project smoothly (above 15 kHz) by 3 decibels. Listen carefully, and you will find many commercial recordings on which this has been done to create a smooth tonal balance.

The sounds below 250 Hz and above 12.5 kHz constitute exceptions to the cautions and guidelines given here. You can play a lot with the extreme upper and lower frequencies without ruining the mix, so feel free to adjust to your acoustical taste, but do still keep in mind that a little bit of EQ goes a long way.

Earlier it was said that there was no chart in the world that could tell you exactly how to equalize or process sound to achieve a specific result and that you had to use your ears and your knowledge of music to make these decisions. The EQ chart included in this chapter can be used as a very basic guideline, but it certainly is not specific (see audio demo, cut 6-13).

Finally, listen critically to your mix and keep in mind that you cannot cut or boost something that is not there in the first place.

Dimension

Every sound you hear is a combination of the original sound and how the sound reacted to the environment it was in. Dimension processing adds perceived size or depth to a recording. Introducing a time delay in the form of echo, reverb, or a chorus effect creates dimension. The artificial delay creates the impression that the listener is in a larger space by simulating natural-sounding repetitions of the original signal that normally occur in a venue like a concert hall or stadium. Short delays are normally associated with small spaces, and longer delays with large spaces. A simple rule of thumb in using a dimensional processor is to imagine yourself in a familiar acoustic space and remember what it sounds like, and then try to re-create the sound you are hearing in your head around you in the studio.

Good candidates for dimensional processing are recordings made in a **dead studio** or location, one that has little or no natural ambience, echo, or delay. Dimension is also used to mask the mixing or blending of two or more audio sources together.

Echo, Slap Delay, and Doubling

Echo, slap delay, and doubling are the simplest forms of introducing depth into a recording. An echo or delay uses one or more repeats of the original signal to create the effect. Delay-effects processors have input- and output-level controls and a control, referred to as the delay time or delay length, to adjust the time delay in milliseconds (see figure 6.11). Such controls may be individual knobs on the face of the equipment or appear as a part of a software menu. Slap delay is generally a single repeat of the sound with a delay of over 35 milliseconds (see audio demo, cut 6-14). When adding this effect to a music bed it is a good idea to adjust the repeat time in relation to the beat, or tempo, of the music. By matching the music's beat, the delay adds to the music rather than fighting with it by being out of tempo with the music.

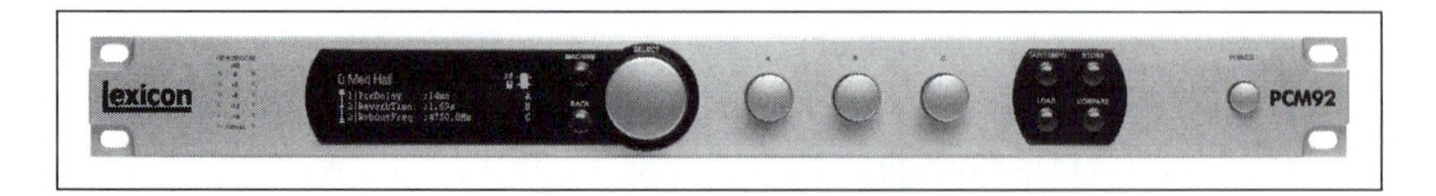

Figure 6.11 An electronic reverberation unit re-creates the acoustic environment of a number of different spaces, from a small room to a large concert hall. Most come with numerous preset effects and also let you create your own new effects and place them into memory.

A feature of many delays is a feedback control. A feedback control does exactly what its name implies; it feeds the delayed signal back into the delay input so that a delay of a delay is created. The higher the feedback setting, the more delays are created (see audio demo, cut 6-15).

A single slap delay of less than 35 milliseconds is called doubling, since it creates the impression that there are two of whatever you are applying the delay to (see audio demo, cut 6-16).

Reverberation

Reverberation, or reverb, is different from delay in that it is a complex combination of multiple, randomly mixed delays designed to simulate a specific acoustical space such as a large concert hall, sports stadium, or small intimate room. A distinguishing characteristic of reverb is that as the repetitions of the signal slowly drop in intensity, the number of repetitions increases.

Reverb has three distinct purposes. The first is to create an aural space, to trick the brain of the listener into believing that he or she is in a particular listening environment. Most reverbs are infinitely adjustable; the space can be adjusted to taste to enhance the production. The second use for reverb is to make a sound appear bigger, fatter, or deeper. The selection of the space and the decay time affect this. And the third purpose for reverb is to be able to use a pan control to position an audio signal between the left and right sides of the stereo image—for example, you could put an instrument in the left channel and let the reverb trail off into the right channel, creating a huge space. Remember that audio being sent to the processor prior to the addition of reverb is called dry, and when the audio is returned from the reverb after processing, it is called wet.

The adjustable parameters for a reverb include an input-level adjustment, delay or space adjustment, and decay time (again, see figure 6.11). The delay or space adjustment regulates the size of the acoustic space that the reverb creates, from a small, tight room to a football stadium (see audio demo, cut 6-17).

The decay time control sets the length of the effect. Smaller reverbs and shorter delays make things sound bigger. To create these effects, reverb decay times should be under a second and delays should be 100 milliseconds or less. These are maximum settings, and you will likely find that shorter decay and delay times work as well or better, particularly if the recording is stereo (see audio demo, cut 6-18).

Likewise, longer reverb decays and longer delays push the sound farther away from the listener. This begins to happen at around 100 milliseconds of decay or delay and really kicks in at over 200 milliseconds. If the decay or delay is too long and too loud, the track just sounds far away and is not very interesting.

Reverb can be used to bring a track to the front, closer to the listener, or blend it into the background, moving it away from the listener. For example, say you have a dry piece of music for a commercial and a dry announcer or vocal track that you want to mix together. To make the voice stand out from the music, leave the announcer dry and wet down the music (add reverb); the announcer moves forward and dominates the mix (see audio demo, cut 6-19).

If the reverse is true and you want the music to stand out and the announcer to be in the background, leave the music dry and wet down the announcer (again, see audio demo, cut 6-19).

A third option is to blend the two together, with equal amounts of reverb on each. This typically happens when a dry announcer track is being laid down over a production music bed that has some reverb. In this case, the voice stands out in front until you add some reverb to the voice, blending it with the music.

Chorus, Phase Shifting, and Flanging

Chorus is an electronic effect that modifies the sound of a voice or instrument to simulate a group of vocalists or instruments. For example, one vocalist is processed to sound like a group of three; a violin becomes a convincing-sounding string section.

The chorus effect is based upon tricking the listener into thinking there are several of something instead of one. When you are seated in front of a musical group at a concert, the acoustic environment modifies the sound that you hear by introducing time delays. Sounds naturally bounce off the floor, ceiling, walls, furniture, and so on, and each arrives at your ears at a slightly different time. If you remember from an earlier chapter, this shift in time is called a phase shift. The acoustical environment also introduces slight shifts in the pitch of the instruments.

Using an adjustable time delay of between 20 and 35 milliseconds creates a chorus effect. The distinguishing difference between the chorus effect and delay or reverb is that in the chorus effect, the length of the delay is randomly varied, introducing phase shifts that simulate what naturally happens in an acoustic environment. The random phase shifts also cause slight variations in the pitch of the sound. This creates the impression of several of the same signals arriving at your ears at various times. This effect is then mixed back into the original signal, causing one of something to sound like a group (see audio demo, cut 6-20).

The same processor that creates a chorus effect can also create a phase-shift effect. A phase shift creates the illusion of a swooshing motion in the music, sounding like the mid and high frequencies are being swept through frequency equalization (more on that later in the chapter). Adjusting the delay time of a chorus processor from 1 to 3 milliseconds creates a phase-shift effect (see audio demo, cut 6-21).

Flanging is a more dramatic phase-shift effect. It is used in the recording industry on vocals, keyboards, guitars, bass, and drums. The flanging effect is created by adjusting the delay time of a chorus processor to 10 to 20 milliseconds of delay. Although flanging is a digital electronic effect today, famed guitarist Les Paul is said to have created the flanging effect using four reel-to-reel tape machines. Listen to the Beatles, Jimi Hendrix, and many other artists of that era to hear examples of the original flanging effect (see audio demo, cut 6-22).

As with most audio-effects processors, the best way to learn the processor is to feed it some audio and start experimenting with the controls to learn firsthand all of the unique effects that you can create.

Re-amping

Re-amping is one of the oldest audio effects that can be created. It is used to simulate a natural-sounding environment. The process is so simple many people overlook it. Take an audio signal and play it back through an amplifier and monitor speaker in the studio. Place a microphone, or a pair of microphones for stereo, in the studio and rerecord the track using the ambience in the studio to create a variety of effects, depending on how far the microphone is placed from the speaker. By experimenting with the placement of the speaker and microphone, you can create some unique acoustical effects. Try placing the speaker facing away from the microphone and recording the sounds that have bounced off a studio wall. There are no rules to re-amping and no expensive audio processor; that is what makes it so much fun to try. The effect that you create is unique to your studio, speaker, and mic placement (see audio demo, cut 6-23).

Broadcast Audio Delay

Broadcast audio delay is not an effect; it is an extremely valuable audio processor used in the radio industry. During live talk shows or telephone conversations with listeners it is critical that the caller not be put on the air live, since the broadcaster is responsible for anything that the caller might say. Indecent and obscene language can bring heavy fines from the FCC. A single word can cost a station thousands of dollars and put your job in jeopardy.

Typically, broadcasters use a 7-second delay on listener calls or talk shows. If someone says something the producer thinks is objectionable, whether it is objectionable language or a libelous remark, the producer has 7 seconds to delete the language by pressing and holding the dump button. When the dump button is released, the processor then automatically plays catch up and restores the 7-second delay (see figure 6.12).

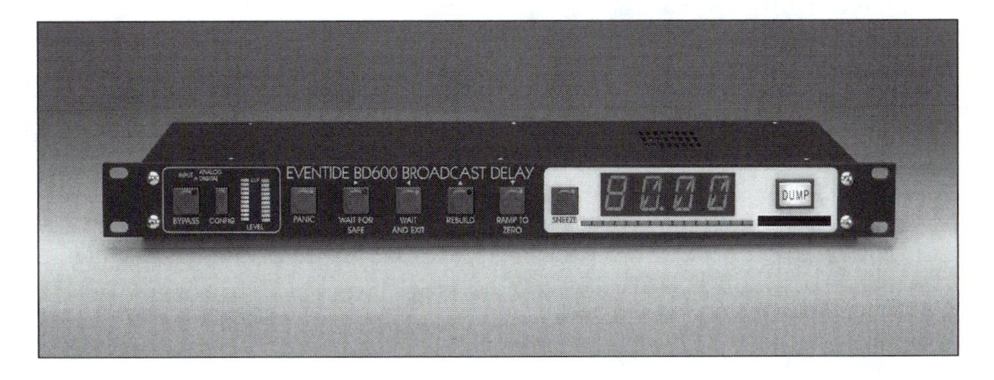

Figure 6.12 A broadcast delay does exactly what the name implies; it takes live audio and delays it a preset time (typically 7 seconds) so that in the event of inappropriate language the producer or operator has a few seconds to dump the offending audio so that it never reaches the air. Note the dump button on the right side of the unit.

Broadcast delays go into and out of delay very smoothly. During the opening of the show, while the host is talking, the digital audio processor begins introducing delay over a few seconds so that by the time the first caller is taken there is a full 7 seconds of delay. This is called ramping up to delay. At the end of the show, when the announcer is wrapping up the program, the producer deactivates the delay and the unit ramps down from a 7-second delay to real time.

The listeners are not aware of what is going on until they call the show and are told by the producer to turn their radio down so that they don't get confused while talking when they hear their own voice over the radio 7 seconds later.

Normalizing

A normalizing processor is included in most digital audio software. It corrects the problems associated with a track that has been recorded with too little amplitude (volume) or too softly. The processor analyzes the audio and searches for the highest and lowest peak levels to determine the dynamic range of the signal. With this data, the processor raises the overall volume level of the track while maintaining the signal's original dynamic range. Normalizing allows a producer to select the percent of modulation to be applied to the signal (see figure 6.13). If 100 percent is selected, the highest audio peak occurs just before hard clipping (see audio demo, cut 6-24). Normalizing, though, is not a substitute for recording at the proper level to begin with.

Figure 6.13 A normalizing processor corrects the problems associated with a track that was recorded with too little amplitude (volume) or too softly. Normalizing allows a producer to select the percent of modulation to be applied to the signal. The audio on the right side of the window has been normalized to a maximum level. Normalizing, though, is not a substitute for recording at the proper level to begin with. (Adobe product screen shot reprinted with permission from Adobe Systems Incorporated.)

Normalizing can also be used to match the audio level between two different audio segments that you want to join together seamlessly. You can match existing audio levels in a track or session by first using the processor's gain-analyzer function to determine the maximum peaks in the existing audio tracks. After analysis, adjust the processor accordingly and normalize the new track to match the original track's level.

Most professional recording engineers prefer a properly recorded track with good levels to normalizing. With that said, the exception to the rule is in radio, where program directors want maximum modulation to achieve a sound on the air that jumps out at the listener. In essence, radio production directors want it loud, and normalizing a commercial or station promotional announcement to 100 percent modulation will certainly do that.

Voice Processors

Voice or microphone processors are designed to be the first stage after the microphone in the audio chain. Designed primarily for broadcasting and voiceover work, voice processors are designed to punch up the talent's voice, giving it a unique quality and making it stand out. The thing to keep in mind is that a voice processor can make a good voice sound dynamic and exciting. A voice processor is not capable, though, of magically turning a mediocre voice into a great voice.

Voice processors are actually a combination of a number of processors in one convenient box. A typical voice processor includes a high-quality mic preamp with phantom power for condenser microphones. Typically, following the mic preamp are a de-esser, expander, compressor, parametric equalizer, and limiter. Some voice processors are even offered with reverb (see figure 6.14).

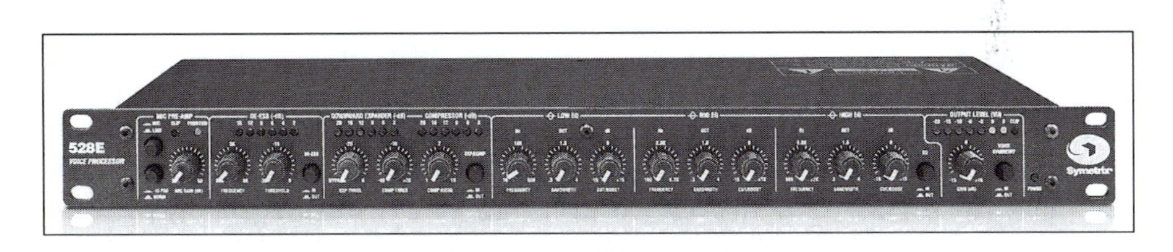

Figure 6.14 Voice processors are actually a combination of a number of processors in one convenient box. A typical voice processor includes a high-quality mic preamp with phantom power for condenser microphones. This processor also includes a de-esser, expander, compressor, and parametric equalizer.

Basic Voice Processor Adjustments

The first stage of a voice processor is typically a microphone preamplifier. A preamplifier features an input-gain control to match the output level of the microphone to the input level of the processor. Additionally, most professional voice processors have 48-volt phantom power for condenser microphones. Start by adjusting the input level so that the preamplifier is seeing an incoming signal of 80 to 100 VU from the microphone (see audio demo, cut 6-25).

The preamp section is usually followed by a compressor (gain reduction) to limit the variations in the levels of the voice that occur in normal speaking. As discussed earlier in the chapter, a good compression ratio to start with is from 3:1 to 6:1. These lower ratios reduce the transients in the voice without reducing the dynamic range of the voice too much (to avoid making it sound flat and without life). After the voice is compressed, use the compressor output control to raise the average level of the voice back up to 80 to 100 VU (see audio demo, cut 6-26).

Keep in mind that when you compress a voice to reduce its dynamic range and then turn up the average level, you also turn up any electrical noise or ambient studio noise. To reduce or eliminate the unwanted noise, most voice processors offer a downward expander. A downward expander works in just the opposite way from a compressor, and instead of making loud sounds louder it makes quiet sounds quieter. Set the threshold of the expander at the point below which you want the expander to close and silence the signal. Then adjust the amount of expansion you want (see audio demo, cut 6-27). A downward expander set to its extreme limits becomes a noise gate, simply turning the audio on and off when the threshold is reached.

A de-esser removes sibilant "ess" sounds that can be quite annoying. De-essers are frequency-specific compressors. By setting a high compression ratio and then adjusting the frequency-range control, the specific frequency causing the problem can be found and completely removed in most cases. However, it does have to be adjusted for each person (see audio demo, cut 6-28).

Most voice processors have a frequency equalizer, and unless you have specific training in EQ it is wise to leave this alone or seek the help of others if there is a serious problem with the frequency response of the person's voice. As discussed earlier, more harm than good can be done to a signal through improper EQ.

Many voice processors offer adjustable high- and low-pass filters. Typically, the low-pass filter is set at 80 Hz, since there really is not that much sound energy in the human voice below this. In addition, this part of the frequency range is where most building rumble and vibration occur, so eliminating the audio below 80 Hz provides a cleaner-sounding bottom end to the audio. Likewise, adjustable high-pass filters allow you to adjust how much audio above 12 kHz is used. There is very little vocal energy above 12 kHz, and eliminating these unused frequencies eliminates hiss and noise in a signal. By rolling off or eliminating these two frequency ranges, the audio appears cleaner (see audio demo, cut 6-29).

The last section in a voice processor is the peak limiter. If you remember from earlier in the chapter, this is the device used to adjust the maximum audio level. The limiter is useful in preventing an overload of the output stage and in setting the maximum audio level (see audio demo, cut 6-30).

The only remaining control on the preamplifier is the output-level adjustment, to match the input level of the control board or recorder the preamp is feeding an audio signal to.

There is no chart or list of the exact settings for voice processors and what works with what kind of voice. By careful experimentation, an announcer can determine the right combination of settings to reinforce his or her radio voice.

A digital voice processor is computer controlled and can store up to 100 different announcer presets, so that each announcer's voice can be maximized. Before announcers go on the air, they select their preset, and the processor recalls all of the settings for that particular person.

Effects Processors

To a production person, an effects processor is the equivalent of a dump-truck load of Legos to a six-year-old. This computer-based audio device is capable of creating any sound or effect you can imagine by manipulating audio through advanced digital algorithms. Effects processors can be used to combine multiple effects and create unique new sounds, storing them in the processor memory to be easily recalled (see figure 6.15).

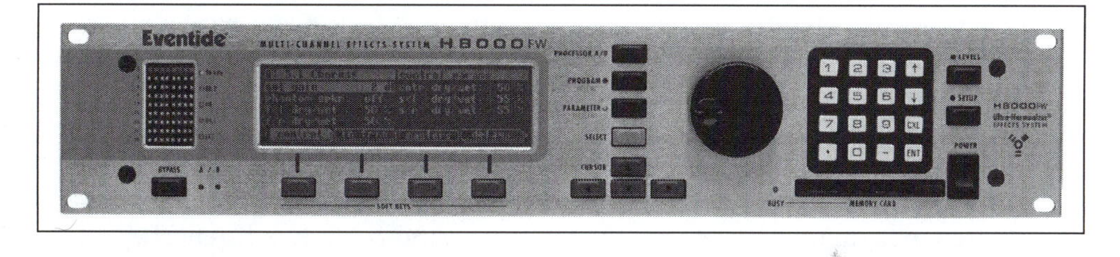

Figure 6.15 An effects processor is capable of creating literally thousands of special audio effects in addition to the hundreds of effects that come preprogrammed with the unit. Newly created effects can be saved for future use. An effects processor can create everything from reverb to sophisticated phase shifts and repeating effects.

Effects include endless variations on reverb, slap delay, echo, chorusing, flanging, phase shifting, pitch shifting, time compression and expansion, tremolo, vocal harmonizing, video-game sounding electronic weapons fire, vibrato, random pitch shifting and panning, vocal stutter, complete reversal of words and phrases, and sweep tones, just to name a few of the thousands of effects available on some processors.

These are very powerful and elaborate tools, and they provide the production person with a valuable edge in creative audio work in commercials and creative productions (see audio demo, cut 6-31).

Broadcast Audio Processing

Listeners are a little smarter and a little more sophisticated than many people realize. The first thing that usually catches listeners' attention as they scan through the frequencies on their radio is the music or the show. Once they have located the show they like, they focus on the station's presentation. Is it loud enough so that it can easily be heard in a car or over the office chatter? Does it sound great on their car or home system? Most cars have sophisticated audio systems. Is it a unique sound that complements the music and is easy to listen to for long periods of time? Of course, listeners are not consciously asking themselves these questions, but programming research has shown that the answer to each question may be a reason why a listener will select another station. Just as important as the music itself is the audio processing the station applies to the music.

Radio stations process all program audio before it is sent to the transmitter. Originally, the reason for on-air processors was to control audio levels so they didn't overload the transmitter and distort the on-air sound. Today's broadcast audio processors offer the program director and engineer an incredible array of multiband digital signal processing options, including sophisticated compressors, expanders, limiters, gates, filters, and EQ to customize the station's sound. Select any song and listen to it on ten different radio stations, and it will sound ten different ways. Audio processing is often the reason a listener chooses one radio station over the other, even though they both play the same music (see figure 6.16).

Broadcast audio processing is based on a psychoacoustic musical image created by the program director in conjunction with the station engineer. As an example, a classic rock station understands that a lot of its listeners turn up their radios and listen at louder-than-average levels. The processing has to be adjusted so that the audio peaks in the music don't cause distortion on a car or portable radio. An urban hip-hop station, to take another example, has to have a clean, loud, distinct bottom end; muddy bass is not tolerable to a serious hip-hop listener with a high-

Figure 6.16 Audio processing often is the reason a listener chooses one radio station over another, even though both play the same music. Digital broadcast audio processors offer the program director and engineer an incredible array of multiband digital signal processing options, including sophisticated compressors, expanders, limiters, gates, filters, and EQ to customize the station's sound.

wattage sound system in his or her car. The bass has to have a solid punch to it. Finally, a light rock station that targets at-work listeners adjusts its processor so that the station sounds good on small, portable radios or computer speakers that are not turned up very loudly.

As a production person, you should consult your program director and station engineer on how far you can take audio processing in your work with programs and commercials. Broadcast engineers prefer that you leave the frequency response as flat (as even throughout the frequency spectrum) as possible on your projects and use only slight amounts of processing to correct minor problems. Because all on-air audio will be processed again when it goes through the broadcast processor, there is the possibility of heavily processed production causing listener fatigue or creating a **tune-out factor**.

Webcast Audio Processing

Putting an audio signal on the web would appear to be a pretty simple task at first. Tap into the on-air audio someplace in the radio station and feed it to a computer hooked up to the Internet. Sending good-sounding audio to the Internet is not that simple, though.

The first problem that broadcasters encounter is the fact that, unlike for the radio station, there are no FCC-mandated technical standards for transmitting the signal to the listener's receiver via the web. Because webcasters don't know what kind of computer, smart phone, tablet, or Internet connection their listeners are using, they typically prepare audio based on "the average site visitor" so that as many people as possible will be able to hear the webcast.

The second problem a webcaster encounters is the fact that it is not easy to jam all of that digital audio data down a very narrow pipeline to the listener. High-quality audio (44.1 kHz, 16 bit) guzzles more than 10 megabytes of data per minute. Either the pipe has to be bigger or the amount of data has to be reduced to stream an on-air signal on the web.

The audio processing required to originate a good-sounding webcast is as technically involved as the on-air broadcast processor. The first step is the use of an Internet audio processor, as they are different from their broadcast counterparts. Many of the companies that produce on-air broadcast audio processors also produce complementary audio processors for the web. Sometimes a broadcaster will elect to use a third-party service that specializes in audio streaming to host its streaming audio and website.

As a general guideline, audio production intended for the Internet needs to have a very good signal-to-noise ratio. The frequency response should be left flat. Heavy compression and limiting should be avoided, as most commonly used web audio processors do not work well with audio that has been heavily processed.

■ SUGGESTED ACTIVITIES

1. Try feeding a plain voice track and a stereo music track through the compressor that is included with Adobe Audition CC. (You can also use a compressor in your radio production studio, if you have one.) Spend 30 minutes adjusting the gain control, threshold, and attack and release times to see how many different sound treatments you can create from the same tracks. Afterward, see if you can adjust your compressor to act as a limiter. Experiment with establishing maximum levels.
2. Using the different dimensional-effects processors that you have access to in Adobe Audition CC, record one track each with echo, reverb, and slap delay. Can you tell the difference among them all by listening? Now try chorus, phase shift, and flanging. Again, can you tell the difference by listening?
3. Record two separate tracks, one a dry voice track and one a dry stereo music track. Using reverb, bring the voice to the front of the recording and then place the voice behind the recording. What challenges or difficulties did you encounter?
4. Record a short musical selection using Adobe Audition CC. Experiment by using the different filters, or bands, on a graphic EQ and see how many ways you can change the spectral balance of the music. Can you make any instruments seem to disappear or to stand out from the rest?
   a. Now experiment specifically with cutting frequencies to make things come to the front.
   b. Repeat the experiments with a parametric EQ to see how it differs in operation from the graphic EQ. Which gives you the most control?

■ WEBSITES FOR MORE INFORMATION

For more information about:

- *broadcast audio processors,* visit Omnia Audio at www.omniaaudio.com and Orban Audio Processors at www.orban.com
- *echo, slap delay, doubling, and reverberation,* visit Lexicon at www.lexicon.com and TC Electronic at www.tcelectronic.com
- *effects processors,* visit Eventide at www.eventide.com
- *streaming audio on the Internet,* visit SHOUTcast at www.shoutcast.com
- *voice processors,* visit Symetrix at www.symetrix.co/products/ or Focusrite Audio Engineering at www.focusrite.com

PRO SPEAK

You should be able to use these terms in your everyday conversations with other professionals.

attack time—The fixed or adjustable length of time it takes for an audio processor to sense the presence of audio and react to the audio, measured in milliseconds.

band-pass filter—An audio filter that has a high and low cutoff frequency, allowing only the frequencies in between to pass unaffected.

bright—A sound rich in high frequencies.

compression ratio—The fixed or adjustable ratio of the input to the output of an audio compressor.

crispy—A sound with extended high-frequency response, like the sizzle of crisp bacon frying.

dark—A sound that is the direct opposite of bright; dull with a weak high-frequency response.

dead studio—A studio or a location that has no natural reverberation or echo.

de-esser—A specialty compressor that has a variable frequency range used to reduce sibilant "ess" sounds in speech between 6 kHz and 8 kHz.

fatter—A sound with a good, solid, low-frequency response with emphasis in the 100 to 300 Hz range.

high-pass filter—A filter that blocks audio frequencies below a specific frequency, allowing those frequencies above the specified frequency to pass; sometimes referred to as a low-cut filter, a term that is more descriptive of what the filter actually does.

hollow—A sound that has a mid-frequency dip similar to when you cup your hands around your mouth and talk.

listener fatigue—A condition brought on by misadjusted audio processing, causing the listener to become consciously, or subconsciously, tired of listening to a musical selection or radio station.

low-pass filter—A filter that blocks audio frequencies above a specific frequency, allowing those frequencies below the specified frequency to pass; sometimes referred to as a high-cut filter, a term that is more descriptive of what the filter actually does.

muddy—Muffled-sounding audio, not clear, with too much reverb at the lower frequencies.

notch filter—An audio filter designed to eliminate a very narrow bandwidth of sound, creating a notch in the frequency. Generally used to eliminate unwanted sounds.

release time—The fixed, or adjustable, length of time it takes for an audio processor, such as a compressor or limiter, to return a signal to its normal level after processing, measured in milliseconds.

sweet spot—The physical location in a concert hall or studio where there is the least phase cancellation of live sound and where the direct-to-ambient sound ratio is in proper proportion.

threshold—When used in relation to an audio processor, a threshold sets the audio level at which the device will become active and begin processing.

tinny—A sound that has an almost telephone-like quality to it. A very narrow-band sound that sounds as if it were coming through a tin can or over a telephone.

tune-out factor—Any psychoacoustic element of a radio station's signal that is annoying to the listener.

7

The Art of 60-Second Storytelling

Highlights

Regardless of the medium—radio, television, or the web—nothing goes on the air without the help of a production person. Production people transform the ideas and concepts that come from clients, ad agencies, and salespeople into commercials. These 60-second messages are really short stories about how the listener benefits from the products and services offered by the advertiser. (Occasionally, advertisers may choose to use a 30- or 15-second commercial, depending on what goals and objectives they have for their advertising campaign.) Ideally, a commercial is so well crafted that it motivates people to take some form of action. There is only one measure of success when it comes to a commercial: Did the client's customer traffic and sales increase?

Radio and television commercials may appear to be two entirely dissimilar media. However, a well-crafted television commercial is simply a radio commercial with pictures. A good television commercial does not depend upon the images alone to deliver the advertising message, it depends on a combination of the audio track and the images. The most effective television commercial has an audio track that can stand alone as a radio commercial. During a commercial break, when the audience gets up to go into the kitchen and get a soda, the commercial is still effective because the audience can hear it. These stand-alone audio tracks are often used on radio because they have the ability to trigger the matching visual images in the radio listener's mind. The visualization of a television commercial while hearing it on the radio is called **image transfer**.

Although many major market operations employ writers and continuity directors (responsible for writing all the commercials), many others do not. There are times when you will need to put together a commercial. Here's a real-life lesson. A salesperson shows up at 4:30 on a Thursday afternoon with a sales order for several thousand dollars from a major client (big car dealership). The copywriter has gone home early because of illness. The commercials need to start running at six o'clock tomorrow morning. What are you going to do? Tell the client, "We're sorry, our copywriter went home sick, so we can't take your thousands of dollars today and you'll have to wait until tomorrow afternoon to start your schedule of ads"? Events like this are the reason you need to be able to assemble a commercial on your own.

A note about commercials: before you go any further in this chapter you must rethink a word that may be in your vocabulary. Production people do not produce **spots**. When you dribble marinara sauce down the front of your shirt or blouse while eating spaghetti, you produce a spot. Dry cleaners remove spots. The fact is, salespeople DO NOT SELL, and clients DO NOT BUY, spots. Clients buy well-crafted solutions to their marketing problems that help them increase customer traffic and sales. You produce ads, or commercials, that are a critical part of the client's sales process and marketing plan.

Production as a Part of the Sales Process

Commercial production is one of the final steps in the sales process. Although the client may have signed a contract for airtime, until a commercial is produced and approved to go on the air, the order is incomplete and cannot start. Often, the client sees the production service as a valuable resource that the radio station offers. This is particularly true when the client is planning on using **dubs** for advertising schedules on other stations.

Why Advertising Works

Advertising is not difficult. But somehow, in all that is written and said, many people forget the simple formula that is the basic premise of advertising. Successful advertising consists of four components: the right product, at the right price, at the right time, aimed at the right target audience. When the four components are brought together, the result is a chain reaction of events culminating in increased sales for the client. Change any one of the four variables in the formula, though, and the advertising will not work nearly as well, if at all.

Here's a story about some hardware stores in Florida that were trying to sell lifetime-guaranteed 25-foot garden hoses in the summertime—the price, $49.95. A chain of three hardware stores had two pallets of these things that had been gathering dust in their warehouse. The stores decided to try radio advertising. They thought that if they put commercials on the radio, people would somehow magically snap up the hoses. The radio salesperson reminded store management of the formula. A garden hose in Florida is a "got-to-have" item—the right product. The summertime is the right time. The radio station's target audience included 35- to 64-year-old homeowners—the right target audience. The price, though, was a problem. Who buys a $50 garden hose? The salesperson convinced hardware store management to do a half-price promotion and sell the hoses at 50 percent off (they were still making a buck). The ads started on Wednesday, and on the following Monday all but two of the hoses were gone. The right product, the right price, the right time, the right target audience; it's not rocket science.

AIDA, an Opera about Success

No matter how well produced a commercial is, just putting an ad on the air will not instantly generate sales increases. Regardless of the medium used, there are four stages in a successful advertising schedule leading to increased sales. There are a number of variations of these stages. You can remember them by thinking of the opera *Aida*. AIDA, in this case, stands for awareness, interest, desire, and action.

Awareness. First, a potential customer has to be made aware of a business. Depending on the schedule of ads, this can take some time. Can a single ad in the Sunday paper accomplish this? Can five radio ads on Friday do the trick? Can one ad each night during the evening news on TV create awareness? Regardless of the medium, the faster a business wants to build awareness, the more ads the business will have to run. As an example, radio salespeople typically recommend a *minimum* of 15 ads per week to accomplish the kind of results and sales increases that an established radio advertiser expects. A new advertiser, though, might have to increase that number of ads proportionately to develop the consumer's awareness of his or her business before potential customers can be moved along in the buying process.

Interest. Before customers can be moved to buy a product, they have to be interested in the product. Their curiosity needs to be aroused. To develop this interest, the advertiser shows or tells the listener what the product does. What problem will the product solve for the customer? What kinds of satisfaction will the customer receive from it? What kind of benefits does the product provide the customer? It is critical to remember that the focus needs to be on the benefits the product provides to the consumer. Until sufficient interest is developed in the product, the consumer is not ready to move to the next stage of the buying process.

Desire. With product awareness and consumer interest established, the advertiser can focus on developing a desire for the product. This is the stage at which the advertiser tells the consumer how good the consumer will look wearing, using, or being seen with the product. The advertiser appeals to many of the consumer's most basic desires to look good, smell good, be safe, be successful, have esteem, prestige, self-fulfillment, and so forth. The advertiser can do this by appealing to the consumer's most basic wants, needs, and desires.

Action. This is the final stage of the buying process. The advertiser asks for some type of action on the part of the consumer. To quote some well-worn clichés, "Don't miss this once-a-year event." "Act now and save thousands." "Send your request before midnight tonight." Customers go where they are invited and do what they are asked. The advertiser must ask for some type of action on the part of the consumer.

What's In It for Me? (WIFM)

Have you ever met someone who did nothing during the first five minutes of a conversation except tell you how great and wonderful they were, the things they owned, and the people they associated with? If you are like most people, you probably started looking for an easy getaway from this egocentric bore.

That is exactly what happens to a commercial when the advertiser spends 60 seconds telling the audience how great his or her business is. Who cares? Customers don't care about a business for its own sake. What customers care about is as old as the ages: "What's in it for me?" (WIFM). If they spend their money with the advertiser, what will they get in return?

Another element of a commercial that can be a real turnoff for the consumer is the use of the advertiser's industry terms and jargon. Although this may impress another business, the consumer has no idea in the world what the business is talking about and, frankly, does not care (see audio demo, cut 7-1).

The focus of a successful commercial is *not* on the advertiser but on how the products offered by the advertiser will solve the consumer's problems and meet the consumer's needs, wants, and desires. This does not mean that the advertiser is not important. It means that the advertiser has to learn to think from the consumer's point of view (see audio demo, cut 7-2). Remember, WIFM: What's in it for me?

The Salesperson/Client Relationship

As a production person, you are going to be working with very valuable property: commercials. A commercial has a real cash value based on the time invested by the salesperson and everyone else who processes the commercial, including the production time you are going to invest in creating the commercial.

As a production person, you need to be sensitive to the fact that a commercial is a collaborative effort among the client, the salesperson, and you. Notice that the salesperson is between you and the client. Beginning with the salesperson, everyone employed at the radio station depends on your work for his or her income. In a small market, a monthly schedule of 100 commercials at $10 per commercial is $1,000. The salesperson earns $150 (a 15 percent commission) on the sale, and the station nets $850. In a larger market, those $10 ads are suddenly $175 each—the client is spending $17,500, the salesperson's commission is $2,625, and the net income for the station is $14,875. You should be aware that people become very sensitive when money is at stake.

As a production person, you should follow your company's policy regarding any direct contact you have with the client. Usually, the salesperson will want to handle all communications with the client, and rightfully so, since you may not be aware of where the client/salesperson is in the buying process. As a general rule, it is not a good idea to get between a salesperson and a client. Channel all your communications with the client through the salesperson. This includes having the salesperson call the client for ad approval, taking the client a copy of the ad, or emailing the client an MP3 of the ad. As simple as it may sound, one of the last steps in the buying process is the client's approval of the ad before it goes on the air. Usually, only the salesperson is prepared to handle the client's questions and concerns at this critical time in the sales process.

There are times when a salesperson may ask you to have direct contact with a client (because he or she has a scheduling conflict, or for some other reason). You should talk with the salesperson beforehand and find out exactly what he or she wants you to do and say. If the client asks you any questions you don't know the answers to, politely refer the client back to the account representative. To sum it up, don't try and go around the salesperson to the client.

Why You Need to Create Great Commercials

A client said yes, a salesperson wrote a contract and electronically submitted a broadcast order, traffic has entered the schedule in the station's logging system, and a production order shows up in your email (see figure 7.1 on the next page). The success of the radio station is now on your shoulders. No kidding.

There are two reasons you need to create great commercials. First, a bad commercial fails to help the client generate anticipated sales increases. Successful radio advertising is based upon solving client marketing and advertising problems through increased customer traffic and sales. Radio's success is dependent on helping a client build his or her business with radio advertising so that the business becomes a regular advertiser. If a commercial fails, it fails the business that bought it and the radio station, and the station has lost a repeat customer.

Second, you need to create a great commercial because bad ones are *dangerous* to the station's audience ratings. During a television show, commercial breaks can be filled with bad commercials. The audience still watches the show, but when the commercials come on viewers have the option to get up and leave the room, returning when the commercials are over to watch their show. Regardless of whether or not the viewer watches the commercials, the television show still gets good ratings. Radio, though, is different. A commuter stuck in a car during morning rush-hour traffic when a bad commercial comes on has only one option. The driver can't get up and leave the car, so he or she changes the station in search of something better. Every listener is valuable. Losing listeners causes ratings to drop. As ratings drop, so does the station's income. Poorly produced commercials are just as responsible for the loss of listeners as are bad on-air talent.

Turning a Sales Order into a Working Commercial

Let's look at the steps involved in creating a great commercial.

Brainstorming

The process of creating an award-winning commercial begins with brainstorming. Brainstorming is a wonderful part of the creative process, because it triggers people's imagination and brings out innovative ideas. All you need is a piece of paper or a marker board and a quiet place to think. As you think, write

WRPC Production Order

104.1 FM *Classic Hits*

Client Number: 88-9679
Date: 5/2
Salesperson: Jim Underwood
Client: Carolina Furniture Outlet
Client Contact: Rick Green
Address: 2617 Hendersonville Road
Phone: 555-5011 **Fax:** 555-5165
E-mail: rickg@spiderlink.net **Website:** www.cfoutlet.net
Start Date: 5/24 **Start Time:** 0600
End Date: 5/30 **End Time:** 1700
Client approval due date: 5/22 **Time:** 3:30 PM
How many different commercials? 2
Length: 60
Who Writes Copy? Continuity _X_ Salesperson ___ Production ___

Specific Production Requirements

(Staple additional information to this order if needed)

Client wants an aggressive car dealer sound to commercial. Really Hard Hitting.

Please see if Jeff will do the voice on this.

Make sure you mention the phone number 4 times, 2 at front of ad and 2 at end.

Client wants to approve music before you do production.

Sofas were $899 now $599

Love Seats were $799 now $499

Save $600 on complete living room sets from only $1,195

All leather recliners are on sale this week only, pick any leather recliner and take 40 percent off

Save nearly $1,000 on complete bedroom sets from only $895

MORE COPY ON ATTACHED SHEET

Produced by: ________________ Date: ________ Time: ________

Salesperson notified on: Date: ________ Time: ________

Music cuts used: ________________

SFX used: ________________

Figure 7.1 A production order is the first step in producing a commercial. Note the amount of information shown here. The production order also tracks who produced the commercial and what music and sound effects were used to create it. Most stations use online or web-based production orders so that everyone connected with the sales and production process can follow the order.

down all your ideas. Nothing is too silly, too stupid, or too off the wall. Another brainstorming style is to think of all of the worst possible things you could say about the product. Often these negatives trigger ideas for some very strong positives that become powerful advertising concepts or market positions. At this stage in the process go for the preposterous, push the limits, be a little daring, and don't be critical or make any judgments regarding your ideas. Just write them down. The goal is quantity—come up with as many ideas as possible for the ad.

Build on the ideas by grabbing anyone nearby and soliciting his or her ideas. People tend to play off one another, and the more people you have, the more imaginative you can be in coming up with creative solutions to meet the client's needs.

Brainstorming does not have to be a long, drawn-out process. Allow 10 to 15 minutes, focused on the project at hand. Focused means *no* interruptions: close your door, turn your phone off, and minimize your computer screens. You're brainstorming.

Refinement

In the refinement stage, you evaluate what you have written down. Start the refinement process by using the guidelines that the salesperson and client have established for the commercial. These guidelines are your filter. Read the ideas you have on paper aloud, since this may trigger more new ideas from you or someone on your team. Group similar ideas together and see if anything begins to take shape. Combining a new idea with one you already have could be the key to an incredible commercial. As you filter the ideas, some may be tossed out as too off the wall, and that is fine, since the purpose of the filtering process is to narrow your options and focus in one direction.

Selecting a Style

As a part of the refinement process, you are likely to come up with an idea about a possible style for the commercial. Some might argue that selecting a style has to come before refining the ideas. But the two depend on each other. Which came first, the chicken or the egg?

In this respect, the process originates with you. Do whatever you are most comfortable with. Refine the ideas and select a style in the order that makes sense to you.

There are five broad categories of commercial styles. Nevertheless, these categories can and do overlap.

Dialogue and Narration, with and without Music. Commercials can use plain language very effectively, whether it's one person talking the whole time or two or more people acting out a scene together. Spoken commercials like this can be kept entirely plain, with no musical effects whatsoever, or they can be enhanced with production music.

The most basic commercial features a single person talking for 60 seconds (narration). On its face, this may seem like a very dull and boring approach. But consider this: When you go into a car dealership to talk to a salesperson, does he or she play the company's theme music and sing jingles while trying to talk to you? Nope!

In spite of this, just about every commercial on the air has music, sound effects, and multiple voices. After a while, with two or three ads to a **stop-set**, the commercials begin to blend together and sound alike to the listener. A narration commercial (one person) or a commercial in the form of a dialogue between two people can be very effective at capturing the audience's attention and standing out from other commercials.

One of the most effective commercial approaches is a business owner voicing his or her own commercial. Not only does the commercial stand out because it is just narration, it stands out because it is not a slick-sounding professional voice. As a real person and a known quantity in the community, the business owner has a great deal of credibility. People will tell the owner, "I heard you on the radio." These are some of the most powerful words a business owner can ever hear.

A dialogue or narration commercial can also use music, in addition to the talking, to reinforce ideas and themes. Music can add a powerful element to a commercial, providing nonverbal cues to many emotions that people experience. The tone of the music provides a cue to the audience ranging from upbeat and positive to the opposite end of the spectrum, very serious. Listen to the music cuts you downloaded from the book's website (http://www.waveland.com/Connelly/) and you will experience a variety of emotions that can be used to enhance a narration-type commercial.

Production music can be purchased in the form of music libraries that are either bought outright or leased. Production libraries feature 30- and 60-second-long cuts of music that are copyright-cleared for commercial broadcast purposes. It is a violation of copyright laws to edit and use popular general-release music behind commercial material, unless you have paid for the right to do so. Aside from the copyright aspect, using popular music behind a commercial causes the audience to focus on the music rather than on the dialogue in the commercial. Several cuts of copyright-cleared music are included in the available downloads (http://www.waveland.com/Connelly/).

Customized music can provide an advantage for a client. Custom music beds and jingles become a client's signature, so that every time the music airs, the audience immediately associates it with the client.

Testimonials. Many business owners feel that the best advertising is word-of-mouth advertising. This is good for radio commercial sales, since a radio station has the biggest mouth in town.

Word-of-mouth advertising is testimonial advertising. Using a business's customers to tell listeners why they should shop at a particular business is very convincing. As an example, restaurants use this commercial style very effectively, featuring customers telling listeners why they like the restaurant and what their favorite dishes are. With the right mix of people, the customers can voice the entire commercial. The only downside to this type of commercial is that it takes more time to produce, since it typically involves location recording and a lot of editing.

Product Comparisons. Many business owners want to go head-to-head with their competition and compare products and prices. One school of thought on this approach says never give the competition your airtime; another says if you have the advantage, take it.

Comparative advertising can be effective with products and services that the listener needs and that have high brand loyalty. If a small independent grocery store wants to challenge a big chain store, they can use price and product, demonstrating that they have the better buys on specific "need" items.

Lifetime Experiences. The lifetime-experience style of commercial involves using talent that mimics people in the listener's life. Real-life situations are used to show how a business can solve problems or make life easier for the listener. People tend to relate to people like themselves, and so this type of commercial can be very effective.

Lifetime-experience commercials must be carefully written and developed so that they are realistic and believable. Otherwise, they appear less than professional and not very credible. The same applies to the talent in the commercial; the speakers have to be believable so that the audience identifies with them. Although announcers have good voices, often they make poor actors. You may have to reach outside the announcing staff to voice a believable lifetime-experience commercial. This is particularly true when you need the voices of older adults or children.

Humor. Humor can be a great selling tool; it can also be a commercial's downfall. Humor has to be well done to be an effective and persuasive advertising style. Just like with anyone telling a joke, the message has to keep the listener's interest and attention until the punch line, and it has to provide some real entertainment for the listener.

Humor requires care so that you entertain, rather than offend, listeners. It is best to stick with generic themes that apply to lifetime situations such as family, children, and friends. All of these people have similar lifetime experiences that you can base humor on. Keep in mind that what you may think is a scream, your audience may not find funny at all (and vice versa).

More than anything, humor is an exaggeration of life. If you stick too close to real life, then it is unlikely that anyone is going to laugh. It is when you step on the line or push the envelope that things begin to get funny.

Always, though, keep in mind that your goal is to entertain and not to offend. Try out your ideas on people before putting them before the client, and listen carefully to the responses.

Assembling a Commercial

This is not a chapter about writing but rather how to assemble a commercial, how to put the pieces and parts together. Here are eight things to keep in mind when assembling a commercial.

The Target Audience. Contrary to popular belief, commercials cannot force anyone to buy anything. Most consumers are motivated to buy something only because the product or service offered solves a problem they have, or at least think they have. A focal point to keep in mind as you write a commercial is WIFM:

What's in it for me? Keep the ad focused on the benefits that the target consumer is going to receive or the value that the consumer is going to get from the product.

Almost every business is number one at something and wants to toot its own horn. And yet, although customers are drawn to businesses they perceive as leaders, that is not the customer's primary concern. Customers are not so interested in the business; they want to know how the business's products and services can solve their problems. Focus the commercial on the target audience, not on the advertiser.

Some consumer problems are easy to identify because they are based on need. As an example, people get hungry and have to shop somewhere for food. Consumer problems that are more difficult to pinpoint are those based on desire or want. For instance: the first major purchase many college students *want* to make after leaving college is a new car. The kind of car they can, or cannot, afford is going to depend on the job they get right out of college.

The client and salesperson must identify the target audience. Who is the business trying to reach? Who are you supposed to be talking to? What kind of problem are you trying to solve for the target audience? The answers to these questions include demographic as well as psychographic information. Demographic information includes quantifiable data, such as the target audience includes married couples that are 25 to 54 years old, with a combined income of $80,000 per year, and have 1.2 children. Psychographic information about this couple includes information such as they prefer country music, eat out two times a week, buy 11 music downloads a month, rent 2 Blu-ray DVDs a month from a video vending machine, have a smart TV with subscriptions to Hulu and Netflix, and like domestic beer. The more information you have, the more effective your commercial can be.

The Objective. A schedule of ads, or a campaign, has to have an objective to succeed. Why is the client running these ads? This information is determined by the salesperson working with the client. The overall campaign objective involves solving a specific marketing problem for the client.

After determining the overall campaign objective, the specific purpose of the ads can be identified. What does the client want each commercial to do? Before you start producing, finish this sentence in clear and concise language: "This commercial's specific purpose is. . . ." A commercial should be focused on solving only one advertising objective or purpose. Trying to accomplish multiple objectives within the same ad dilutes the message and makes it difficult for the listener to follow. If necessary, create more than one ad and rotate the ads in an appropriate schedule so that each ad gets the proper **frequency**.

As a part of the commercial creation process, write the overall campaign objective and the specific commercial objective, or purpose, on the copy. Some stations use copy forms to define objectives and strategy more clearly and to help keep the writer "on message" (see figure 7.2).

One purpose that every commercial has in common is to establish the name of the business in the listener's mind. This is done through a three-step process: repetition, repetition, and repetition. The client's name needs to be mentioned at least

WRPC Production Copy
104.1 FM *Classic Hits*

Client Number: 88-9679
Date: 5/2
Salesperson: Jim Underwood
Client: Carolina Furniture Outlet
Client Contact: Rick Green
Address: 2617 Hendersonville Road
Phone: 555-5011
Website: www.cfoutlet.net
Fax: 555-5165
E-mail: rickg@spiderlink.net

Start Date: 5/24
Start Time: 0600
End Date: 5/30
End Time: 1700
Client approval due date: 5/22
Time: 3:30 PM
How many different commercials? 2
Length: 60

Overall campaign objective: To establish Carolina Furniture Outlet as THE only store in Western North Carolina that is a true manufacturer's outlet. The only direct-from-the-factory manufacturer's furniture outlet.

Specific commercial objective: Move overstock dining room sets in a 12-hour one-day only monster sale.

COPY

Attach additional copy sheets as needed.

Figure 7.2 Excellent commercial copy is important to the success of an advertising campaign. This copy form repeats much of the information on the production order so that it is always in front of either the writer or the producer, making it easy to check vital client information and due dates. Like the production order, most stations prepare copy online for easy access and editing.

four to six times in a 60-second commercial. Avoid using words like "we," "us," or "our" when referring to the business, use the client's name, and use it often, throughout the ad (see audio demo, cut 7-3).

A Few Phrases to Eliminate from Your Vocabulary. Before you ever start to write commercials, you need to eliminate the following words and phrases from your vocabulary. They are overused, trite, clichéd, and meaningless. They are junk words with no real meaning, which wastes your client's valuable time. These are just a few of the junk phrases that show up over and over in commercials.

- That's right. . . .
- You, the customer. . . .
- So hurry on down to. . . .
- So you don't forget. . . .
- Call right now. . . .
- This is a limited edition that won't last long. . . .
- For storewide savings. . . .
- We service what we sell. . . .
- Our friendly and knowledgeable staff is here to serve you. . . .
- Where our service makes the difference. . . .
- Plus, you'll save. . . .
- With prices too low to mention. . . .
- Come see us today. . . .
- For all your [insert product here] needs, see. . . .
- Conveniently located at. . . .
- They're going fast and won't last long. . . .
- Drive a little and save a lot. . . .

Start listening critically to radio commercials. You will quickly find out that there are even more useless phrases that need to be trashed. Sixty seconds is not much time; learn to edit your copy, and don't waste your client's money. Remember, the goal is to create copy that increases customer traffic and sales.

Opening the Commercial. Time is money when it comes to creating a commercial. Although the entire length of the ad may be 60 seconds, the reality is that you have only 8 to 10 seconds to catch and hold the listener's attention at the beginning of the commercial. This can be done through a variety of ways, including definitive statements such as:

- This weekend, say good-bye to your old clunker and drive off in a new Subaru for only $259.95 a month, at Ridgemont Subaru.
- This week, Wild Bill's Thriftway Store has over $30 in special in-store coupons to help you save on your groceries.

- Would you like to own a new home for what you're currently paying in monthly rent? The First Bank of Cullowhee can show you how, with First Bank's low-interest, no-down-payment home loan.

Notice that each of the above statements targets a listener's needs, wants, and desires. An ad can also be opened with a startling question, an unusual statistic, an example, a comparison, a quotation, a survey result, or a simple story. Do something different, be creative, and get the listener's attention. In opening the commercial, remember the short amount of time you have and stay focused on consumer benefits.

The Price and the Product. If you take nothing else from this chapter, remember this: above all else, every commercial must mention price and product. This is not an option. If the commercial does not have a price and a product, you are not writing a commercial. A commercial without price and product will not produce increased customer traffic or sales for the client.

Stop and think about it for a moment. Every car dealership in the world can tell you it has the lowest prices, easiest financing, an award-winning service department, friendly salespeople, no-hassle pricing, a convenient location, years in business, prices too low to advertise, and so forth. So what! The *only* thing that distinguishes one car dealership from another, or any business from any other business, is price and product.

Marketing research has shown time and time again that although consumers are loyal to a particular brand, they hold little loyalty to a particular retailer. When it comes time to buy, the retailer with the low price wins. If you don't think this is the case, ask yourself this question: "Where do I shop?" Complicating matters, the Internet makes it easy for a consumer to determine not only a fair market value for a product, but also its wholesale price as well. If the product is not available locally at the right price, consumers will drive to a nearby city or purchase the product online.

When presenting the price and product in a commercial, the price needs to be as consumer friendly as possible. Ask yourself, "What would attract me?" For example, an ad for a cable television service provider featured the company's premium package of channels, which were regularly $89.95 a month, on sale for only $59.95 a month, plus free installation. The savings was over $30 per month. Pretty impressive, huh? But the company wasn't selling any packages because the thought of paying $60 per month ($720 per year) for television was still enough to scare customers away.

To solve the problem, the ads were rewritten using a price-comparison technique. The $59.95 monthly cost breaks down to about $1.99 per day. The comparison could have been made with anything, but the choice was made to compare this to the cost of food. The ad copy read:

> A buck ninety nine a day won't even buy your lunch at a fast-food restaurant, but it *will* buy you Acme CableVision's Gold Deluxe package, with 132 channels of incredible television *and* 12 channels of HBO and Cinemax. Which would you rather have every day, a greasy hamburger or quality entertainment for the whole family?

Based on the results from this one campaign, the local cable company changed its whole marketing strategy. This type of approach is called baby steps. Just like a baby, you walk the consumer through the buying process, one little step at a time. Car dealerships use this technique every day. Advertise a car for "only $26,995" and it will sit on the lot. Advertise the same car for "zero down and only $227.51 a month" and it will sell. A consumer cannot afford to spend $27,000 for a new car, but he or she can afford $227.51 a month.

With respect to percentage-off sales, they mean nothing to a consumer. As an example, a client wanted to advertise lawn furniture at "thirty percent off their already low, low, price." In effect, the client was saying "blah blah blah blah blah" to the listener. Instead, tell the listeners they will "save $50 this weekend on an outdoor table and four chairs." In another instance, a 105-year-old family-owned department store had long experience with an annual Founder's Day sale. Each year, the store closed for the afternoon before the start of the sale to mark down the prices in the store. Their radio and print ads said things like "save $14 on men's Dockers slacks." The store owner said that his family had learned quite a long time ago that a percentage of nothing is nothing, and that customers understand money, not percentages.

Here are two sample pieces of copy for a furniture store:

> This Saturday is super savings day at Hallmark Furniture. Save 15 percent on select Hallmark leather sofas and love seats! Leather three-way recliners are 20 percent off! Relax tonight in your new Hallmark sofa or love seat with Hallmark's interest-free financing!

> This Saturday is super savings day at Hallmark Furniture. Save $200 on select Hallmark leather sofas and love seats, starting at only $799. Save $175 on Hallmark three-way leather recliners, starting at only $599. Relax tonight in your new Hallmark sofa and love seat for just $28.95 a month, with Hallmark Furniture's interest-free financing.

Which of the above clearly communicates price and product? A critical part of creating a successful commercial is considering how the consumer evaluates the cost of the client's product. You have to present the price to consumers in terms they understand, making it appear more affordable from the consumers' point of view (remember: WIFM).

The Call to Action. Ask and you get, don't and you won't. A business has to ask, or invite, customers to shop with them. As a part of the commercial, the client has to tell consumers what the client would like them to do or ask consumers to take some action that will result in a sale for the client. This is the stage where it's easy for the clichés highlighted earlier in this chapter to show up if you're not on the alert. Avoid them!

Combining a product benefit with a call to action works well. Be creative. Here's an example: "Don't fight bugs in your home any longer; call Acme Pest Control today at. . . ." Another example could be: "Schedule your free Acme Pest Control home inspection at acmepestcontrol.com."

An Alternative Call to Action. Many stations create alternative ads for clients to run on the station website. The client's ads run simultaneously on air and on the web. The website ad is basically the same as the on-air ad with one exception, the call to action. For example, the on-air ad for a car dealership's service department might say, "Call XXX-XXXX to schedule your service appointment today." The call to action in the ad on the website might say, "To schedule your service appointment now click on the Davies Chevrolet ad on the right side of this page."

The call to action in the website ad directs the listener to do something on the web page that is displayed while the ad is running. Essentially, the website radio ad becomes interactive.

Closing the Commercial. A good wrap to a commercial is very important. In just a few words all of the elements of the commercial need to be tied together into an easy-to-remember message for the audience. The wrap, or closure, should reinforce the purpose and advertising objective of the client (see audio demo, cut 7-4).

Most businesses want to end their commercials with their address, phone number, and web address. This block of information can take up a significant amount of time and needs to be very well written if the audience is going to remember it.

A business address has to be reduced to the simplest of terms so that it can be easily remembered; not many people can write well while they are driving, jogging, or listening in the shower. Eliminate long strings of numbers and use landmarks instead to create an address that can be remembered. Which is easier to remember: "5-6-8-7-9 West State Road 107" or "Highway 107, across from Danny Lenn Subaru"? Other ways to give an address include using an intersection as a reference point or saying that the business is just past, or near, a major landmark. People expect a business to have a large sign they can see if they get close enough, so generic addresses work well. The goal is to keep the address as short and simple as possible.

Phone numbers take up a lot of valuable time, 5 to 7 seconds for a local number if it's repeated and longer for 10- or 11-digit toll-free numbers. A business often includes its phone number just because all the other businesses have included theirs. The reality is that most phone numbers serve no real purpose in an ad and just take up valuable time. Before including a phone number, ask if it is really necessary. Do you ever hear a phone number in a McDonald's commercial?

On the other hand, a florist who has Teleflora services or a beauty shop that takes appointments needs to include their phone number. If a phone number is used, it should be easy to remember. Have the announcer combine the last four numbers in the telephone number into two groupings, such as 555-53-86. This makes it simpler for someone hearing the ad to remember the number. Fifty-three, eighty-six is easier to remember than 5-3-8-6 (see audio demo, cut 7-5).

Businesses that have e-commerce websites need to include their web address in their ad. As an example, car dealerships encourage customers to make online service appointments, golf courses let players schedule tee times, and so on. Just like the phone number, the web address must be simplified. Long web addresses with underscores, hyphens, and dashes *do not* work verbally. Since there is only one World Wide Web, drop the www from the web address (it takes a couple of seconds just to say w-w-w). People understand what to do when they hear "schedule your service appointment on the web at mountainchevy.com" in a commercial.

However you simplify the business's contact info at the end of the commercial, remember to keep the closing strong, memorable, and structured to reinforce the client's objectives.

Record Keeping and Archiving

Throughout the commercial production process, it is important to keep accurate records and copies of all of the information related to the commercial, including the production order, original script, and an archive data copy of the digital audio file.

The production order is a record of what you've done and how you did it. Usually there is a place on the order to record when the client approves the written copy and when the client approves the final production before it goes on the air. It is important to keep production notes as to what you did and how you did it, either on the production order or on the copy. These notes should include the track numbers and content along with a listing of the music bed and sound effects sources used to create the commercial.

Clients often develop a liking for a specific ad because it significantly increases their sales. It is not unusual for a client to ask a salesperson to update last year's big blowout bonanza ad because it worked so well. The client won't be able to remember the exact wording or the music that was used, just that it was the best ad the business had ever run and management wants to run it again. With complete records, it is easy to re-create the ad or make changes to the original. Most stations have policies regarding how long a commercial should be stored on a hard drive and, when it's time, how to archive creative materials. Such materials are kept in active client files, sometimes for as long as two years.

The best rule of thumb when it comes to record keeping and archiving is to leave an information trail that is accurate enough so that a complete stranger can pick up the file and re-create the ad. Not that a stranger is going to do that, but the chances of your remembering exactly what you did to produce an ad, a year after the fact, are very small.

Suggested Activities

1. Look up a listing for a pet store online. Round up two friends and brainstorm a commercial for the pet store based on the online information. Spend no longer than 15 minutes writing down your thoughts. See if you can come up with more than 20 different ideas.
2. Select and record five radio or television commercials off the air. Carefully listen to the commercials and see if you can pinpoint the target audience, advertising objective, and specific purpose of each commercial.

a. What is the style of each commercial?

b. Describe the wrap to each commercial and its effectiveness.

Websites for More Information

For more information about:

- *books on radio advertising,* visit the National Association of Broadcasters at www.nab.org
- *creativity,* from one of the most creative people on the face of the planet, visit www.danoday.com
- *why radio advertising works,* visit the Radio Advertising Bureau at www.rab.com

Pro Speak

You should be able to use these terms in your everyday conversations with other professionals.

dubs—Slang for the word duplicate; a term used for a copy or copies of an original.

frequency—The number of times a radio ad airs during a given period of time.

image transfer—When an audio track from a television commercial is used as a stand-alone radio commercial, causing the radio listener to visualize the images of the matching television commercial.

spot—Slang term for a radio or television commercial. Don't use this word. Although advertising salespeople call them spots, clients typically think in terms of ads or commercials.

stop-set—A break in a station's music programming, composed of a jock talking, commercials, and promotional announcements.

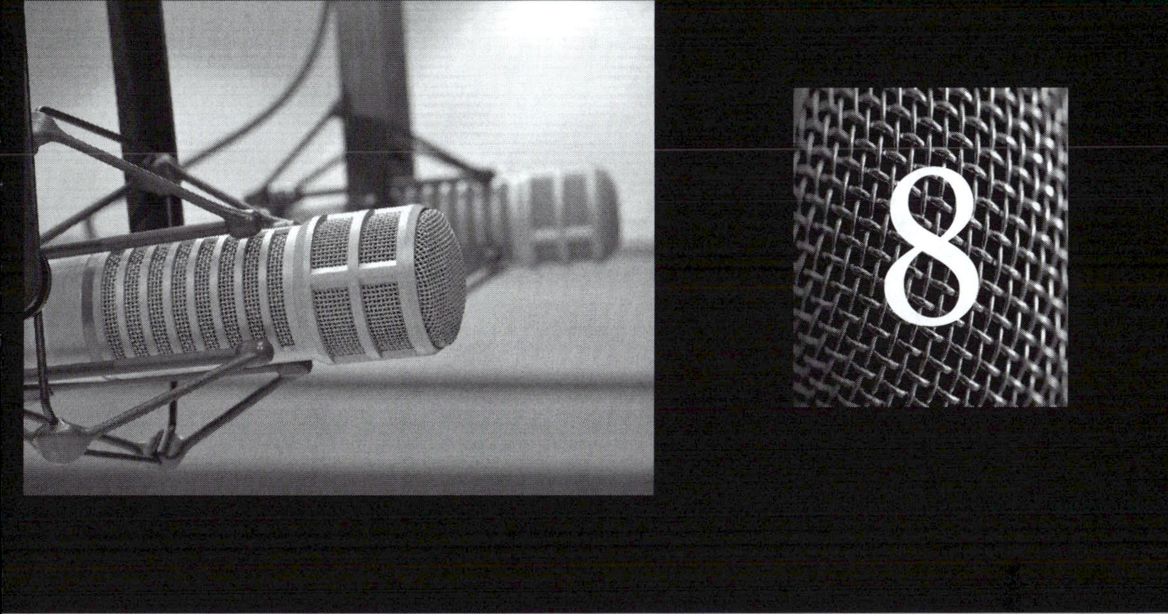

Producing Commercials, Promos, and News

Highlights

Talent
Production Preplanning
  Script
  Technical Considerations and Track Layout
  Music Selection
  Sound Effects Selection
  Talent Selection
  Scheduling Studio Time
Production
  Recording
  Editing
  Mixing and Processing
  Agency- and Manufacturer-Supplied Commercial Material
  Client Approval
  Preparing the Commercial to Go On Air
  Preparing the Commercial Archive: Back Up
  Duplication for Other Stations
  Documentation
News Production
  Newscasts
  Long-Form Programming
Suggested Activities ■ Websites for More Information ■ Pro Speak

Typically, when we think of production in a radio station we think of commercials, station promotional announcements, and the imaging elements that give a radio station its on-air and web personality. However, digitally integrated stations are turning out audio clips of the station's most popular shows for social media distribution. Stations are posting a variety of local audio and video content ranging from newscasts to sports and entertainment news. On-demand podcasts of complete shows and events are also popular.

From a production standpoint, commercials, station promotional announcements, and public service announcements are all basically the same. A commercial is produced for a station client, a promotional announcement is produced for the station, and a public service announcement (PSA) is produced on behalf of a non-profit organization, but the core production elements of recording, editing, and mixing are the same for each. When we refer to commercials throughout this chapter, think of any on-air and web content, promos, and PSAs.

News is a different critter and gets its own special section at the end of the chapter. But even there, keep in mind that many of the core elements of production with regard to recording, editing, and mixing are all the same basic production skills, just used in a slightly different manner.

Talent

What in the world is talent? Simply, talent is a person's inherent, distinctive ability to do certain things a little better than other people do. Some people explain talent as God given; others attribute talent to learned behavioral patterns or skills. It really doesn't matter how a person comes by talent. Some people appear to be naturals, whereas others work hard to build skills in a particular area. Talent is also a generic term used to describe announcers.

Whether a production person is talented or skilled, one of the pitfalls he or she will face in their day-to-day work is turning out "formula" production. Call it falling into a rut, or turning out the "same old, same old." This is especially true in consolidated markets, where a production person typically has a more diverse and heavier workload. The individual may be producing work for five to seven different radio stations. Give any experienced production person the name, address, and phone number of a business and three items with price and product, and he or she can likely ad-lib a 60-second commercial for you. It won't be pretty, but it is a commercial, and if scheduled properly it stands a chance of working.

Of course, the time may come when you are backed into a corner and are forced to resort to a formulaic work product. When this author was a general manager, I had a standing house rule ("house" being the radio station). In an emergency, we would process a sales order and have the commercial on the air in an hour or less, using formula-type production. I can't tell you how many times a car dealership called in near cardiac arrest because a newspaper had screwed up a big weekend ad or a competitor had underpriced the dealership, and the client wanted to use the flexibility of radio to correct the situation *now.* Once the emergency commercial was on the air, though, the salesperson and production had to revamp it and bring the commercial up to station production standards.

To remain fresh, talented, and skilled, production people need to stop from time to time to evaluate their work and the production process, from the creation of the script forward. This evaluation can involve anything from a lunch away from the station to attending an industry seminar. Anywhere you can talk freely and objectively evaluate your production process with others will work. Evaluation should be a constant, ongoing process. If nothing else, use a simple form of self-evaluation on a daily basis. Stop at the end of each day and ask yourself what you did that day that was really good. Likewise, ask yourself if there was anything you could have improved upon, and what you could have done to make it better. Regular evaluation is a valuable resource for news and public affairs producers as well.

Production Preplanning

Before going into the studio, take a moment to plan out what you are going to do and how you are going to do it. This does not have to be a lengthy process. Keep in mind that studio time is valuable, especially in a multistation group.

Script

Regardless of whether you or the continuity coordinator wrote the script for a commercial, review it before you go into the studio. Believe it or not, people do make mistakes, and before you spend time on a project you should make sure you are working with good copy. Do the names, telephone numbers, and addresses in the script match up with the production order? A simple error caught at this point saves everyone time and trouble at some point in the future.

Check pronunciation and look for any new or unusual words you, or the talent, may not be 100 percent sure how to pronounce. If you find something you are not familiar with, ask the continuity coordinator or salesperson how the word should be pronounced. Sometimes it's the simple words that will jump up and bite you. As an example, a pronunciation that comes up frequently in the furniture business is a "suite" of furniture. Is the word "suite" pronounced like a man's suit of clothes, or is it pronounced "sweet"? The dictionary's preferred pronunciation and what the furniture store wants may be two different things. In this case, after confirming with the salesperson, you go with what the client wants.

Look for any phrases that require special attention, such as the client's slogan. Sometimes the client has a particular way he or she likes the slogan to be read, with emphasis placed on certain words. A car dealership, for example, uses the slogan "Nobody, but nobody, beats a Mountain Jeep Deal." In this case, the client insists that the talent read the first "nobody" quickly, and that "but nobody" be stretched out, with a lot of emphasis on the second "nobody."

Check the script's timing by reading the script aloud and timing it with a stopwatch. How long is a 60-second commercial? Sixty seconds. When it comes to timing, you also need to be aware that most stations use live-assist computer systems to play back the ads in the stop-sets. These systems are adjusted so that the ads in the stop-set play back to back, and often the system overlaps the ads by a fraction of a second. Check with the production director or the station's engineering staff to see what the overlap is. If a station is using a common overlap of 750 milliseconds, that means that during the last three-fourths of a second, the ad is being faded down and the next ad is starting. The end result is that your voice track must be finished at 59 seconds or the end of the voice track will be clipped off by the live-assist system as it fades the ad.

Technical Considerations and Track Layout

After previewing the script, you should have a good idea of what is going to be required to create the commercial. The first question to get out of the way is whether or not you have everything in-house to complete the project. Have you got the voice talent, sound effects, and music to produce the commercial? Are there any production elements, like a client-supplied jingle or music bed, that you still need in order to complete the project? If the commercial is a simple dialogue (voice-only) commercial, then all you need is the voice talent and you're ready to go.

Of course, if you are producing an ad for a car dealership you might be asked to do a little more. For example, the dealership may be required to use a manufacturer-supplied music bed, to which the dealership wants to add synthesizer rips, called laser blasts, to emphasize price points. The dealership might also like to overlap the announcer's voice on top of itself with reverb and close the commercial with the eight-second tag line from a custom jingle the dealership has purchased.

As a part of studio prep, make a quick track count and see where you stand. For the car dealership's commercial, eight tracks are required to complete the project in stereo—two tracks for the manufacturer's music bed, two tracks for the client's eight-second custom jingle tag line, two tracks for laser blasts, and two tracks for the voice talent so that the voice can be overlapped. Having an idea of how you want to lay out the tracks will speed things up in the studio.

As a part of preparing to go to the studio, figure out how you want to group or lay out your tracks. There really is no set rule, except to try and keep it as simple as possible. The best track layout is the one that works for you. Don't ask me why, but this author's personal habit is to lay down (record) the voice tracks to the left, or top (depending on software), of the computer's monitor, starting with track one. Music tracks get set to the far right, or bottom (again, depending on software), of the computer's monitor. The sound effects (SFX) usually end up on the tracks in between the voice and the music. Regardless of how you lay out your tracks, though, you need to have a plan in mind before you go into the studio. It is at this stage that you can take a production/track layout form and begin designing the commercial (see figure 8.1). Over time you will become proficient at doing this and will likely be able to do it in your head.

Music Selection

The time to make general music and SFX decisions is before you go to the studio. To avoid tying up a studio, most people use the computer on their desk to audition music and SFX.

Before considering music, consider the lack of it. In chapter 7, plain voice and dialogue commercials were discussed as being very powerful at communicating ideas. Some commercials really do not need, nor should they have, music behind them. When a friend is talking to you about the big sale at the electronics store, does he play music in the background? When a couple is talking about wedding plans, does the wedding planner stop and play music in the background so they can talk? Nope.

If you think you just have to have music in the background of a straight voice, or dialogue, commercial, try this test. Go ahead and mix a track of music you like with the voice track. Play it back and listen critically to the commercial, then immediately play back the commercial without the music bed. Does the commercial message suddenly sound like it's missing something? If it sounds like it lacks something without the music, then you were right: you know the music is adding something to the commercial. But if the commercial message still sounds good without the music, drop the music.

The best music for a commercial is something written and scored specifically for the commercial, something that appeals to the commercial's target audience,

WRPC Production Creative Concepts

104.1 FM *Classic Hits*

Client Number: 88-9679
Date: 5/2
Client: Carolina Furniture Outlet
Salesperson: Jim Underwood
Length: 60 __X__ **30** _____
Date due for approval: 5/22

Production music bed used: ________________

SFX used: ________________

____

Any special music or SFX you created for this ad? ________________

Track 1: ________________
Track 2: ________________
Track 3: ________________
Track 4: ________________
Track 5: ________________
Track 6: ________________
Track 7: ________________
Track 8: ________________
Track 9: ________________
Track 10: ________________
Track 11: ________________
Track 12: ________________

Notes:

Use reverse side for more tracks.
Complete form online or attach to completed production order.

Figure 8.1 A production track-layout form helps you to preplan your work before you go into the studio. It also serves as a record of what you did, should you have to re-create a similar commercial for the client at some time in the future.

and that matches the tempo, mood, and style of the commercial. If you are fortunate enough to be in a larger metropolitan area, there are likely to be a number of talented professionals who can quickly score 60 seconds of music for a very reasonable fee. Such fees will depend on a number of factors. For example, who owns the music and retains the copyright: the composer, client, or the radio station? Is the fee a one-time-use fee, with additional fees to be paid whenever the client uses the music again? Or is the music an outright one-time purchase, where the radio station or client pays for it and it is the property of one or the other in perpetuity? Some radio stations negotiate quantity discount rates from these independent producers, making the cost even more affordable for the client. (Typically, the client pays for any additional costs associated with a custom music bed, even if the costs are folded into an annual contract or commercial rates.)

The next best thing to custom music is a music library. Very likely, most of the music for your commercials and station promotional announcements is going to be selected from a production music library. Production music libraries consist of 60- and 30-second cuts of music that are copyright cleared for commercial use. Some libraries include 10- and 15-second cuts as well. Production libraries come in a variety of music formats, tempos, and styles, fitting just about every business and mood. Most libraries include news, sports, and public affairs theme music as well. A small production music library is available for download (http://www.waveland.com/Connelly/) for your class projects, including news and sports themes (see the appendix for a list of options).

Music libraries are created and sold by music production companies that specialize in production music and jingles. These libraries offer music to fit numerous radio formats. A music library can be acquired on a market-exclusive basis, meaning that the production company will only sell a format-specific library, such as classic rock, to one station in the market. However, the company will sell that same library to a number of stations across the country on a market-by-market basis.

There are three ways a station acquires the rights to production music. First, a station can buy the library outright, with market exclusivity, for a one-time payment. Prices for libraries purchased in this manner range from the hundreds to the thousands of dollars, depending on the number of cuts and the quality of the library. Second, a station can lease a library for a monthly fee, never owning the music in their possession. The advantage of a leased library is that it is updated on a regular basis, therefore you get fresh music to work with, exchanging older music for new cuts. A third way, and the least common way to get a library, is on a **needle-drop basis**. This is a pay-as-you-go situation in which the station initially pays nothing for the library. Each time the station uses a cut from the library, it has to report the usage to the music production company and pay the fee for that cut. (The term needle-drop comes from the early days of radio, when turntables had a replaceable phonograph "needle," or stylus, and a payment was made for each drop of the "needle" onto the record.)

Clients sometimes request a popular song as the background to a commercial. These requests should not be encouraged or readily pursued. In virtually all cases, it is a violation of the copyright held by the music's owner. When you hear and see national and regional ads with popular songs used behind them, additional rights fees have been paid to the music's copyright holder. (An exception to this rule is when your station is promoting a music concert and you use the band's own music behind the ad for their concert.)

As far as selecting the music for the commercial, there are a few more considerations to take into account. First, your program and production director are going to require you to use music that complements the station's format. You can't use music that goes counter to the format. For example, if your station were a light rock station, you would not select something with a heavy metal sound to it. Conflicts like this are usually avoided through the careful selection of the production music library in the first place, but don't be surprised if you occasionally see a production library with some of the cuts labeled "not for air."

Second, consider the commercial's pacing and energy level. You are looking for a piece of music that will appeal to the commercial's target audience and complement the announcer's timing and energy level. The music is not the star of the commercial. The music is backing up the announcer and needs to remain in the background, assisting the announcer to reach the target. Often, the music influences how the talent reads the copy. If you select something too fast and intense, it appears as though you are trying to push, or rush, the announcer along. Likewise, selecting something too slow with a weak energy level drags the announcer down.

Third, consider the emotional level, or appeal, of the commercial. Music is a very powerful emotional tool; use it wisely. Select just the right track for the commercial and you can greatly enhance the commercial's impact. Choosing music is often a trial-and-error process as you look for the track that fits just right with the commercial. What you are looking for is a piece of music that complements and reinforces the commercial to help reach the target audience, all without calling too much attention to itself. You want to affect listeners, but you don't want them to know it is the music doing it. You want the focus to be on the dialogue of the commercial message; the music stays in the background.

Fourth, consider the client's business type. Try to match the music style and tempo, or pacing, to the business. A bank is going to get an entirely different music bed than a car dealership or a fast-food restaurant will. A commercial for an Italian restaurant needs something that sounds like an Italian restaurant. Again, the trial-and-error process comes into play, trying cuts of music until you hear the one that causes you to say, "That's it!" Of course, for effect, you can also try going 180 degrees from what you think the music should be. For example, try putting classical music behind a local Dairy Queen Halloween commercial for hot dogs. Bach's Toccata and Fugue in D Minor, which is rather scary, makes a convincing background for Dairy Queen's Hallow-weenies!

Fifth, you have to consider the client's request, or lack of request. Clients sometimes ask for a specific style or type of music and, just as often, clients leave it up to the salesperson to select something that fits with the commercial. The ultimate music decision is going to rest with the client, and it's a good idea, if possible, to run the music by the client first for approval before producing the commercial.

Sound Effects Selection

The purpose of a sound effect in a commercial is to give the listener an audible cue to what is going on. Depending on how sound effects are used, they can either add to your commercial and enhance the message, or they can detract from your work and make the message difficult or impossible to understand. It's a balancing act. Use too few sound effects, and the listener misses the point of the commercial. Use too many, and the listener misses the commercial message entirely (see audio demo, cut 8-1). When it comes to the use of sound effects, less is generally more.

There are three ways to use sound effects to enhance a commercial. The first is as background continuity, to set the stage for a commercial. For example, if you have a dialogue between two people who are supposed to be hiking in the mountains, an outdoor background with birds would set the stage for the ad. Other examples for different contexts include street sounds, sports crowd noise, babbling brooks, and chirping crickets. Background continuity is just that, nondescript background sounds. Background continuity is sometimes referred to as a secondary sound effect. Nothing in the background should stand out and draw attention.

A second use for sound effects is as foreground continuity. Foreground continuity is a sound effect that supports the dialogue or the event that is being depicted in the commercial. For example, if an ad for a tire store talks about you having a flat tire, you need a tire blowout with the flop, flop, flop of the tire as the car comes to a stop. Foreground sound effects are sometimes called primary sound effects, as they give the listener important clues as to what is happening.

The third use of a sound effect is as punctuation to the dialogue or action in the ad. This is an effect that has to occur at precisely the right moment or the ad does not work. For example, an ad for a cell phone company uses a single cell phone ring to punctuate each price point in the ad. A beer company depends on the sound of a pop-top and the fizz of opening a beer at just the right moment. Timing is absolutely key for sound effects used as punctuation.

Of course, there are also commercials with very little dialogue and are composed mostly of sound effects. (These are just the opposite of the ones described above, in which sound effects merely support the dialogue.) In this case, the sound effects have to carry the commercial; they have to be clear, simple, and definitive. This is not easy to do, since you are depending on the listener to figure out what is going on. Such ads require a lot of extra work and trial-and-error experimentation to find just the right mix of sound and dialogue. It is always a good idea to test this type of commercial on a number of people to make sure they understand the ad, before it goes on the air.

Specific Sound Effects. One class of sound effects that you should avoid entirely is realistic-sounding police and fire sirens, or for that matter any realistic-sounding emergency alert effect, as a primary sound effect. Although this can draw immediate attention to your commercial, it can potentially cause confusion on the part of car and truck drivers and distract them from what they are doing. It is a violation of the FCC Rules and Regulations to use any of the Emergency Alert System

(EAS) tones or announcements as a sound effect in any commercial, promotional announcement, or program. The FCC has issued fines of over $100,000 per occurrence for the inappropriate use of the EAS alert signals.

There is one sound effect that scares everyone to death. It is so stark, startling, and abrupt, that many are afraid to use it in a commercial. Silence. The late radio commentator Paul Harvey will hold the title "Master of Silence" forever. His pauses while delivering commercials and news copy dramatically punctuated and enhanced his stories. Two to three seconds of silence will grab an audience's attention and sounds like an eternity on radio; it can really make a point.

Talent Selection

Voice talent is what makes a commercial. The talent becomes the actor in the theater-of-the-mind story you are creating. As a production person, you need to develop a sense of what talent you have available within the radio station and what they are capable of doing with their voices. Most production people, as they review a script, are mentally casting station staff members in the various roles in the commercial. However, it is important to note that jocks with great voices are not necessarily going to be good actors for your commercial projects.

When casting the voices for your commercials, go for character and style first and voice quality second. Network with everyone you meet, because someday you are going to have to reach outside the radio station for a child, an elderly person, or an ethnic personality to do an accent or role in a commercial. Remember, too, that there is a distinct difference between an announcer and an actor. It is not how the person's voice sounds but what he or she can do with that voice that makes a person an actor.

Every commercial has a personality, character, or attitude to it. Your goal in selecting the talent is to reflect and enhance the commercial's character. As simple as a plain voice, or dialogue, commercial might seem, it can be the most difficult to cast because you have to match the speaker's personality to the commercial. There are male and female speakers who specialize in a style that sounds warm and fuzzy, firm and authoritative, relaxed and easygoing, full of energy and upbeat, or even a touch goofy (but still with some credibility), just to name a few. A key to the commercial's success is your ability to match the right vocal character and qualities to the commercial.

In the preplanning stage, select the talent you need to complete the commercial and set a time for them to record, or cut, the tracks. The talent can cut their tracks anytime they are available and you can come back to the tracks later to produce the commercial. As an example, a station produced a series of commercials for a high-end jewelry store chain that involved a 6-year-old girl and the chain's owner talking about Christmas and Hanukkah gifts. The little girl was brought in first and recorded her lines. The owner of the store came in a week later and, listening to the little girl, recorded his responses to the little girl's lines. The voice tracks were recorded in September and mixed and edited in October; no one believed the two had never met.

However, be aware that you don't always get to pick the talent for your projects. There will be many times that station staff are assigned or outside talent is brought in to work on a project. Most production directors rotate station talent on commercials so that voices don't get overused or scheduled back-to-back in stopsets. Sometimes a client even provides the talent for the commercial, as in the case of the previously mentioned jewelry ad, and in most cases the salesperson works to try and accommodate the client's wishes.

Another aspect of talent selection is a talent fee. Who gets a talent fee and who does not varies from market to market and even within stations. Most often, a station's on-air announcers do not receive talent fees for in-house work, but they do receive a fee when a commercial is used on another station in the market at the client's request. In larger markets, it is not unusual for the talent to belong to a labor union. The Screen Actors Guild and the American Federation of Television and Radio Artists (SAG-AFTRA) represents approximately 160,000 media professionals. It is the most prominent union in the broadcasting field. SAG-AFTRA members work under local and national contracts and must be paid accordingly.

As a word of caution, any time people are brought into the station as outside talent, be sure they sign a talent release authorizing the station to use their voice or likeness. This is particularly important when working with children. In most states you will need the parent or guardian's legal permission to use the child's voice or likeness in a commercial, even if the child is not identified by name.

Scheduling Studio Time

The last stage of preplanning is to schedule your studio time to complete the project. Each facility has its own policy about studio time; don't be a studio hog. In simplest terms, use the studio for the actual work of creating the ad. The key to making maximum use of your studio time is: (1) know what you want before you go into the studio and (2) know how to create it with the equipment you have to work with. Do not be afraid to ask questions of others or to observe others while they work. Most people are more than happy to assist you when you ask for their help, and watching others is a great way to pick up tips and software shortcuts.

Production

It is time to gather the script, music, sound effects, and talent and head to the studio. When you stop to think about it, radio production is absolutely incredible. Where else can you create something so powerful that it motivates people to do something they would not otherwise do? Radio is particularly satisfying because it is such a cost-effective and creative medium. Radio's cost per thousand (cost to reach 1,000 people) is lower than the traditional mediums. Furthermore, the cost to produce a radio ad is much less than producing a television commercial or creating the content for a billboard.

Recording

This is really the easiest step in the commercial-production process. You pre-planned what was going onto each track; now it's time to record each element of the ad as cleanly as possible on a separate track (or tracks, in the case of stereo music and some sound effects). You don't have to do things in real time because you can edit the project later. You are working in a medium that allows you to rearrange everything, including time, so the order in which you record the tracks is immaterial.

Recording Talent. We are in the communication business, but sometimes we forget to communicate with those who are right around us. We just assume they know. In working with talent, let the announcer have the script early so that you can answer any questions he or she may have. Give your talent a general feel for what it is you are looking for. If you want an upbeat read with excitement in the voice, tell the announcer. He or she cannot deliver what you want without your guidance. As an example, one producer of an ad for a Florida retirement community had an almost 20-minute long discussion with the announcer doing the voice-over ahead of time about the read. The producer emphatically explained he did *not* want a typical commercial read on the script. He wanted from-the-heart sincerity, just as if the announcer were talking to her own parents, trying to convince them that this retirement community was the right place for them. Tell your talent what you want; it saves a lot of time in the studio.

Ideally, you would like the talent to give you a clean read, without errors, since this saves editing time later. Don't be afraid, though, to ask the talent for two or three reads of the copy, even when they have given you the read you want (see audio demo, cut 8-2). In fact, try asking the announcer to read the commercial however he or she would like to read it, and encourage the talent just to have fun with it. Not every paragraph of every read is going to be perfect. As the talent gives you that second and third read, you may hear individual sections from the various reads that you want to mix and match in the final edit. Be aware, though, that some talent find it hard to read more than one paragraph at a time without a screw up. That's fine—that's why digital editing was invented.

Don't ever let anyone you are working with gloss over screw ups with the infamous words that production people hate: "Don't worry; you can fix it in post" (meaning postproduction). Even with digital editing, there are some things that cannot be easily changed. Listen carefully to the talent as he or she reads your copy take to take, making sure that the announcer maintains the same distance to the mic, vocal level, intensity, and pacing. Also, follow the announcer's read with the script, making sure it is word for word. By doing so you make it easier to edit the voice track later.

This should go without saying, but after each track is recorded to your satisfaction, save your work and check playback to make sure it is OK. When you have completed recording all of the tracks, back up the project. Editing and mixing of the project can always be redone, but if you inadvertently lose a voice track, the talent is going to charge you to return and cut it again.

Recording Music. The music for your commercial has likely been recorded by a composer/producer, or has come from the station's production music library. Depending on your digital audio software, you can either rip the music track into the project or drag and drop the audio file from the music source into your project. If you are using multiple cuts of music, say different cuts for the open and close of the commercial, you'll need a stereo pair of tracks for each cut of music you are recording. Following your multitrack layout pattern allows you more easily to edit and mix the music later.

Recording Sound Effects. There are three sources for sound effects: a sound effects library, "Foley," or go out and get it yourself. A sound effects library is just like a music library. It is a huge collection of copyright-cleared sound effects that is catalogued and easy to find. Sound effects libraries range from a simple set with basic door creaking and dog barking kinds of sounds to the BBC Complete Sound Effects Library, which includes just about any sound you can imagine from around the world. Often, sound effects are combined or mixed to produce just the right effect that the production person is looking for.

The second source for sound effects is Foley. Jack Foley was a movie sound effects editor at Universal Studios for over 30 years who pioneered the creation of synchronous sound effects in the studio for film. The process of creating sound effects and matching them to dialogue is named after him. (Ironically, after all those years on the sound stage, there is no known recording of his voice.) A true Foley effect is created in a studio and recorded. For example, you are producing a commercial about a wedding. A couple is celebrating a wedding announcement with a bottle of champagne. You need the sound of the cork popping out of the bottle. You have a choice: you can either open a real bottle of champagne (not always practical), or you can place your index finger in your mouth and pop it out to make the sound. Foley is about using things and objects to create sound effects in the studio.

The third source for sound effects is to go out and record them. Sometimes the needed sound effect is so unique that there is no substitute for getting the sound itself. For example, a station was producing a commercial for a tourist railroad that featured a historic steam locomotive. The ad called for the sounds of *the* steam engine in the background. A generic steam engine from a sound effects library was out of the question because railroad enthusiasts *know* what the various types of steam engines sound like. A bogus-sounding steam engine would have cost the railroad customers. Armed with a digital recorder and condenser microphone, the production person spent about an hour with the locomotive, even recording the train's historically correct steam whistle (see audio demo, cut 8-3).

Sound effects are like talent in that they can be recorded at basically any time and edited later. The goal is to get a good, clean recording. Most primary sound effects, like a phone ringing, are recorded on a single channel (monaural) to be close to the dialogue in the sound field. Background sound effects, such as street sounds, are often recorded in stereo to broaden the sound field.

Editing

Editing is the art of cutting, pasting, and assembling the elements of a commercial in the specific order, or time line, that you desire. It is often referred to simply as cutting.

Editing Talent. When you work with good talent, you often get a great take in which every word is pronounced properly, the inflection in the voice is right, and the timing is dead on. Listen to the entire take carefully. Listen for the quality of the presentation, how the dialogue flows, and the impact and overall impression the track creates. If you are satisfied with the take, then leave it alone. If not, move on to editing the voice track.

Editing allows you to move lines around within a commercial and to cut and paste dialogue within the ad. When working with multiple voices, you have the control to edit voices as tight (close together) or loose (far apart) as you want. By adjusting the space between sentences, and sometimes between words, you have the power to make the commercial sound very natural. You can even overlap voice tracks and make the talent sound as if they are conversationally overlapping the ends of the sentences, as people do when they are talking to one another (see figure 8.2).

It is a good idea, as previously suggested, to record several takes of the talent while they are in the studio. Assuming that you did not get one perfect take, select what you think is the best take and use that cut as your basis for building a better

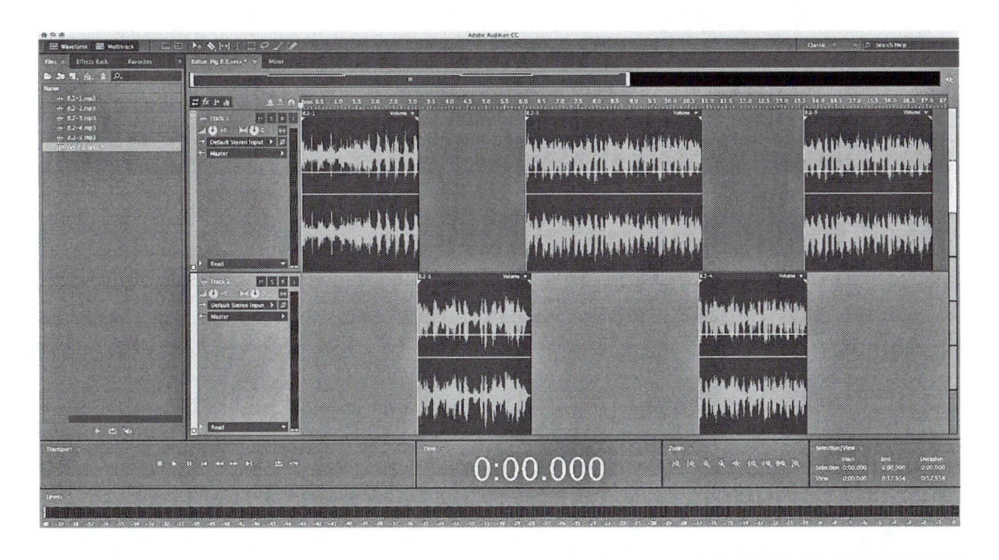

Figure 8.2 Editing allows you to move the announcer's lines around within a commercial and to cut and paste dialogue within the ad. In this example, voice tracks 1 and 2 are very slightly overlapped to make the talent sound as if they are conversationally overlapping the ends of their sentences, just as people do when they are talking to one another. (Adobe product screen shot reprinted with permission from Adobe Systems Incorporated.)

track. With the track selected, listen carefully to all of your other takes (see audio demo, cut 8-4). You may find a sentence, paragraph, or even a word that is better in one of the other takes and can be edited into your original.

When editing dialogue, you have to use great care in making sure to match the vocal intensity, tonal qualities, audio level, and pacing of the original track you are cutting to; otherwise the audio edit won't match, which is quite noticeable. You also need to make sure that when a talent's phrases are edited together, the person's speech flows in a natural pattern.

Editing voice tracks is not too much of a problem. However, even with nondestructive editing, if you spend 30 minutes editing only to have to start over again, you have wasted valuable time. It is possible to do a perfect job of editing and still screw up a commercial. A person may cut and paste a voice track until each individual element within the commercial sounds just right, only to listen to the entire 60-second voice track and have the overall impact and presentation fall flat. Early in the author's broadcasting career I edited a person who spoke broken English. I did a perfect job of taking out every pause, "uh," and stammer. One problem, though: in my zeal to make this person speak perfect English I forgot to leave in any breath sounds. The person spoke for 60 seconds without taking a breath! The words were in perfect sequence, but the natural flow of the person's speech pattern was completely off and it sounded very unnatural.

While editing, another thing to keep in the back of your mind is that the final voice track has to be slightly shorter than 60 seconds. If you are cutting a dialogue commercial, then you can take it right up to your station's time limit. If you are using a music background, try for a voice track that is 57 to 58 seconds long. The two to three seconds left over gives you a second or so at the start of the commercial and a second or so at the end of the commercial for the music to establish and to fade out, respectively.

Editing Music. Music editing is an art, and the interesting thing is that when it is done well, no one even notices the editor's work. There are four basic ways to approach the music track of your commercial. The first is to write a commercial and send it to a composer and have him or her score and produce a custom piece of music to appeal to the target audience and to fit the commercial. If you have the time and the budget to work with, this produces outstanding results because the music is written to match the tempo and style of the ad by surrounding and reinforcing the dialogue and sound effects. The music can even punctuate lines and elements in the commercial.

If you don't have the budget for a custom music cut, there is a way you can make your commercial sound as if the music were custom-scored for it. Select the music first and write the commercial to fit the music. You will need a stopwatch to time the breaks in the music where you want to put copy. Your talent will have to hit those timing marks as he or she reads each segment of the voice track to make it fit with the music. In essence, you are working backwards. If this is done with just a little bit of care, no one will ever know the music wasn't custom-scored for the ad.

The third way to add music to your commercial is to create all of the elements of the commercial as previously discussed. Then select a cut, or cuts, of music from your production library that fits the commercial. You can also select something that is close enough that it can be edited to fit. Depending on how familiar you are with your music library, selecting an appropriate music cut could take a few minutes. In any case, don't spend more than 20 minutes trying to select music. If you have not found anything after this amount of time, you need to ask yourself whether the ad really needs the music or not. (Another option is to select a couple of cuts of music and edit them together to fit the ad.)

A fourth option is to use a commercially available music loop or loop service to create your own musical track. Looping, as it is called, can produce some interesting pieces of music (see audio demo, cut 8-5).

Even when you have found just the right piece of music, you are going to want to clean up the cut. A commercial is 60 seconds long. How long is your music? Chances are it is not exactly 60 seconds long; even in production libraries, you find cuts that range from 57 to 62 seconds. You can easily correct this by loading the cut into your digital audio software and using the time compression and expansion feature to adjust the music track length to exactly 60 seconds.

When working with production music, you will discover that much of it is written so that it can be easily edited. Production music typically has well-defined intros and extros with a bed in between. You may be happy with a cut but want to chop off the first two seconds before the downbeat. You may want to clip the end of the music, ending it on a more definitive downbeat instead of a fade. Then again, there may be two cuts of music you want to blend together, creating a third cut of music that fits the commercial.

Digitally editing music allows you to use hard cuts (butt cuts), butting the ends together, or overlapping cross fades to make an edit (see figures 8.3 and 8.4 on the following page). A trick to mask, or hide, edits is to apply a small amount of reverb or delay to the music track. Reverb is also handy for smoothing out the end of a music cut that has been cut to a hard downbeat. The reverb adds a few milliseconds of fade, making the music appear to trail off naturally.

The simplest form of music editing is to use two cuts of music in an ad. One typically leads into the ad, and the other leads out of the ad. Keep things simple: use the first cut of music to open the ad and slowly fade it out at the appropriate spot. If this is done properly, the listener is unaware that the music is disappearing. Likewise, for the end of the ad, slowly fade up the second music cut and time it so that the cut ends when the commercial ends. Don't try to overlap the music. Leave a couple of seconds, or more, of silence between the two cuts.

When cutting music, it is easiest if you can butt cut on the beat and match the two ends of the edit together as you cut seconds from or add seconds to a selection (again, see figure 8.3). The beats are easy to see on a monitor, which makes the job fairly simple. Only whole bars of music should be cut from or added to a selection (see figure 8.5 on p. 247). Refrain from cutting or adding individual beats. Cutting or adding just a few individual beats to or from the music can really bother listen-

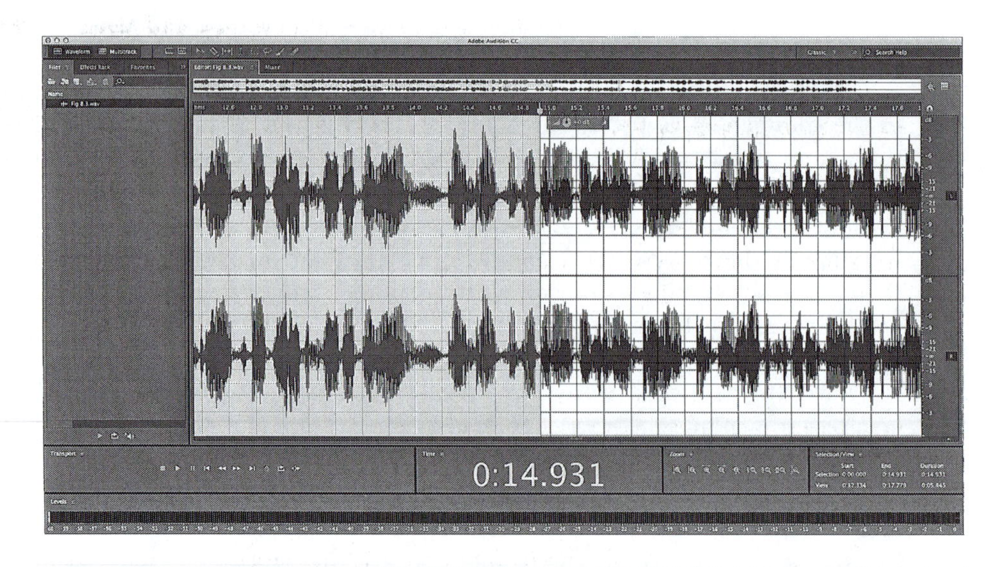

Figure 8.3 A hard-cut, or butt-cut, edit is when two audio clips are joined together end-to-end. This is fairly easy when there are clear indicators of where to cut within the audio waveform, such as at the ends of words, sentences, and beats of music. The edit here is at 14.931 seconds into a 30 second commercial. (Adobe product screen shot reprinted with permission from Adobe Systems Incorporated.)

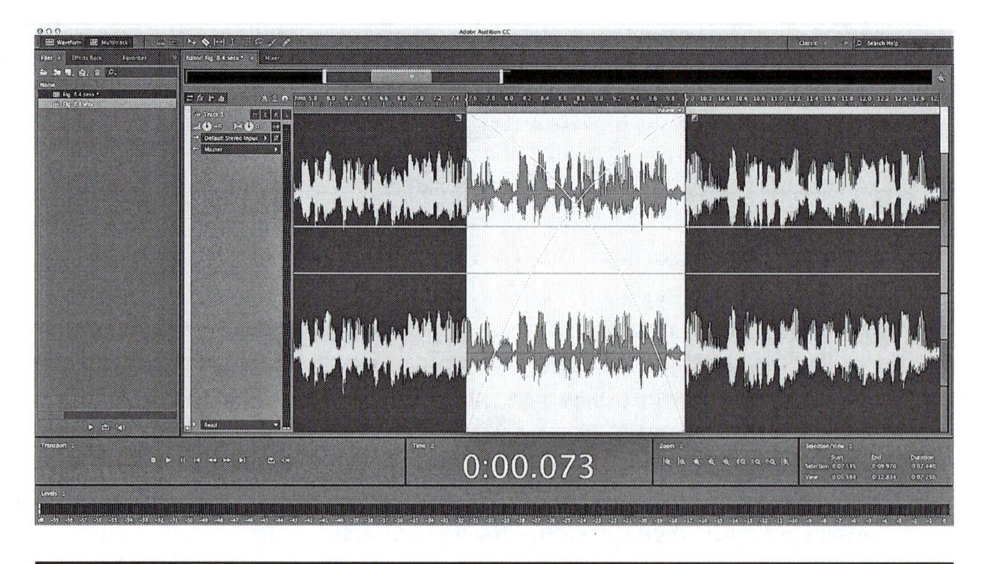

Figure 8.4 When a butt-cut edit sounds too abrupt, it can often be smoothed out with a cross-fade edit. Put the two audio cuts to be joined in the same track opposite of each other. With the cursor, slowly slide one of the tracks over the other at the edit point. You will see the cross fade created as you slide the one edit point over the other. In this example, the cross fade has been greatly exaggerated for illustration purposes. This is an effective way of making a smooth-sounding edit or transition. (Adobe product screen shot reprinted with permission from Adobe Systems Incorporated.)

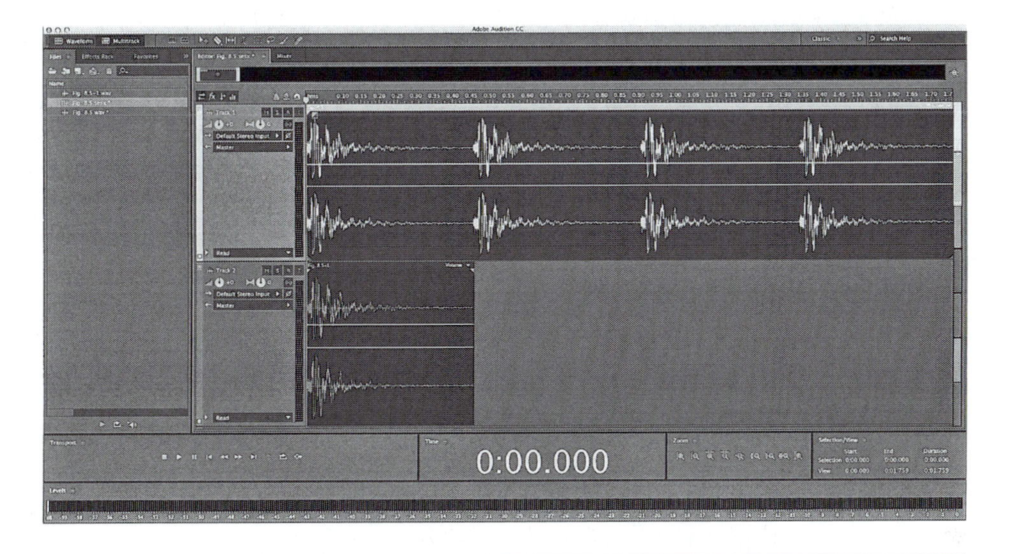

Figure 8.5 It is best musically to edit whole bars of music. Track 1 shows a complete bar of music consisting of four beats. Track 2 shows just one beat of a drum track. However, adding or cutting just one beat can really throw off the flow of the music. (Adobe product screen shot reprinted with permission from Adobe Systems Incorporated.)

ers, even though they probably won't know why the music sounds a little off. (If you have little or no knowledge about music, then these suggestions may not make much sense to you. If this is the case, seek help from someone with more musical knowledge before attempting to edit music.)

Finally, when you are cutting two pieces of music together, they must be in the same key, or very close. If they aren't, the abrupt change in the musical key will be distracting to listeners and draw attention away from the message and to the music.

Editing Sound Effects. With the voice and music tracks cut to length, the next step is to add any secondary (background continuity) and primary (foreground continuity) sound effects to the ad. In editing the background continuity, keep in mind you don't want anything in the effect that stands out and draws attention to the background. For example, if you are using an outdoor background that suddenly has a blue jay screeching loudly several times, the blue jay has to go. The background is just to give the listener an audible cue as to where the talent is located; it should not draw the listener's attention.

After the background continuity is established, record the primary sound effects, which are the foreground continuity. These are the effects that you want noticed; they establish action or movement in the commercial. For example, in a commercial for a sale on lawn mowers the typical homeowner is trying unsuccessfully to start his or her old lawn mower. The sound of a lawn mower failing to start is needed to support the dialogue. The purpose of the effect is to support the voice track and the action in the commercial, but it should not overpower the voice track.

Sound effects that you are going to use to punctuate the commercial are added last. These are effects that are critical to the success of the ad, such as a phone ringing at the precise moment, a car starting, or the sound of a cash register to match the dialogue. These are effects whose placement are critical and should dominate all other effects. Again, the effect should complement the voice track but not overpower it.

Experiment with different placements of a sound effect just before, during, or after dialogue to see how they work. Also learn to experiment with the length of the effect and the effect's strength or intensity. Don't be afraid to shorten or stretch the length of an effect. What you are trying to create is an effect that complements and supports the dialogue in the commercial. Although it might be a great oversimplification, play with the effects until they sound like they belong in the commercial and sound real in your "theater of the mind."

Mixing and Processing

Until this point in the creation process, the concern has been recording good clean tracks and editing each segment of the commercial to fit within the 60-second time frame. We have not been concerned with adjusting levels or using audio processing to enhance the audio. During the mixing process, audio levels are adjusted. The tracks are blended together by assigning the various elements to particular channels so that a stereo sound field is created. Mixing and processing is when the commercial starts taking shape and suddenly comes to life.

Mixing. As you learned in chapter 4, mixing involves track assignments and audio-level adjustments to create a sound field or stereo image. For commercials, music and background sound effects should be mixed in stereo and everything else should be assigned to the center channel (monaural). Stereo music and background sound effects give the commercial a bigger, wider, fuller, stereo sound field. Primary sound effects and any music used as a sound effect, such as a radio on in the background, are assigned to the center channel to make them stand above the background with the voice track.

On a two-voice commercial, as mentioned before, don't even think about putting one voice in the left channel and one in the right. The reason you don't do this is because, although it may sound cool in the studio, it will not sound that way on most car stereos. In fact, depending on where the listener is sitting in the car, it very likely will sound as if one of the voices is disappearing into the background. This acoustic effect occurs because the interior of a car does not provide the same acoustic listening environment that your studio does. Likewise, you need to consider the home or office listener who just wants background music. In these contexts, since a stereo has two speakers, one speaker might easily get put in one room and the other speaker in another room or cubicle.

After making the track assignments, start mixing with the voice track first, since it is the most important element. Use the solo feature of your audio software to isolate the voice track and adjust the track's level to near maximum. After the voice track, bring up the background and then the primary sound effects, one by one, listening to the mix after each is added. The background sound effects should be very low behind everything else. The primary sound effects are more important to the ad, and their levels should be adjusted to stand out more. If you are using a music bed, at the start of the commercial the music should establish at full volume for a second or so before dropping or fading down under the commercial copy. At the end of the commercial the music bed comes back up during the last second or two and ends the commercial.

Processing. As you learned in chapter 6, less is more when it comes to equalization and audio processing. Unless there is a problem that needs correcting, there is really very little processing to do to a commercial to make it sound better. There are some basics, though, that work on just about every commercial.

For the voice track, there are two things you can do to make it stand out. First, mildly compress the announcer's voice track with three to five decibels of compression. This gives the announcer's voice a consistent level and allows you to raise the overall level of the voice track, making it appear louder. Second, apply a small amount of reverb to the voice track and match the reverb length of the music or the background sound effects. This takes the **dead-studio sound** off the voice, giving it a more natural sound (see audio demo, cut 8-6).

Unless there is a problem with the voice track, avoid the temptation to equalize it. On occasion, a voice blends with the background sound effects. As discussed in chapter 6, rather than boosting a frequency in the voice to make it stand out, cut a frequency from the background sounds and carve an equalization hole for the voice to sit in. If needed, apply this same technique to any other problem you want to try to fix. With regard to equalization, always try to lower background levels to bring out a sound effect or voice track first, and when equalization is the only option, cut before you boost.

As a final note on processing, keep in mind that most engineering staffs have spent a lot of time adjusting the station's audio processor to obtain maximum modulation (loudness) and on-air presence for the station. Their adjustments are based on average program levels and frequency curves. Most engineering staffs request that you do not "over process" a commercial in an attempt to dramatically alter the overall equalization, intensity, or tonal quality of the ad. Such attempts often go counter to the station's audio processing and only result in your commercial not sounding as good as it should.

Monitoring. How you monitor, or listen, to the commercial is important to the mix and audio-processing choices you make. If you are working in a nice studio, you may be listening to a big pair of high-end monitors at a pretty good volume level. But how many of your listeners are going to be listening on similar speakers at similar levels? How you monitor a mix can make a big difference in what the listener hears when the commercial is played back on the air or over a website.

First, learn to mix at normal listening levels; only turn the monitors up loud when you need to listen critically to a quiet passage. Second, when you have completed the mix, listen to it at a normal listening level on a sound cube or a portable boom box with the tone controls adjusted flat. (A sound cube is a small monitor speaker designed to emulate a portable radio or car speaker.) As an alternative, make a copy of the commercial and run out to the parking lot and listen to the ad on a couple of different factory car stereos. Can you understand the voice track? Is it clear and above the background? Can you hear all of the sound effects clearly? If the commercial does not pass the sound cube/boom box test, then a remix is in order.

Agency- and Manufacturer-Supplied Commercial Material

Advertising agencies often supply radio stations and local clients with commercial material. This material can be one of two varieties. The first type of ad that an agency may provide a station is a commercial that is already produced and ready to air. This material is supplied by some means of electronic delivery, such as posting the audio file on a website for you to download. When a station receives an agency ad in this format, all the station does is dub, or copy, the ad into the station's server system. The second variety of agency-supplied material is a script that still needs to be produced and must be produced by the station exactly to the agency's specifications.

A common variation on the preproduced, agency-supplied ad is a commercial that is almost complete but requires the addition of a local tag or insert to complete it. The agency may send an ad that has a blank space of 10 to 15 seconds in which the station is to insert local copy. The first of this type of ad is one that has the blank space at the end of the ad. At about 50 seconds, the agency voice is out, leaving 10 seconds of background music, or sound effect, over which you put the local information the agency has specified (see audio demo, cut 8-7). This is called a **tag**, or adding a tag.

The second type of local insert commercial is one that has the hole for the local copy in the middle of the commercial. This hole can range from 10 to 30 seconds. Because the copy is going in a hole in the middle of the ad, this type of ad is known as a **donut** (see audio demo, cut 8-8).

Producing a tag or donut-type agency commercial is simply a matter of laying down the commercial on two tracks (stereo) and adding a third track for the local voice. The mix involves matching the local voice level and intensity to the already-produced agency material.

Always check the timing when you are working with agency materials. Agencies sometimes forget their stopwatches in hopes of getting just a little bit extra for their client. Preproduced agency materials sometimes run a couple of seconds long—a couple of seconds they have not paid for. Agency-recorded materials are easily time adjusted with the time-compression or -expansion feature of your audio software.

In the case of scripts, it is not out of the ordinary for a "60-second" script to read longer than 60 seconds. When written scripts read over the time allotted, the salesperson needs to let the agency know that the script is long and needs a rewrite.

Client Approval

After the ad is produced and you are satisfied with the mix, it is time to play the ad for the salesperson and get his or her approval. Don't be surprised if the salesperson has questions or makes a couple of suggestions for changes. After all, he or she is representing the client and is more familiar with the client's wants and needs.

The ad-approval process can take a couple of different forms. The salesperson may choose to call the client and, with your help, let the client hear the ad over the phone. This is how most ads get approved. On the other hand, the salesperson may choose to make an electronic copy of the ad and send it to the client attached to an email, or take the client a copy of the ad and play it in person.

However the ad is delivered, until the client approves it the ad cannot go on the air. Typically, clients who are more experienced with buying advertising are more likely to listen to their ad on the phone, whereas clients newer to the advertising-buying experience like to have a physical copy of the ad that they can play back for themselves and friends. As previously mentioned, if the salesperson is leaving the client-approval stage of the buying process up to you, ask the salesperson exactly what he or she wants you to do and what you should say to the client when you call on the phone.

Preparing the Commercial to Go On Air

Once the client has approved the ad, the commercial is ready to go on the air. Depending on your station's physical layout, this generally involves moving the ad, as an audio file, from the computer in production to the main on-air server. Usually, this is a drag-and-drop procedure. In many station systems, the new file is placed in a folder on the production computer that the on-air server automatically checks every few minutes, picking up new production and transferring it to the on-air server. Or you might drag the commercial audio file from the production computer across a local area network (LAN) to the on-air server folder where the commercial is to be stored for on-air playback.

If you are working with a local group of stations, this likely involves a LAN. In some cases it involves putting the commercial on a wide area network (WAN) server so that a number of regional stations within your company can access the commercial.

A less-sophisticated method for transferring audio from one radio station to another, but one that certainly works, is through standard Internet FTP (file transfer protocol). This involves logging into a computer via the Internet and either taking files from, or sending files to, the destination computer. FTP is discussed in detail in chapter 13.

Preparing the Commercial Archive: Back Up

In chapter 5, backing up files was discussed in detail, and the points we made there certainly apply to commercial material. Files get corrupted, and computers crash. Each station has different procedures to follow. The best backup is one that is on a different medium from the primary work. If you created a commercial on a hard drive, then back it up on a personal flash drive and store it in a safe place.

You should do two kinds of backup, if possible. First, back up an audio copy of the final mix as the commercial aired. Second, back up a session file or data copy of the individual audio files used to create the ad. These are the software files that hold the data that created your commercial. With a session file or data copy, you can return months from now and open the commercial to change a price or edit an individual track.

Always back up your work. There is nothing more frustrating than to have a file corrupted or have someone accidentally delete a file, and it will happen. With a backup file, you grin like a Cheshire cat and save the day for everyone.

Duplication for Other Stations

As previously stated in this chapter, many clients value a radio station for the quality of its production. Clients often bump up their advertising buy on your station since you are the one producing a great ad and sending dubs to the other stations in the market. When a client asks you to make dubs of a commercial, it is a compliment to your work.

Typically, a copy of an ad for another station is posted on a website for the requesting station to download, or the audio file is attached to an email and sent to the station. Occasionally, a salesperson may want to drop a hard copy of an ad at another station as a courtesy to the client or to the other station. This is especially true in a market in which one owner has several stations that may be located at different sites. Make a quality copy of the ad, or ads, on a flash drive and give it to the salesperson. Likewise, you will retrieve dubs from other stations and ad agencies electronically to prepare to go on the air. The key to a good, professional, working relationship with other stations and ad agencies is to deliver a quality product on time.

Documentation

As discussed in chapter 7, it is critical that you leave a trail of information regarding the work you have done on a project. Each station has different procedures to follow. Usually, though, a client file contains a completed production order with clear notes on anything special you did. In particular, note the sources for all of your music and sound effects. Better yet, a track layout for the ad helps the next person who may be assigned to work on the account.

Many production people keep personal backup copies of all of their work. These can be handy not only for station purposes, but also when it comes time to produce a personal audition.

News Production

It is not that often that production people produce newscasts or sound bites for the news department. Newspeople are pretty independent and prefer to do their own thing when it comes to creating a newscast. However, newspeople do need good basic production skills in today's consolidated market structure. The days of the newsperson just writing copy and delivering the news are long gone. Today, newspeople write their own copy, gather their own video and sound, edit and produce their own newscasts, and post their video clips and work to the station's social media and website. And to some degree, technology is to blame for this.

Using software such as the WireReady system, newspeople can sit at a single computer terminal and receive their news service, such as the Associated Press, write stories, record and edit audio cuts, and embed those audio cuts into the copy (see figure 8.6). They can retrieve data services and audio cuts from major news networks and can handle their email, all on one terminal (see figure 8.7 on the following page). When they step into the studio to do the news, newscasters read from a computer screen and play back the audio cuts embedded in the story by pressing a key on a computer keyboard. It is a paperless news operation. With a WAN, news stories can be shared with any other station in a market, region, or nationally, all with a mouse click. Further, such systems often push the news content up to the station's website.

Generally, about the only time production people are directly involved in the newsroom is when they are asked to assist in producing long-form public affairs shows or to help clean up a poorly recorded sound bite.

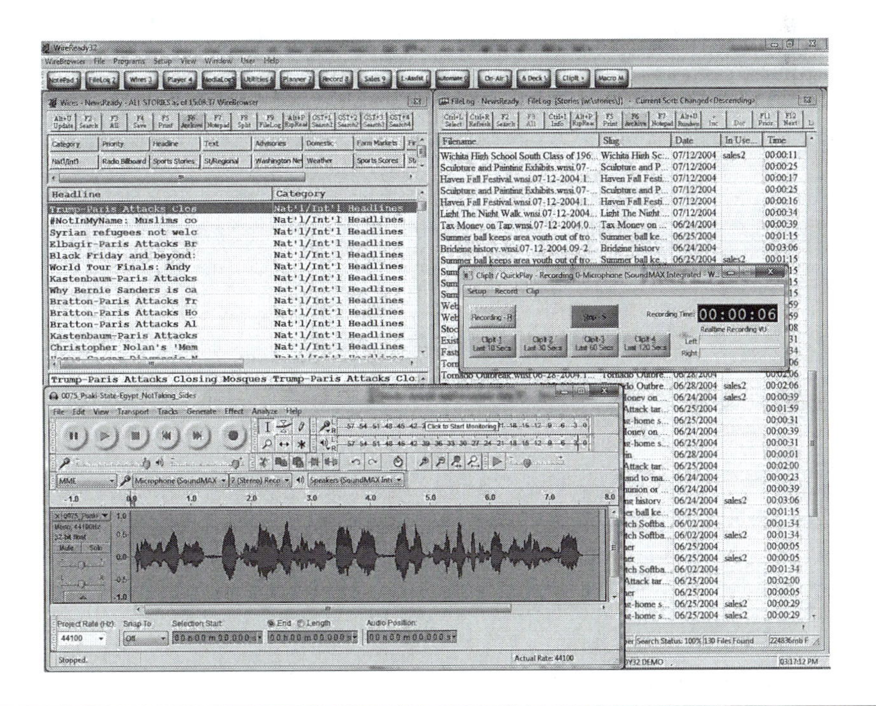

Figure 8.6 Digital workstations make it possible for newspeople to sit at a single computer terminal and receive their news service, such as the Associated Press, write stories, record and edit audio cuts, and embed those audio cuts into the copy. They can retrieve data services and audio cuts from major news networks and can handle their email, all on one terminal. (Courtesy of WireReady.)

Figure 8.7 In a digital newsroom, the sound bites (audio files) are embedded into the copy so that as a newscaster reads the story from a computer screen he or she can play the cuts by simply pressing a key on the keyboard. It is a paperless news operation.

Newscasts

As previously stated, newspeople do need good, solid production skills. Typically, they are recording and editing content for on-air and online use such as video clips, pulling audio from the video clips, and editing sound bites that have been received over the phone, from a television audio source for which they have clearance to use the audio, the Internet, or location content recorded in the field. The key element here is that a lot of their work is comprised of recording and editing voice tracks.

Digital audio and video production has raised some interesting ethical questions with regard to news reporting. How much do we edit and mix when it comes to news? With digital editing it is easy to lift sentences or quotes out of speeches or news events, completely out of context. It is easy to rearrange digitally the time line of an event or the order of a presentation. Likewise, it is possible to reduce the background noise behind a newsmaker to make him or her easier to understand, or conversely, it is possible to inject more background noise to make it sound the way you think the event should. A newsperson could use a recorded background to enhance a report (see audio demo, cut 8-9).

In addition to editing and mixing, there is the ability to use audio processors to enhance or detract from a person's voice or likeness. How far are you permitted to go with audio processing devices such as compressors, limiters, or EQ to make a sound bite more intelligible?

You may ask why these questions are being raised here in a production text, and the answer is simple: they have already been raised in newsrooms across the country. Your station should have its own set of ethical standards, which you are required to follow. Keep in mind, if the *New York Times* has had to deal with someone altering the news, the potential is there for it to happen anywhere.

Basic news production is recording sound and then editing that sound to length, so that it fits in a newscast. The same recording and editing skills required for producing a commercial apply to news. Although these skills make up the core of the newsperson's job, often considered just as important is the presentation style. Most stations design their news format and style of their newscasts around the station's target audience. Announcing style and newscast presentation are discussed in detail in chapter 9.

Long-Form Programming

Long-form news programming is usually associated with public affairs programs or a major breaking story, such as a weather disaster. The newsperson most often needs assistance with the volume of work and the length of the project. Radio stations are required by the FCC to produce public affairs programs that serve the community interest. Most news departments rely on production staff to assist them with recording, editing, and mastering projects. Usually, though, the newsperson wants to assist with editing the program to retain control of the program content.

Once produced, such shows are placed on a server for airing. A backup copy should be made and given to the newsperson for safekeeping. Newspeople like to go back and pull archival cuts from such programs. In short, as a production person you are going to be more an audio engineer than a producer when it comes to the news department, since they have the final judgment with regard to content, just as a commercial client does with an ad.

■ Suggested Activities

1. Using audio demo cuts 8 to 10, sharpen your dialogue-editing skills by editing the file so that the randomly spoken numbers 1 through 10 are spoken in the proper sequence, allowing for a natural pause, or breath, between each number.
2. Using audio demo cut 8-4, edit the 60-second voice track to create a good, clean read. Remember to try to maintain the natural flow of the announcer's voice.
3. Using audio demo cut 8-3, edit the raw sounds of the steam engine to make a 60-second secondary sound effect or background continuity bed to go behind a commercial.
4. Select a 30-second music cut from the production music provided for download (http://www.waveland.com/Connelly/). Identify the intro, extro, and bed of the

music. Try your hand at turning the 30-second cut of music into a 60-second cut of music by repeating measures of music from the music bed.

5. Select a 60-second music cut from the production music provided for download (http://www.waveland.com/Connelly/). Identify the intro, extro, and bed of the music. Try your hand at turning the 60-second cut of music into a 30-second cut of music by removing measures of music from the music bed.
6. Using three tracks in Adobe Audition CC, create a commercial with a monaural voice track and stereo music bed; experiment with mixing the dialogue to fit the music bed so that the dialogue is dominant in the mix.
7. Record a two-person dialogue commercial with a background sound effects bed and three primary sound effects. Add mild compression to the voice tracks to make them stand out and some reverb to help them blend with the background sound effects. Mix the project.

Websites for More Information

For information about:

- *custom jingles and production music,* visit TM Studios, Inc. at www.tmstudios.com
- *Foley,* visit the Motion Picture Sound Editors organization at www.mpse.org
- *newsroom software,* visit WireReady, Inc. at www.wireready.com
- *production music and sound effects,* visit Sound Ideas at www.sound-ideas.com

Pro Speak

You should be able to use these terms in your everyday conversations with other professionals.

dead-studio sound—A studio is said to be a dead studio when there is little or no natural reverberation—the less reverberation, the more dead the studio.

donut—A type of agency-supplied commercial that has a hole in the middle of the ad for the insertion of local copy.

needle-drop basis—An old radio term used to describe paying for production music on a per cut used basis.

tag—A 10- to 15-second space at the end of an agency ad over which the local station puts copy.

Communicating with the Listener

Announcing

HIGHLIGHTS

The success or failure of a radio station is based on the ability of its announcers to communicate with one person—the listener. Not "all you guys out there," not "all our listeners," and certainly not "everyone out there in radio land." As it has been said many times, "That's not a microphone, that's someone's ear." Regardless of whether it is production work or an on-air show, radio is about one-to-one communication between the announcer and the listener (see figure 9.1). An on-air announcer's relationship with listeners is unique because, even though the two may never meet, a good announcer builds a lasting bond with the listener. As an extension of the on-air contact you have with the listener, you will be expected to maintain a social media presence with the listener across multiple platforms.

Entire semester-length announcing courses are offered by many universities. The focus of this chapter is not so much about announcing as it is about good communication skills and professional announcing practices. Many of the things discussed in this chapter relate to on-air work, production, and social media presence.

Communicating On Air

Very few people know the amount of work that goes into creating a great radio show. Most people have this image of a jock standing in a studio playing music and having a good time for four hours. On-air announcers just work four hours a day, don't they? Nothing could be farther from the truth. Creating a good show takes a lot of work before, during, and after the show. Additionally, you very likely will be

Figure 9.1 An on-air announcer's relationship with listeners is unique because, even though the two will probably never meet, a good announcer builds a lasting bond with the listener. One of the simplest things an announcer can do to improve his or her delivery is to smile.

required to maintain a social media presence across a number of social media platforms both during and outside regular work hours. In fact, many stations have daily and weekly quotas for posting to various social media.

Preplanning

A good show starts with preplanning. Beginning jocks often ask, "When should I plan my show, the day before or the day of my show?" The answer is: both and neither.

You should plan your show 24/7. Every waking moment, if you see something, hear something, or think of something that might work well on your show, write it down, take a picture of it with your cell phone and email it to yourself, or find some other way to remember it. Keep a file of these things. As crazy as it sounds, you may think of a great New Year's Eve bit for your show, but in August. File it away in your brilliant idea file (BIF).

Another often-misunderstood aspect of preplanning a show for new jocks has to do with keeping their show local. A lot of general managers and program directors instruct talent to "keep it local." Beginning jocks often think this means that everything in their show has to be from, or related to, the people and things in the station's physical coverage area. That's not the case: if something is important to your listeners, it's local. For example, every summer there are deadly tornadoes that occur across our country that receive national attention. Every rational person in your community fears tornadoes, therefore, the event becomes local. Could your radio station and community help with relief efforts? Could your station feature local disaster planning professionals talking about emergency plans for your community in the event of such a disaster? Likewise, if a unique video or something on social media goes viral, your listeners are likely interested. Anything that affects or touches your listeners can become "local."

A Talk Clock. Formal planning of your show starts the day before, prior to going home. The main thing you'll have to plan is what to say and do in the stop-sets during your show. A **stop-set** is a break in the music or other programming during which you talk and play back commercials, promos, and so on. As the show host, you will need to know when your stop-sets appear and the length of each stop-set (see figure 9.2 on the next page). The stop-sets are usually set up as a computer form, sometimes called a show planner or talk clock, depending on your station. News, weather, traffic reports, contest promotional mentions, giveaways you have to conduct, and vertical promotional mentions are some of the common examples of things you will schedule. Some of your stop-sets will have to be taken up with things mandated by the program director (see figure 9.3 on p. 263), and these are the first things to put on your talk clock, along with items that always happen at scheduled times such as weather, traffic, etc. When all of the mandatory items are taken care of, you will know exactly how much time you will need to plan.

Read Your Local Newspaper's Website and Social Media. If you remember how to write a commercial, you already know how to plan the remainder of your stop-sets. Focus on the listeners (your customers) and their wants, needs, and desires. What interests them? What is happening in their lives? It might be anything from a local Little League baseball team's regional tournament to traffic snarls due to road construction. You cannot overlook the local things that are important to your listeners. That is why you read (no, you CONSUME) the local newspaper online each day before your show. Notice that I say the *local* newspaper's website. It is imperative that you know what is going on in your listeners' lives, so although you may want to read something like *USA Today* for ideas, read your local paper *first* and then check social media sites. The local paper and social media sites are a great source of show-prep ideas. Remember, it's not about you, it's about your listeners.

BIF: Brilliant Idea File. After you have checked out the papers and social media, pull together the items in your BIF (brilliant idea file) and see if anything fits for tomorrow. It might be something that is timely or something going on somewhere else in the media, such as a major star or political figure who is going to appear on a talk show. What question would you ask that person if you were the host? Do you know a little-known fact about that person that you can share with your listeners?

Another kind of item you might have stored in your BIF is something from your life or the life of one of your listeners. For example,

> "I flew home to see my parents this last weekend and you would not believe what was in this lady's suitcase in front of me at the security checkpoint. Talk about embarrassing! What is the most embarrassing thing you have ever done

in an airport? Give me a call at XXX-XXXX or log on to our website at themorningshow.com and tell us what made you the center of attention at the airport!"

or

"Have you ever sent a text message to someone only to have the auto correct feature change a word or two and really weird out the person you're sending it to? Give me a call or hit me up on Twitter or Facebook with the craziest message you never intended to send."

WRPC Talk Clock
104.1 FM *Classic Hits*

	Morning Drive (6:00–7:00 AM) *All breaks must include live promo for station's social media.*
:03	Traffic and weather (:60)
:06	*Stop-set #1 (3:00)* • Promo next 20 minutes • Promo contest in second half of hour
:13	Traffic and weather (:60)
:21	*Stop-set #2 (3:00)* • Promo tomorrow's show and guests
:23	Traffic and weather (:60)
:33	Traffic and weather (:60)
:37	*Stop-set #3 (3:00)* • Contest: Morning Breakfast Brunch Giveaway • Promo afternoon drive • Dinner for Two and a Movie Giveaway
:43	Traffic and weather (:60)
:53	Traffic and weather (:60)
:54	*Stop-set #4 (4:00)* • Promo next 20 minutes • Station ID at end of stop-set
	All breaks must include live promo for station's social media.

Figure 9.2 A talk clock is a listing of the mandatory things you are required to do and is a part of your show planning. Items from the talk clock are merged into your show planner.

Often you can tie listeners' experiences into your show through current events or something that everyone has been through in his or her life. Holidays and seasonal topics are a natural tie-in for a show: "Do you know where valentines came from?"

One Stop-Set, One Thought. While you are putting your show together, there is something you always need to keep in mind, and that is: one stop-set, one thought. Cover only one topic or thought per stop-set. The reason is that you want

WRPC Show Planner

104.1 FM *Classic Hits*

Date: ____________

Announcer: ________________________

	HOUR 1
Contest promo	
Horizontal promo	
Vertical promo	
Giveaways	
Top of hour (:55)	
Break 1	
Break 2	
Break 3	
Break 4	
	HOUR 2
Contest promo	
Horizontal promo	
Vertical promo	
Giveaways	
Top of hour (:55)	
Break 1	
Break 2	
Break 3	
Break 4	

Use reverse for additional hours. Retain this sheet for 5 working days.

Figure 9.3 A show planner lets you plan what you are going to do in advance and when you are going to do it. Planning helps give structure to your show and helps avoid repetition or forgetting a critical promo or contest promo.

listeners to be able to remember your really good material, and they can't do that if you are rambling all over the place from one topic to the next during a stop-set. Pick one topic and stick to it.

Live Delivery

A live show is a lot of fun if you know what you are doing and where you are going. In other words, it is well planned. Live delivery has a unique spontaneity and excitement to it that keeps listeners intrigued, and careful planning enhances that and allows it to work. A live show lets you do a lot of things to attract the listener's attention, such as interacting with listeners by taking their calls, participating in station promotions, and pulling off stunts (see figure 9.4) or engaging them through social media. The more interactive you are with your listener, the better. Quite bluntly, there is no substitute for the energy and enthusiasm of a creative, live show.

Part of what makes a live show so attractive to listeners is the unpredictability of the show. Where is it going today? Hopefully, you are not doing the same bits at the same time every day; that would be boring for you and boring for your audience. That's why you have a written show planner that directs the show and provides you with a record of what you did, what worked, and what didn't.

Figure 9.4 A live show is a lot of fun if you know what you are doing and where you are going. In other words, it is planned out. Live delivery has a certain spontaneity and excitement to it that keeps listeners intrigued.

Making the Human Connection. As a show host, one of the unique aspects of radio is that you are trying to develop a lasting relationship with someone you have never met, using only your voice. Not an easy task! Listeners have to know something about you before they can come to like you. The things you reveal about yourself on the air and through social media are a way to let listeners get to know you.

Listeners would like to feel that you are "one of them" at heart, and this can be something as simple as an on-air birthday party on your birthday, letting your listeners know your likes and dislikes when it comes to food, or talking about your relatives or pets on the air. Your listeners would like to know that there is some commonality between you and them, and the only way this can happen is through what you reveal about yourself.

In revealing this information you have to be careful not to put yourself above your listeners; you want to relate to them on a one-to-one level. I don't think I would brag about my big house, fancy car, house at the lake, or the lavish vacation I took. As a mentor of this author once said, you have to have some "humble." There are also some basic pieces of information I would not reveal about myself, such as my real name and where I live. Most air personalities in larger markets work with assumed names on the air for security purposes. I would also not reveal my religious beliefs or political affiliation. Although religion and politics make for some good discussion topics, they are also a great way to alienate listeners.

Knowing Your Audience. At some point before you start doing an on-air show or production work you need to sit down with the program director and become thoroughly familiar with your station's target audience. Who is the station trying to reach? Production and social media need to be aimed at the station's target audience, just like the announcer's air shifts are. Two types of information can help you to learn more about who it is you're trying to reach.

Demographic information is quantifiable data, such as a target audience's age range, the group's average income, whether or not people are married or single, how many children they have, or what geographic area they live in. Psychographic information is data about a group's lifestyle, such as eating out two times a week, taking two short vacations per year, subscribing to Netflix and Hulu, and liking craft beer. The more you know, the better you can relate to your listeners in your production work and on the air.

Another aspect of getting to know your audience is becoming a local. If you have just moved to an area, you need to become a "local" as fast as possible. You can do this through talking with other station employees and engaging people on social media. The first thing to do is use your smart phone and pull up a map of the area and drive around, getting familiar with all of the local landmarks and roads. Find out how people get around town. Is there a local public transit system? What do the locals use? Are there any vehicles that are popular for one reason or another (e.g., four-wheel drive vehicles in a snowy mountain area)? While you are driving around, locate the top industries and shopping areas in your community.

Find out about the local government and the school system. What kind of city and county government is there, and what is the hot button with the schools? Is it sports, the marching band, or academics? What are the local athletic teams? Who supports what sports?

Explore local foods. What are the regional favorites when it comes to restaurants, soft drinks, beers, and specialty items? Buffalo is known for wings, Memphis for barbecue, and Philadelphia for cheese-steak sandwiches. What does your new area like, or what is it famous for?

Another interesting area to find out about is the local critters. What kinds of animals are in the area and what kind of bugs drive everyone crazy? Florida, for example, is home to alligators and crocodiles. What is the difference between the two, and where do they live? What kind of snakes and spiders live in your area? Some areas are known for mosquitoes the size of Volkswagens. Is the area known for anything special like the annual Possum-Trot Festival? Critters can make for some interesting conversation.

Another topic you should find out about is who the local heroes are. Who are the people that locals hold in high regard and to whom should you show proper reverence?

These are just a few of the many things you need to know to become a local. Don't plan on it just happening by osmosis over a period of time. The faster you become a local, the better you can communicate with your listeners. Some of this knowledge to become a local will come from social media and some will come from face-to-face conversations with people.

Voice-Tracking Delivery

You must learn to voice track a radio show. Voice tracking is recording your stop-sets and letting a computer run your show while you do something else. It is the opposite of live delivery. Voice tracking has been around for some time. It is often used in lower-rated **dayparts** as a way to save money and make a station more profitable (see figure 9.5). Many broadcasters see voice tracking as an opportunity to maximize staff and to give listeners better on-air talent, particularly in the smaller markets. In fact, it was the small-market operators who first embraced voice tracking in order to save money and be able to continue to operate their stations profitably.

It takes about 30 minutes to voice track a 4-hour show. The problem with voice tracking is the safety net. When you voice track a show, there is no fear of failure—you are recording onto a hard drive. Make a mistake, and you can re-record the stop-set. Many announcers look at voice tracking as the way to do a perfect show with no mistakes. But in whose eyes is that really a perfect show? Record a stop-set two or three times and just about every bit of whatever spontaneity there was is stripped away. The result is a lifeless announcer in a can.

Listeners turn on a radio for fun and excitement and for the spur-of-the-moment things that happen. They want to hear someone on the morning team make a goof every now and then, and how the others on the team react to it. There is a certain level of danger in doing a live radio show, an edge that excites listeners, which can easily be lost in voice tracking a show.

Another related concern in voice tracking a show is that announcers have to be very careful not to let the show become just something they do every day before

Figure 9.5 Voice tracking is recording your stop-sets and letting a computer run your show while you do something else. It is often used in lower-rated dayparts as a way to save money and make a station more profitable.

they work in promotions or some other department. Bluntly put, voice tracking is not as much fun as doing a live show. Sadly, some announcers see voice tracking a show as just another job they have to do before they get to go home.

The best way to combat the pitfalls of voice tracking and to do a really decent-sounding, voice-tracked show is to do it live to disc. In other words, cut the safety net. Do each stop-set as if you were on the air and *do not* go back and re-cut a stop-set unless you made a serious technical mistake that will affect later playback. This way, you retain some of the freshness, excitement, and liveliness that listeners seek in a good show. It is not enough for listeners not to be able to figure out that you are voice tracking a show; that's not the point. The point is that listeners have to hear something in your voice and personality that keeps them coming back to hear you every day.

Even more than a regular live daily show, voice tracking requires good, solid preparation and a commitment on the part of the announcer to take possession of the show and be proud of his or her work.

Localizing the Delivery. Voice tracking a radio show, whether it is for your station or another, requires you to be "local." You have to become a local in the ways we have already discussed, even if you live hundreds of miles from the station you are tracking for. You can become a local through the information you have about the community and through the language you use. The moment your listeners suspect you are not real—not one of them, an outsider—you're toast.

Localization for a remote station can be accomplished through general knowledge of an area (maybe you worked or visited there once), emails, or a website the station sets up for the voice-tracking talent so the station can post community information it wants to focus on. An announcer should also maximize the use of social media to keep tabs on a community. It's not the technology that is important, it is the information the jock needs to be local that is.

News Delivery

Unless you are working at a news/talk radio station, news does not get the airtime it once did, particularly on FM stations. The traditional five-minute newscast at the top of the hour and the minute-and-a-half or two-minute news headlines at the bottom of the hour are long gone on most stations. The concentration today is on short, FM-style newscasts of two minutes or less, with four or five brief stories, a couple of which have sound bites. Even news/talk stations have changed their delivery and approach to one that presents a full range of news in a minimal amount of time. Typically, during morning or afternoon drive times the news is even shorter and more upbeat in tempo.

Another aspect of news on a music-based station is that often the morning- or afternoon-drive sidekick is the person responsible for the "news." Usually, the news is little more than some quick rewrites out of the newspaper or off the web. Quite bluntly, it is not really news, since the stories are nothing new and the jock delivering them has very little news authority, believability, or credibility.

Although this might not be the first thing you think of, a good reporter or newscaster has to have a lot of energy and excitement in his or her voice to get and keep our attention. On most stations news needs to be upbeat, quick, and to the point to grab and keep a listener's attention. Just as with production or an on-air show, newscasters have to communicate one-on-one with listeners and, more than anything, tell them the story.

Credibility and Authority. As already mentioned, a good, clean, two-minute newscast can have as many as five stories, with three or four of them including short sound bites. Your delivery style has to be upbeat and authoritative to convey the importance of the information you are telling listeners, although not so authoritative as to sound pompous. A newscast is something special; it has to be, to interrupt the music. It does not need to be sugar-coated. To develop a credible and authoritative news delivery, eliminate the minute or two of happy talk that newscasters sometimes get caught up in with jocks just before or after the news. This drags down the pace, quality, and credibility of the news. On some stations, the chat-up almost sounds like the jock is apologizing to the audience for having to break for news. In effect, the jock is telling listeners, "We are going to do the news now, so if you want to try another station you have a few minutes to spend station surfing."

News should get a neat, clean introduction and start. If the stories are good and the presentation is upbeat with an air of importance, you can hook the listeners and they will hang with the station because they know that the news moves fast and gives them what they need to know. Following the news, jocks and newscasters often go through a typical thank you routine. Like happy talk, it is not necessary, and it distracts from the importance and quality of the news. When the news is over, it is over. Play the commercial or sweeper and move on.

In many respects, a news story is like a commercial, and the announcer has to sell it to the audience. This goes right along with the old adage that a commercial is a news story about a business. All of the basic communication tools discussed later in this chapter are used in a newscast. Reading ability is paramount. Pronunciation is critical. You have to articulate each word during a two-minute newscast or no one will understand you. Good newscasters and reporters read with a lot of vitality in their voices to let the listener know that they are on top of things. The vocal inflection you choose often varies from story to story, depending on the seriousness or importance of the story. The tempo of the newscast needs to be upbeat and is accomplished by reading copy at 15 to 18 lines per minute.

Nevertheless, you have to keep in mind that each station has its own style guidelines when it comes to the news, and it's your job to follow them. Regardless of the guidelines or style, though, the basics we've covered here still apply and can be used to develop a strong, credible presentation.

Weather Delivery

Weather is one of the reasons people listen to the radio. There is any one of a number of ways for a station to produce a weather forecast. The three most common ways (in order of least expensive to most expensive) are (1) for the station staff to use the forecast from the National Weather Service; (2) have a local TV weather forecaster do the forecasts in exchange for the TV station getting credit on the radio; or (3) hire a weather service, such as AccuWeather or another private company. In some areas, such as in the mountains of Colorado or North Carolina, you might need to find a local company that specializes in weather for that particular region's unique weather patterns, which requires extensive local knowledge to accurately make forecasts.

One of the problems with delivering a good weather forecast is the numerous clichés associated with weather that have developed over the years. Every location and region has unique weather conditions. What you have to focus on is what the listener wants. The listener wants a concise, quick forecast that is easily remembered. Most normal weather forecasts can be done in 20 seconds. Obviously, anything other than normal weather warrants a more in-depth forecast that will likely include how the weather is going to impact the area. Probably the best way to explain how to do a good forecast is to explain what not to do or say.

First things first. Never, never, never use the word degrees, or Celsius for that matter. Temperatures are given by their number only. Adding the word degrees is wasted time and fluff since degrees are the only way we measure temperature in this country. This is a good example of something you think you should say but in reality it is the first sign of a rookie.

Eliminate the phrase "the weatherman says" from your vocabulary. Why? First, you are stating that you lack the intelligence to give the weather and that only another person can do it for you. Second, you are stating that all forecasters are men. Third, the weatherman is not saying anything; you're the one reading the forecast.

Never say, "Outside it's 72." Do you normally give the temperature inside as well? Again, wasted time and fluff. Likewise, never say, "It's 72 at our studios." People don't care what the temperature is at the radio station; they want to know what *the* temperature is. Just say, "It's 72." To give the temperature in the morning hours simply say, "It's 72 heading for a high of 82." To give the temperature in the evening hours say, "It's 72 heading for an overnight low of 65."

When it comes to chances of precipitation, most people have no idea what the "chance of rain percentages" mean. Does a 30 percent chance of rain mean it is going to rain over 30 percent of the area, or is there a 30 percent chance it might rain? According to the National Weather Service, the probability of precipitation (PoP) describes the chance of precipitation occurring at *any* location you select in the forecast's geographic area. A simple way to handle chances of rain is as follows: a 70 to 100 percent chance of rain is translated as "plan on showers today" or "it is going to rain today"; a 50 to 70 percent chance of rain is translated as "a good chance of a shower today" or "scattered showers likely today"; a 30 to 50 percent chance of rain is a "slight chance of a scattered shower today"; and a less than 30 percent chance of rain does not get mentioned. When dealing with rain avoid the phrase "take along an umbrella today" since most people can figure this one out for themselves. Besides, a lot of people don't give a hoot about an umbrella.

Each station will have its own definitions for the time of day. Some generic time frames include: morning forecasts are from 5:00 AM to 11:00 AM and talk about this morning, this afternoon, and this evening. Afternoon forecasts are from 11:00 AM to 4:00 PM and talk about this afternoon, this evening, and tomorrow. Evening forecasts are from 4:00 PM to 11:00 PM and talk about this evening, overnight, and tomorrow. Overnight forecasts are from 11:00 PM to 5:00 AM and talk about overnight, this morning, and this afternoon.

An example of the worst possible weather forecast might be:

> "Rock one oh five nine weather this morning. The weatherman says mostly sunny for today with a high of 78 degrees. This evening it will be partly cloudy with an overnight low of 60 to 70 degrees on the dark side. Tomorrow it will be partly cloudy with a 70 percent chance of a thunder boomer. Tomorrow you can pretty well plan on some scattered showers and rain so you will need to take that umbrella along. High temperature is going to be right around 70 or so tomorrow. Right now at the Rock one oh five nine studios it's 70 degrees or 21 Celsius outside in our town. Weather is heard twice an hour on Rock one oh five nine."

Here's an example of a "clean" weather forecast:

> "Rock one oh five nine weather. Mostly sunny with a high of 78 today. This evening partly cloudy, an overnight low of 65. Tomorrow partly cloudy with scat-

tered showers and rain likely, a high of only 68. It's 70 heading for a high of 78. Weather twice an hour on Rock one oh five nine."

Weather often becomes the news. In situations such as tornadoes, blizzards, sand storms, and hurricanes, the delivery style changes to the news style we discussed earlier in the chapter. Usually, such severe weather coverage is handled by the news department and the station's weather service, if the station employs one.

Sports Delivery

A sportscast is something that typically requires a lot of vitality and excitement. Just as you were instructed to look for the passion in a commercial, sportscasters need to look for the passion in sports. Sports fans are passionate about their teams and have clear likes and dislikes. Chances are, nothing you say is going to change a sports fan's biases toward one team or another, but you can certainly share a sports fan's excitement about sports and give that fan something to talk about with friends and coworkers the rest of the day.

Energy and Enthusiasm. More than anything, sports has to be upbeat and forward moving. You have to be enthusiastic. An interesting aspect of sportscasting is something called home-town advantage or bias. It is not at all unusual for sportscasters to give an edge to the hometown team in their coverage. They didn't lose the game last night—it was a hard-fought battle. Our team didn't just win the game—they steamrolled right over the Mud Hens.

One of the ways you can show your energy and enthusiasm is by the language you use. You need to be speaking the same language your listeners are. Often words and terms develop locally that you need to incorporate into your sportscasts. You need to speak in an informed, yet conversational, manner. Your choice of words can play a big part in this. Get away from more formal words and use simple terms. For example:

- Sports figures don't express concern—they're worried.
- Games are not lengthy—they're long.
- Things are not similar—they're alike.
- A sports legend has never expired—he or she died.
- Sports are not timely—they're fast, quick, or speedy.
- There are no sports physicians—they're doctors.

Nothing in simplifying the language is meant to dumb down the message; it is meant to make it clearer and easier to read and understand.

Inflection and tempo play big roles in sportscasting. Your presentation needs to move at about 18 lines of copy per minute and you need to learn to use the full range of inflection in your voice. There is an international soccer broadcaster who became famous by the way he said "gooooooooooaaaaaaaaaallllllllll" when his team scored, stretching the word out for at least 6 to 10 seconds. That's inflection!

With that said, there are going to be times in sports when things get very serious for one reason or another. It is during those times that a sportscaster needs to assume some of the persona of a newscaster. This occurs when a stock car racing driver is killed or an athlete is seriously injured or charged with a crime. And as previously mentioned with news, each station is going to have its own set of guidelines regarding the overall style of the sportscast it wants.

Working with Producers, Consultants, and Production Directors

On-air staff and production people often work with a producer, consultant, or production director who provides guidance and a sense of direction to a show or commercial project. On occasion, talent resists such input since "we know how to do our jobs."

The only problem with knowing how to do our jobs is that sometimes we get too close to the work and don't step back to take a critical listen. Producers and production directors can provide valuable input to a show or project (see figure 9.6). Really good producers and directors assume the role of the listener or client and see your work from the consumer's point of view.

Figure 9.6 Often it is a good idea to involve an extra set of ears on a project. Producers and production directors can provide valuable input to a project. Really good producers and directors assume the role of the listener or client and critique your work from the consumer's point of view.

Generally, when someone takes the time to give you input regarding your work, it's because they want to help you improve the quality of the product. Take the time to listen critically to your work and keep something in mind: you can always learn something from just about everyone.

An Announcer's Basic Tools of Communication

We've already discussed many of the tools that radio talent use to put together their message to their listeners. In this section our discussion encompasses more of the tools of *how* you will get your message across, not *what* you wish to convey. Here are some basics on your voice, how to develop it, basic communication skills, and basic communication tools.

Your Voice

Of all the qualities it takes to be a good announcer, the most important has to be a personable-sounding, natural voice. Whether you realize it or not, your voice is one of the key elements that you have used in building and shaping your personality, long before you considered a career in radio. One of the interesting things about radio is that listeners never get to see the nonverbal cues they otherwise depend on while communicating with others. Your listeners will never get to see your eyes, your facial expressions, or your body language as you do your job each day as they would if you were face-to-face. Your voice is the only aspect of your personality that your listeners have to judge you by.

The voice you have is the one you were born with. You didn't get to choose it. Although a few people still think that all male announcers have to have big, deep, rumbling voices and women have to have low, husky, sexy voices, that just isn't the case. Listeners are looking for a friend and, quite frankly, very few people have friends that sound like that. Your natural voice, the one you use every day, is the one your listeners are looking for in a friend. And just as you can spot someone who lacks sincerity and is a fake, so can your listeners. Radio is not about being someone else, it's about being you, and this is one of the most difficult things for a new announcer to master.

Although delivery styles will vary among a regular daily show, news and sportscasts, and production work, one thing remains constant: your voice. Just like someone who trains to be better at golf, tennis, or some other sports activity, you can train and improve your vocal performance for a variety of presentations and styles.

Developing Your Voice

Before we continue let's get something straight. There is an old radio legend that says if you want a deep voice, drink whiskey and smoke a lot of cigarettes. Nothing could be farther from the truth. Smoking irritates the vocal folds and does give a gravel-like quality to your voice, similar to what happens when you have a cold and your voice becomes deeper. Alcohol dries the vocal folds, causing a similar raspy sound. The reality, though, is that you don't develop a great-sounding voice by routinely damaging your vocal folds, which is exactly what smoking and alcohol do. Politely put, this is about the stupidest thing you can do.

Your Natural Pitch. When it comes to your speaking voice, everyone has a natural pitch. You have had it all your life; you just may not have known what it was. Try forcing your voice above or below your natural pitch for any length of time and your vocal folds become tired; you start sounding raspy. Your vocal folds are complex muscles, and just like other muscles in your body, if pushed too far, they give out. To determine your natural pitch, without thinking about it, spontaneously say "mm-hmm" as if you are indicating something is OK, or as if you agree with something. You don't really say it as much as you hum it with your lips closed. From the mm-hmm, begin speaking with a phrase you use often (such as, "How are you?") at a normal volume level. The two should be very close in pitch because the pitch you hum in and the pitch of your speaking voice should match. If you are having problems hearing whether they match, try recording your voice and playing it back to yourself, or get someone else to listen to you.

If there is a significant difference in the pitch of your hum and your speaking voice, then chances are very good that you are pushing your voice either up or down, and that can strain the vocal folds. This often results in discomfort in the chest and throat. Anytime you experience such discomfort, you are exceeding the comfortable range of your voice.

Breathing and Breath Control. Your voice is powered by air being pushed past your vocal folds by your lungs and diaphragm. To be a good announcer, you need a steady supply of air so that you are not gasping for breath at the end of each sentence.

There is no magic trick for instantly learning breath control; however, with a regular training regimen you can build the muscle control necessary and learn to control your breath. It is not that you don't have enough air in your lungs; you just don't have the muscle control necessary for good breath control. Learning to breathe properly is the first step in expanding your versatility and vocal range.

How to Breathe. Everyone knows how to breathe, right? Not really. Some of us breathe pretty shallowly, meaning we don't use the full capacity of our lungs.

Here is an exercise to help you learn to use your available lung capacity. Standing upright, relax and place your hands on your abdomen (belly) and breathe in slowly and deeply, filling your lungs. You should feel your *lower* ribs and abdomen move out against your hands. The object is not to push your stomach out but to fill your lower lungs with air. As you slowly exhale, your lower abdomen and ribs should relax. During this process, keep your focus on your lower abdomen.

You want the power to come from the abdomen, where you have lots of muscles to do the job, not the chest. Do this deep-breathing exercise for one minute several times a day.

Breath-Control Exercises. There is a very simple exercise you can do to build your breath control and improve your speaking voice: "oohs" and "ahs." Six to eight times a day, while standing upright, relax and breathe out, eliminating as much air from your lungs as possible. Pause a few seconds with your lungs at rest and then allow your lungs to take in as much air as you are comfortable with. You don't really have to think about breathing in a lot of air; your body automatically does that all by itself.

With your lungs full of air, tilt your head back slightly and make the ooh sound (as in "Ooh, isn't that delicious") until you empty your lungs of air. Your goal is to make the ooh sound for 30 seconds. If you are like most people, you are not going to be able to do this the first time you try. The goal is to be able to do this for 30 seconds with ease and still have a comfortable reserve of air in your lungs. Thirty-five seconds is even better. Use a watch or clock, and no cheating.

For the basic exercise, do your first ooh until you empty your lungs, pause a moment, let your lungs refill with air, and repeat the exercise. Beginners should repeat this five times. As you develop muscle control, extend it to 10 repetitions. Doing these 10 repetitions five or six times a day, over 30 to 60 days, will help you improve your breath control. You are training yourself to regulate the flow of air from your lungs so that you have plenty of air to speak with. Each time you do the exercise, alternate from ooh to ah.

When you can do the ooh and ah exercise for 30 seconds, try repeating the exercise and varying the volume level of your voice up and down, from soft to loud. This takes even more breath control and gives you more flexibility in your delivery.

By making the ooh and ah sounds, you are forcing all of the air from your lungs out through your mouth. Another benefit of this exercise is that it helps train the muscles in your throat to route the air that powers your voice through your mouth, not through your nose. Air that escapes through your nose creates a honking nasal sound or overtone to your normal speaking voice that most people find annoying. (An easy way to check for nasality is to place your index finger lightly on the tip of your nose. Slowly say the alphabet. If you feel the tip of your nose vibrating on any letter, those sounds are coming out of your nose. The only two letters you *should* feel a vibration on are *m* and *n*.)

A final benefit of the ooh and ah exercise is that after all of those oohs and ahs, you will have given your vocal folds a great workout. Over time, don't be surprised if your voice mellows or lowers a bit.

Another breath-management exercise is to take a small breath of air and say the number "one" at a normal volume. Take another breath of air and say "one," pause one second, and then say "two." Take in another breath and say "one," pause, "two," pause, "three," and so forth. Adding one number at a time, see how far you can count using one breath. Over 60 to 90 days, build up until you can count to 30 or 35 on a single breath with a slight pause between each number.

Basic Communication Skills

It is one thing to have a great voice, and an entirely different thing to be able to communicate with others. Communicating with others using only your voice requires that you be able to read and that the words you read be audible, clear, and distinct. The basic communication skills include reading proficiency, articulation, and pronunciation. You should already be familiar with reading and pronunciation. Articulation is the way that you move your tongue, palate, teeth, and lips to form the different speech sounds.

Reading. The first aspect of reading is the process of comprehending a written message. Although we typically think of reading as requiring sight, those with impaired vision can read just as well using a number of methods, including finger reading to pick up raised Braille symbols. Comprehending written messages is at the core of reading, but that is just the first step in the radio communication process. Not only do you have to be able to read, you have to be able to translate the written message into spoken words so that someone else can understand the written message.

A large part of reading well for radio comes down to being able to read ahead of the words you are saying, that is, reading several words ahead of your spoken words so that you have time to interpret the words, the punctuation, and their meanings. Each punctuation symbol translates into a vocal cue for you to pass on to the listener. Commas translate into brief pauses; periods indicate the end of a sentence and are about twice as long as a comma. Colons get a pause similar to a period, as do semicolons.

Most punctuation marks also provide a cue as to what you should do vocally to convey the punctuation to the listener. For example, a period indicates the end of something, or finality. Typically, sentences are finished with a slight downturn in the tone and inflection of the voice to indicate the end. A question mark often causes a slight rise in the voice at the end of the sentence. Exclamation points are a signal to be expressive, with some astonishment. Each of the punctuation marks can be a cue to one of the many effects that can be created by adjusting the tone and inflection of the voice up or down.

Reading and speaking simultaneously is a complex process and it is beyond the scope of this text to diagnose and offer a solution to all of the possible reading problems that each person might have. As a university student, if you are not an excellent reader, seek out help through your student support services department. Most universities have programs that can help you become a better reader.

If you are going to be in radio, learn right now to read copy out loud. Typically, when we read silently to ourselves, we leave out a lot of the expressiveness found in a vocal interpretation. This is why it is so important for those who desire a career in radio to read aloud. Reading also involves the pronunciation and articulation of each individual word. Slow reading is an old radio exercise to help you learn to read and pronounce each word separately and to give the proper vocal cues to the punctuation.

Slow reading works like this. Twice a day, block out half an hour and pick up some kind of reading material. It is not really important what you read; it could be a book, a magazine, a newspaper, or something else. In fact, reading different source materials each time provides a good variety of different writing styles. For 30 minutes, read the copy out loud, pausing for one second between each word. Pronounce each word clearly and distinctly. Learn to read four to six words in advance of what you are saying.

This drives most people pretty crazy the first time or two they do it, but under no circumstances are you to speed up the tempo. The benefit is that you hear each and every word, and as you do this in "slow motion" you are programming your mind to see the individual words, pronounce the individual words correctly, and to read the punctuation and interpret its meaning so that you can provide verbal cues to your listeners.

The benefit of slow reading for 60 to 90 days is that you become a better reader, pronouncing every word clearly and distinctly. This is called clipping your words. Over time, you will find that you can read copy with greater ease, since you have trained yourself to read ahead, recognize individual words, and pronounce each word separately. You will discover that you make fewer mistakes reading copy for the first time. Also, the overall quality of your presentation improves tremendously, as words that were once run together are now separated and are easily understood.

The first time you do this, you might want to find a quiet, out-of-the-way place to do your slow reading, as it is quite annoying to other people. As you become proficient at slow reading, try reading different types of materials. For example, read the classifieds in the Sunday newspaper and add some style and emphasis to various words; experiment and have fun while building your reading skills.

Articulation. Articulation is the way that you move your tongue, palate, teeth, and lips to form the different speech sounds. Each of the vowels and consonants is formed using different muscular and skeletal combinations of the throat and oral cavity. Some typical potential problem areas include a tight or stiff jaw, a lazy tongue, and lazy lips. Another problem is a person who talks too fast. Here are some basic exercises that can help you articulate words better.

Start by limbering up. Spread your lips in a big, cheesy, ear-to-ear smile and say "eeeee" several times (see figure 9.7 on the following page). Then open your mouth as wide as an airplane hangar and say "aaaahhh" several times. Finally, close your lips and make the "ooh" sound. Repeat each of these several times to stretch and limber up your mouth.

To limber up your jaw say a series of words that requires you to open your mouth. Try saying the following words while exaggerating your jaw, opening it as wide as you can:

hah	yacht	dog	paw	laugh	yard	tater tots
tab	huh	darts	why	dark	load	wiggle

A stiff or tense upper lip can cause articulation problems. Try saying "pity, Patty" several times in a row to loosen up, making sure to clearly pronounce the *p*'s and *t*'s. When you can do the exercise cleanly at a slow pace, try going faster, but be sure you are clearly saying the "p" and "t" sounds.

Your tongue needs articulation exercise to be directed to its proper place in your mouth for particular sounds. Try saying each of the following words with your tongue forward in your mouth, just behind your upper front teeth:

disc	tea	nope	tabletop	
tale	deal	tick-tock	naughty	nice

Figure 9.7 Open your mouth in a big, cheesy, ear-to-ear smile for vocal exercises and on-air delivery. Listeners can't see you so your voice has to carry the whole load when it comes to inflection and emotion. A big smile adds a whole lot of personality and friendliness to your voice.

Another exercise to limber up your tongue is to place your teeth fairly close together and make the sound of a rattlesnake by trilling your tongue.

Probably the most popular way to exercise your articulators is through tongue twisters. However, the goal with tongue twisters is not to see how fast you can say them but how well you can articulate each of the individual words. Some that I like are:

- The skunk thought the stump stunk; the stump thought the skunk stunk.
- The big black bug bled blue-black blood, and the other big bug bled blue blood.
- Sarah sells seashells by the seashore sitting in her stick-shift Chevrolet. The shells she sells are surely seashells. As she sits she shifts and sits. Sarah, Sarah, sitting in her stick-shift Chevrolet.
- Toy boat, toy boat, toy boat.

Pronunciation. Pronunciation is the manner in which someone says a word. It is based upon the phonetic sounds of vowels and consonants. Pronunciation is also based upon regional peculiarities, accents, and dialects. Regional pronunciations of words are often the result of tradition rather than correct phonetic syntax. For example, go to the city of Cairo, Illinois, and pronounce the name of the city like the one in Egypt and you will get a funny look from residents, because the one in Illinois is pronounced kay-row, with a long "a" sound. And in Missouri, there is a river called the Osceola (oh-see-oh-la) River. In Florida, however, that same Seminole Indian name is pronounced ah-see-oh-la. These are just two of thousands of examples of regional pronunciations.

When you arrive in a new area, be sure to enlist the help of the other station staff to find out what local expressions are used; which local community names, if mispronounced, are dead giveaways that you are an outsider; and any other regionalisms that are important for you to understand. Be sure you can pronounce all the high school sports teams' names, the names of local government officials, and the different neighborhoods or communities in the area. Even though news, as well as commercial copy, often comes with a pronunciation guide, ask a knowledgeable person whenever there is any doubt about a pronunciation.

Basic Communication Tools

With all of the previously discussed elements of your voice in place, including breathing, reading, pronunciation, and articulation, you are ready to begin to have some fun with your voice. As mentioned earlier, it is one thing to have a great-sounding voice, and it is another thing entirely to know how to use it. What follows are some basic tools and techniques you can use to develop more than just a good-sounding voice. Inflection, tempo, vitality, phrasing, and volume are all elements of the creativity used to "sell the copy" to a listener.

Inflection. Speaking in the same tone of voice all the time and showing no expression or emotion is speaking in a monotone, meaning one tone. A monotone presentation does not work on the radio, unless it is for a character in a comedy bit. The opposite of someone speaking in a monotone is someone speaking with inflection in his or her voice. Inflection involves changing the voice by adding emphasis to words through simple things like a change in pitch, pronunciation, the force with which a word is said, and variations in how you say words. It involves subtle changes in your voice to express emotion.

For example, take the phrase "I want you to come here." How many ways can you find to say it? You could say it in a deep, sexy whisper to attract a partner, or at the opposite end of the spectrum, you could say it like an angry father ordering his child to come to him, pronouncing each word loudly, separately, and with a lot of force and emphasis to give a feeling of power. In between, the possibilities are up to you.

It may sound crazy, but all the electronics and audio processing of radio strip your voice of some of its subtle, unique qualities. A good announcer works at being very expressive with his or her voice through inflection, since the listener can't see any of the announcer's body language or facial expressions to get visual emotional cues. Although we use our ears to hear, interestingly, we use our eyes just as much to help us "hear" by interpreting body language and facial expressions. But in radio, your voice is the only real tool you have to reach your listener.

There are three basic things to remember when you are doing an air shift or newscast or producing a commercial. The first is that you are talking to one person and one person only. Pick a particular person and speak to that person. Imagine a close friend, your best buddy, or a close family member. Speak directly to that person, as you would if he or she were sitting in front of you. By focusing on speaking to that one person, you naturally pick up some inflection in your voice.

The second thing that you have to get past is that you are not reading copy or lecturing to this person; you are telling your best friend a story. Concentrate and focus on telling that person the story. Communicate with the person as you would if he or she were sitting right there with you. Don't be afraid to use facial expressions or body movement to emphasize the story you are telling (see figure 9.8). This method works just as well for news as it does for on-air work or recording commercials. Again, by focusing on telling a story to the person who is there with you, your voice changes and you naturally add inflection to your voice.

Finally, above all else, smile. I don't mean a little smirk; as I've said, I mean a big, cheesy, ear-to-ear smile (again, see figure 9.7). And about the time you think you sound friendly, add just a little more. However, if you are reading news copy, you have to use your best judgment as to how to present each story. Some stories are deadly serious, and others are sometimes quite ironic in an almost humorous way. No matter what you are reading, though, keep in mind that your listener can't see you so your voice has to carry the whole load when it comes to inflection and emotion. A big smile adds a whole lot of personality and friendliness to your voice. As a result, a lot of your natural inflections make it through to the listener.

Figure 9.8 Reading a commercial is like telling a story to your best friend. Don't be afraid to use facial expressions or body movement to emphasize the story you are telling. When you focus on telling a story, your voice changes and you naturally add inflection and character.

Here is a fun inflection exercise you can try. Take any sentence and imagine five or six different situations in which you could say that same sentence; imagine the emotion that would go along with each situation. For example, imagine saying to someone, "Is it time to eat yet?" Say this as you would in the following situations:

- You are a six-year-old child saying it to your mother while she is preparing dinner.
- You are on a first date and have been waiting at a restaurant for nearly 45 minutes for your food; you are asking the server if your meal is about ready.
- You have spent the entire day hiking in the woods and it is after 6:00 PM.
- You are a teenager who missed lunch and you have to wait for your sister to get home before the rest of the family eats dinner.
- You are on a cell phone with your lover who is fixing a special dinner for your birthday, and you are already 25 minutes late getting there.
- Your partner bought your favorite chocolate-covered strawberries at the mall and you don't want to wait until you get home to try them.

Announcers often give a client two or three different reads of a commercial, trying out various inflections to see how it changes the commercial's meaning. Sometimes a producer may tell the talent just to have fun with the copy and try a couple of different reads. Overall, inflection adds personality and character to written words.

Tempo. Tempo is the speed at which you deliver the read on a project. Different projects require different tempos. A commercial for a romantic five-star restaurant gets an entirely different read than one for a car dealer's 12-hour mega-sale. Even news and public affairs are presented at different tempos, depending on the situation. Learning to use tempo requires an overall sense of timing. Asked to do so, most experienced announcers can talk for exactly 60 seconds without looking at a clock. Likewise, he or she can look at a piece of copy and tell you if it is too long or short. This kind of timing comes with experience.

Another factor that contributes to tempo are the announcers themselves. Each announcer has a range of speed at which he or she can comfortably read and still maintain inflection and emotion in their voice. Some announcers can't even begin to approach anything closely related to a speed read, and, similarly, others have problems doing slow, intimate reads. As with inflection, you should try to expand your versatility and the range of tempos that you are comfortable delivering at.

The two extremes of tempo each come with built-in difficulties. A slow, intimate read means that every word you say can be heard and the articulation has to be razor sharp. Any defect, no matter how slight, is emphasized by a slow, intimate read. Often, mouth noise or the noise made by your lips and the saliva in your mouth become evident. An old radio trick to help with this is to eat an apple or sip some apple juice before you start. Apple juice does a good job of cleansing the mouth and reducing the excess saliva.

At the opposite end of the spectrum is the high-pressure car commercial in which the announcer is asked to cram 68 seconds of copy into a 60-second commercial by reading at a blazing speed. The difficulty with a speed read is getting each word out of your mouth in the first place. You have to do it so someone can understand what it is you are saying. Adding to the difficulty of a speed read is trying to keep inflection or feeling in the voice.

One of the things you can do to help you learn to control tempo is the slow-reading exercise explained earlier in the chapter. Slow reading lets you hear every word so that you can build pronunciation and articulation skills for those slow reads. It also helps to slow down those announcers who already have a fast natural tempo. By saying just one word at a time, you learn to pronounce the entire word, making sure you say things such as the "-ing" or the "-ed" on the end of words. As you practice slow reading, though, you may notice that over time you can speak faster if you want to, since each word comes out of your mouth as a separate, distinct word.

Vitality. When applied to announcing, vitality refers to the energy in your voice or the animation of your personality through your voice. Vitality is closely tied to inflection and, to some degree, tempo. Vitality is also a component in acting. Really good announcers can deliver a wide range of energy levels in their voice, from something soft and caring to a screaming NASCAR race commercial. Vitality is also a component in doing news. A newscaster needs a certain level of energy to the voice to let listeners know that he or she is on top of things, that there is an interest in making sure listeners are kept up-to-date. Sportscasters need a much higher level of energy, because that is what sports broadcasting is all about—energy, excitement, and enthusiasm.

Radio announcers are similar to actors when it comes to delivering their personality to listeners. One of the interesting comparisons between announcers and actors can be made when actors get the opportunity to voice a character in an animated feature. Most have remarked that creating and carrying a character with just their voice is just as difficult as acting before a camera or on a stage, if not more so. Again, this is because listeners don't get to see any of the body language that actors normally depend on to give viewers so many of the cues to their character.

The best way to develop vitality in your voice is to listen to yourself and to have others listen and critique you. Often you hear producers telling announcers to "kick it up a couple of notches." What they are asking for is a little more personality, a little more of you to come through. Another thing that helps an announcer to improve vitality is to find the emotion in a piece of copy. One of my mentors taught me that to sell the copy, you had to find the passion in it. You had to find something to love and bring it to the audience. In other words, find something in the copy that you have a love for or that the listener has a love for. Or pick something in the copy that you want to really move the listener with. For example, take a soft and soothing commercial for a health spa. Can you be passionate about a health spa? Let me put it this way: Can you be passionate about the incredible cuisine in the four-star restaurant, the total relaxation of a full-body massage, soaking in the hot tub, or the fact that you are going to be pampered for a three-day weekend? What about just getting away from the rat race for three days? Now can you find something in these ideas that you could be passionate about or express a love for? Easily!

From the standpoint of presentation, the flip side is a raucous commercial for a NASCAR race. Can you be passionate about noisy stock cars? The smell of race gas, burning rubber, the roar of the engines, the great junk food, the excitement of the race itself, and the friendship of being with others who understand racing?

If you look, you can find some kind of passion in just about every piece of copy you are asked to read. You just have to take a moment to find it.

If you want to have some fun, go online to a retail website and pick out some ads. Try reading aloud ads for cars or homes. Find the passion in each ad. What point does the seller really think is important? Read aloud from a jobs website and pick out what each employer thinks is important; get passionate about it. Search the situations-wanted section of the website and read the ads there, looking for why each person thinks an employer should hire them. What are these job seekers passionate about?

Phrasing. There is an old adage that says you need to be able to walk and chew gum at the same time. The same thing applies to phrasing the words in a piece of copy. You need to be able to read copy and breathe at the same time. Breathing and phrasing go hand-in-hand. Ideally, you breathe during the natural breaks in the copy that are often indicated by punctuation. A copywriter affects your phrasing from the standpoint that he or she creates the sentence structure and the paragraphs. Different styles of writing present some interesting challenges. The flip side is when you write your own copy and you tailor it to your natural breathing and phrasing patterns.

Copy that is written for a newspaper is entirely different from copy that is written for radio, as newspaper copy adheres to a completely different style of writing. A newspaper is informational, whereas radio is conversational. Try reading a front-page story from your local newspaper aloud. Chances are it does not read well at all because the sentence structure and paragraphs are written to convey a lot of detailed information. A newspaper's journalistic style includes longer sentences and small details.

Radio copy is written from the perspective that we only have a short time to tell a listener a story, whether it is a news story or a commercial. The copy is written to convey the essential details or the big picture, similar to the way you might tell a friend about something. Radio copy typically avoids long, multisyllabic words and run-on sentences. It is purposely written to be read aloud to someone so that the announcer can tell listeners a story.

To phrase copy properly, you have to have a good supply of air to begin with, and you have to learn how to replenish that supply of air without gasping or making loud sucking noises. The way to practice this is to pick a favorite book or magazine and read the first 20 words or so from it. Look first for the punctuation or other natural places to pause in the copy; mark those places with a forward slash mark (/). Every time you reach one of those pauses while reading, allow your lungs to take in air naturally rather than sucking in air consciously to fill your lungs. Pace yourself, and don't speak so long that you allow yourself to run out of breath. It takes some time to develop the ability to pre-read copy—figure out the phrasing that works best for you, and deliver the copy's message. As an example, try reading and phrasing the following paragraph and see how you do:

> It was a very cool, foggy, and breezy spring morning as the sun crested the mountain top and the hikers walked slowly along the Appalachian Trail, their packs heavily laden with provisions for the next two weeks of their solitary journey.

Just as you did to practice vitality, try reading aloud from a variety of sources, including a newspaper, magazine, and a book—even this textbook, for that matter. The more you practice, the better you will become at breaking a piece of copy down into phrasing that complements your style and accents the meaning of the copy.

Volume. How loud do you have to talk to be heard on the radio? Not very loud. We often associate the loudness of a person's voice with a number of emotions, ranging from joy and excitement to extreme anger. Many think that the louder you are, the more forceful you are; the louder you laugh, the happier you are; and so forth. Sadness is often associated with a soft, weak voice. Although these are all typical reactions, actors learn early on that the secret to conveying many of these emotions is not actually loudness or softness, but the emotion itself. Although volume plays a role in expressing the emotion, your control over the emotion is what is important.

An announcer uses a microphone in a studio. Rather than thinking of it as a microphone, think of it as your best friend's ear. Learning to control the volume of your voice is an important aspect of being an announcer. The only thing that happens when you shout into a microphone is that the microphone and associated equipment overload and distort. The end result is that the listener does not get the message. On the other hand, speaking too softly or whispering into the microphone causes the audio processing to automatically turn up the audio levels to the point that background and room noises become very evident.

Good announcers learn to express themselves with their regular speaking voice. You can make it sound as though you are shouting by changing your vitality and the inflection you are using. Likewise, you can make it sound as though you are whispering while still speaking at your regular vocal level.

Your Personality

Communicating with your listeners involves both the technical skill of announcing and learning to convey your personality and express yourself through your voice. As said earlier in the chapter, your voice is the primary means you have to communicate with listeners; they will likely never see you in the studio. To be successful, it is vitally important that you learn to practice vocal skills and at the same time experiment and have fun with your voice. Learn to listen to yourself critically and to depend on others you trust for critical evaluation.

Developing a voice and personality does not happen overnight; it might take months of work to get to the professional entry level you want to achieve. One final thing to keep in mind is that developing your voice and personality is a continuing process that will follow you throughout your career.

Suggested Activities

These questions assume you already have a show of your own. If you don't, just imagine that you do, or put yourself in the place of a show host you are familiar with.

1. Listening to other stations is a great way to improve your show. Listen critically to any local radio morning show and do a talk clock in reverse. In other words, track the station and chart what the jock or jocks are doing during each stop-set. When are the stop-sets each hour? How long are the stop-sets? What do they contain? Do this for the length of the show. Track it over several days. See any predictable patterns or routines?
2. Log onto the web and locate a major-market radio station streaming its signal. Carefully listen to the jocks and how they present and do things. See if you can find one or two stylistic things you could incorporate into your show.
3. Listen to a radio station, not for the music but for the commercials and promos. Record them if you can and then listen to them again and take each one apart. Take notes of your observations. What is different about their work from yours? See if you can find two things you like that they are doing and that you can incorporate into your commercial work.
4. Record yourself every time you are on the air and listen to yourself after you are off the air. Likewise, keep a copy of every commercial and promo you do. Then, listen to them a week later. Are they still as good as you thought they were when you produced them? What could you do to improve your work? In both cases, have someone you respect critique your work.

Website for More Information

For more information about:

- *announcing, production, and being a great jock*, visit Dan O'Day's website at www.danoday.com

Pro Speak

You should be able to use these terms in your everyday conversations with other professionals.

daypart—A broadcast industry term that classifies a specific time period of radio listening. For example, 6:00 AM to 10:00 AM, Monday through Friday is generally accepted as the morning-drive daypart.

stop-set—A break in a station's music programming, composed of a jock talking, commercials, and promotional announcements.

Promotion and Station Imaging

HIGHLIGHTS

Creating a great-sounding radio station is much more than just playing a lot of music back-to-back. In fact, listeners don't really want to hear back-to-back music; they listen to radio to be entertained and to be a part of something. A really great radio station is based around creating an original **sound design** that sets the station apart from the rest of the competition. The purpose of the sound design is to give the radio station its charisma and a personality—the same kind of charisma and personality that a listener seeks in a good friend. These traits will also carry over to the station's website and social media marketing strategy.

The production person is at the heart of any station promotion, imaging, or contest whether it is on air or online. You are the person challenged to bring the concept to life for the listener, to create a theater-of-the-mind event. In this chapter you are going to learn about two of the most important elements of creating a station's sound design: promotion and imaging. Promotions and imaging are also critical to the development of unique web and social media promotions and imaging that tie back to the on-air product. It is imperative that you understand the broad scope and purpose of station promotion and imaging so that you can do your job more effectively. Station promotion and imaging is on air, online, and on demand across multiple platforms. These key elements play a significant role in programming and sales.

Station Promotion

There is an old media saying that if something is worth doing, then it is worth promoting. The key to a successful radio station is how the station promotes itself. An outstanding radio station has a continuing stream of promotions and promotional announcements, one after the other. Promotions include station events, contests, and public service. Promotions help create the positive, forward-moving image in the listener's mind that is required for a station to be successful. Such forward movement also carries over to the station's website and social media, where listener interactivity can be maximized.

The Purpose of Promotion

Stations use promotions to (1) build **Nielsen Audio** audience ratings by increasing the number of time segments (**dayparts**) listeners tune in to and (2) extend the amount of time those listeners spend listening to the station (**time spent listening**). A station also uses promotions to direct listeners to a specific daypart the station wants to build an audience for, such as for the morning or afternoon drive time. Nielsen Audio ratings are discussed in detail in chapter 14.

A well-designed promotion creates an awareness of the station and interest in the community beyond the station's **core audience**. Promotions help build a bond with listeners and enhance the station's image. A second major motivating factor behind station promotion is to generate sales revenue; 70 percent or more of station promotions are sales-driven (see figure 10.1). Even a station's public-service events are often sponsored by businesses (for public relations purposes) on behalf of a charity or community-service organization. Building an audience and generating revenue are two key components of promotion.

Successful radio stations tie their on-air promotions into their websites and social media. Promotion directors often use the station's on-air signal as a **barker** to drive listeners to the station website and social media platforms for maximum interactivity with the listener. There is almost an endless list of things a station can do on its website and social media, including such things as contest registrations, registering listeners for special advertiser email coupons and discounts, and providing links to social networking sites. A promotion that ties the on-air product to a station website and social media (and vice versa) can create very powerful results for the station and its advertisers. The effectiveness of a promotion can be determined using ratings services like Nielsen Audio, as well as Nielsen Online, which provides analysis of online audiences, consumer behavior, and advertising effectiveness.

Promotions require preplanning for proper execution and to generate the visibility, publicity, and revenue the station seeks. A promotions-driven station plans major promotions as much as a year in advance. Concert promotions require three to four months of work. A simple appearance by a musical artist or concert ticket giveaway should be planned three to six weeks in advance. An effective promotion campaign includes a charted time line from brainstorming the promotion concept, preplanning the event, executing the promotion, up to the follow-up. In effect,

Figure 10.1 A well-designed promotion creates an awareness of the station and interest in the community beyond the station's core audience. This remote was from a college football pregame event. Promotions help build a bond with listeners and enhance the station's image.

promotion planning is preparing to go into a ratings war on every front with your competitors. The promotion office and the production studio become the ratings war rooms.

A well-designed promotion plan of attack includes clearly stated overall objectives. You must be able to finish the following sentence: "Our promotion will. . . ." To achieve the objectives, the station must develop a strategy, or plan, that includes a detailed time line on a calendar. Finally, the station must develop specific tactics or methods to carry out the strategy. Promotion requires objectives, strategies, and tactics to have a maximum impact on the station's listeners and other radio stations.

A station committed to success works with a promotional marketing plan that targets and highlights the key elements in the station's programming that set it apart from the competition. The marketing plan includes positioning and branding the station.

Positioning. Each station needs a clearly defined **positioning statement**. A positioning statement is a unique selling proposition, or slogan, used by the station to separate itself and its entertainment products from the competition. Only one station in a market can own or dominate a particular position. Positioning requires the station to focus and specialize on a target audience both on air and online.

Figure 10.2 Although we wouldn't consider "99.9 KISS Country" to be a statement in normal language, it qualifies as a positioning statement on this T-shirt and hat. The radio station's dial position, music format, and signature logo are all easily identifiable at a glance.

A positioning statement is clear and includes definitive language about the station (see figure 10.2). Broad, undefined positioning statements fail to differentiate a station from its competitors. "Your FM weather station," for example, is a particularly weak statement because every FM radio station reports the weather in one form or another. The other weakness is that a station does not need to remind people they are listening to FM. Maybe in 1965 it was necessary, but not today. Similarly, "The Valley's best music" does not tell the listener what valley the station is a part of or what kind of music is "best." Equally bad is "Less talk, more music"; less of what talk and more of what kind of music, from what radio station?

The best positioning statement includes the station's dial position, the station's location, and the kind of music it plays. Short and simple is best. Examples of strong positioning statements include:

- "Fun times and great oldies, Kansas City's one oh four one."
- "Key West's home for rock 'n' roll—ninety-nine five."
- "Live ninety-five five, Waynesville's classic rock superstation."

Each of these short positioning statements lays claim to a specific target audience in a specific geographical area. A casual listener scanning the dial instantly knows where the station is located, what kind of music it plays, and what the station's frequency is so that he or she can return or tell a friend.

Notice that two things are missing in each of the preceding positioning statements. First, the letters AM or FM. They are not needed. Second, the station dial position is simplified by removing the "point" from the sentence. No one outside the industry tells a friend, "I listen to 104 point 1 FM." Instead, they say, "I listen to one oh four one." So why not speak to your listeners in the language they are using?

Instead of saying 104 point 1, make it simply one oh four one. There are some cases where this will not work. For example, 90 point 5 would become ninety five, and this is confusing, since there is a 95 on the dial. In this case, something creative should be done, such as calling the point a dot; ninety dot five has an Internet-related tech sound (see audio demo, cut 10-1).

Positioning statements evolve over time as stations seek to maintain their current market position, expand their listener base, and promote their website and social media. This often involves a refinement of the station format or a change in the musical trends within a format. The goal is to maintain the current target audience and to add listeners through programming revision and updating of the positioning statement.

Branding. **Branding** is a part of the overall marketing plan and works in conjunction with a station's positioning statement. Whereas a positioning statement is tightly focused, branding encompasses the entire marketing effort of the station. Branding provides a clear differentiation from the other stations in the marketplace based on the station's overall identity and attitude, as defined by the listener. The station's brand should be aimed at making the listener feel as if he or she is part of the radio station.

A station's branding concept is not something the station dreams up and throws at the listener, hoping it will stick. Branding comes from the consumer; in this case, the listener. A radio station has to go out and walk through the mall and talk to its listeners; it has to become involved with its listeners to learn their likes and dislikes about the station; it has to track website and social media visitors to find out what causes people to spend more time on the sites. A station needs to find out what the key elements are that listeners associate with the radio station.

In effect, branding is a promise by the radio station to provide a solution to a problem that the listener has. For example, if listeners seek a station with an upbeat, lite rock format and a super morning team so that they do not fall asleep while driving to work, that is a need, or problem, that a station can solve. If there is a large base of sports fans in a community, there is an opportunity for a sports-talk station to brand itself as "the sports authority" and serve the needs of rabid sports fans.

Branding comes down to two things: listener awareness and attachment. Everything, from the music on the station, the station's logo, and the station's website and social media, should be aimed at making listeners and potential listeners aware of the station's positioning and branding. Attachment comes when the station provides a solution to the listener's needs and the listener adopts the station as "his or her radio station." In an ideal situation, listeners feel like they are a part of the station. The radio station needs to find a place in each listener's life. Once this occurs, an emotional bond is built.

Using listener-based information, branding should be applied to every form of "packaging" the radio station, including the station's positioning statement, logo, music, on-air personalities, contests, liners, sweepers, promotions, and public service (see figure 10.3). Branding is all about top-of-mind awareness, so that when the ratings are being taken, the listener remembers the station. By using all of a station's unique qualities and variables, successful branding turns the generic product of "a radio station" into "my radio station."

Ultimately, it is not so much the branding as it is the relationship that develops between the listener and the radio station that is the most powerful. Branding creates a mental image or identity for a station that makes it instantly recognizable as unique or one of a kind. In a best-case scenario, the radio station reaches out beyond the station to form an emotional bond with the listener and becomes a part of his or her life. An even more powerful bond is formed when radio's reach is combined with the web and social media.

Figure 10.3 Branding is all about top-of-mind awareness, so that when the ratings are being taken, the listener remembers the station. This station uses the same logo and trademarks consistently, on everything from billboards to temporary tattoos.

Program Promotion

Every successful station seeks to promote itself and the uniqueness of what it is doing. Good station promos follow the station's positioning statements and branding, tying all of the elements together. Promos highlight the station's music, contests, personalities, website and social media, public service, and all of the things that the station does that benefits the listener.

Promos should be forward moving (promoting the future) and make maximum use of the theater-of-the-mind effect so that listeners develop visual images in their heads of what it is you are doing. They should be exciting, fun, and attention-grabbing 60-second stories about your station aimed at the station's target audience or demographic. A demographic group of people is identified based upon age, gender, income, and other quantifiable statistical information. Even if a station does not have a huge promotion budget, if it has a creative production person it can unleash a firestorm of promos that build the station's image, create excitement, and attract and help hold listeners (see audio demo, cut 10-2).

The fact is a creative thinker with a good imagination and a production room can run circles around a station with a big promotions budget and little creativity. Promos are valuable, just as valuable as a commercial. Salespeople ask clients to sell their products on the radio station and its website. Likewise, the radio station needs to sell itself, and it should be its own biggest client.

Cross-Promotion. Promotions must have a purpose, with the most basic being to cross-promote (i.e., to promote another event or program on the radio station) to help build audience ratings. For example, every radio station should promote the next day's morning show 24 hours a day on air and online since it is usually a station's most-listened-to show. Each day, within an hour after the morning show is over, there should be a promo running about tomorrow's show, its features, and how great it's going to be. Sometimes cross-promotion extends to promoting an event such as special news or sports coverage of a major event on a sister station in the same market group (see audio demo, cut 10-3).

All jocks should be cross-promoting ahead to other dayparts on the station to build interest and maintain listenership. A station that is constantly cross-promoting ahead holds the listeners it has and develops a steady base of listeners across a number of dayparts throughout the week. There are two basic types of cross-promotion and both should be extended to the station's website and social media platforms.

Horizontal Promotion. Horizontal promos are announcements that promote events, contests, and station public service on a day-to-day basis (see figure 10.4). They also include morning-show promos. You should always start promoting at least a day in advance. For example, promos for the weekend should start on Wednesday and promos for the morning-drive show on Monday should start Friday, within an hour after that day's morning show is over, and run up to the start of Monday's show at 5:00 or 6:00 AM. Like all other promos, horizontal promos should promote ahead on a day-to-day basis to a station event.

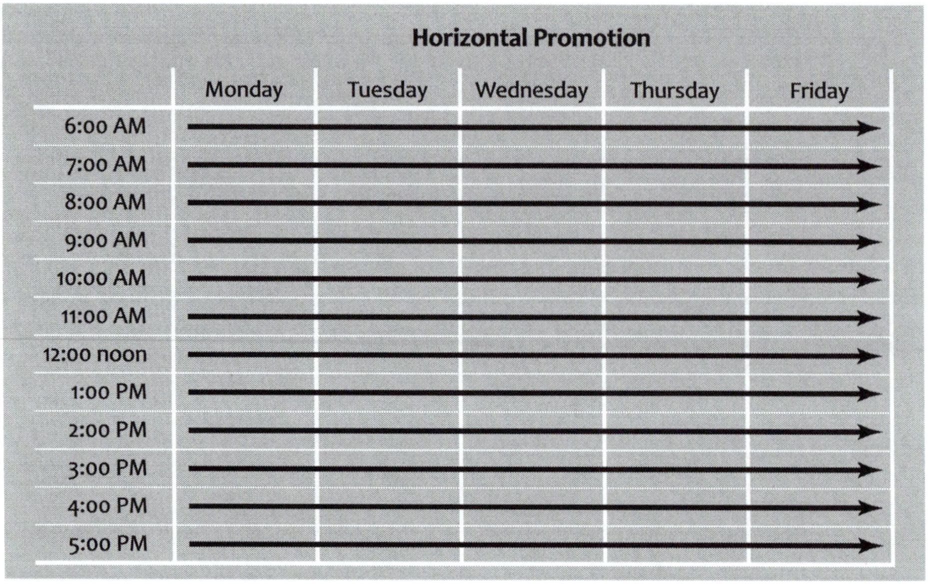

Horizontal Promotion

	Monday	Tuesday	Wednesday	Thursday	Friday
6:00 AM					
7:00 AM					
8:00 AM					
9:00 AM					
10:00 AM					
11:00 AM					
12:00 noon					
1:00 PM					
2:00 PM					
3:00 PM					
4:00 PM					
5:00 PM					

Figure 10.4 Horizontal promos are announcements that promote events, contests, and station public service on a day-to-day basis. Announcers should always cross-promote other dayparts on the station to build interest and maintain listenership.

Vertical Promotion. Vertical promos are announcements that give a listener a reason to keep listening for the next 20 minutes, the next hour, or the next daypart or show (see figure 10.5). Vertical promos create a lot of urgency when it comes to events or contests. When a jock promises to do something in the next 20 minutes, chances are a listener sticks around to see what happens, because 20 minutes is not that long. The reason jocks work with "the next 20 minutes" is that Nielsen Audio ratings are based on quarter-hour segments. If a jock can get you to stay five minutes into the next quarter hour, he or she gets credit for that quarter hour as well.

Jocks should constantly be promoting the next hour and the next daypart as well, and to some degree, there needs to be repetition of theme within vertical promos. As listeners join the station at various times, they need to be constantly brought up-to-date as to what is coming up in the next few minutes.

The Bad Promo: A Dangerous Situation

A station promo is as valuable as a commercial and should be given the same creativity and attention to detail in its production as the station's best commercial client gets. Sloppy production, at any level, is dangerous to a radio station, and here's why.

When people are watching a television show and the commercials come on, they have a choice. Viewers can either watch the commercials, or they can get up and go get another beer and some chips and then come back and collapse in their

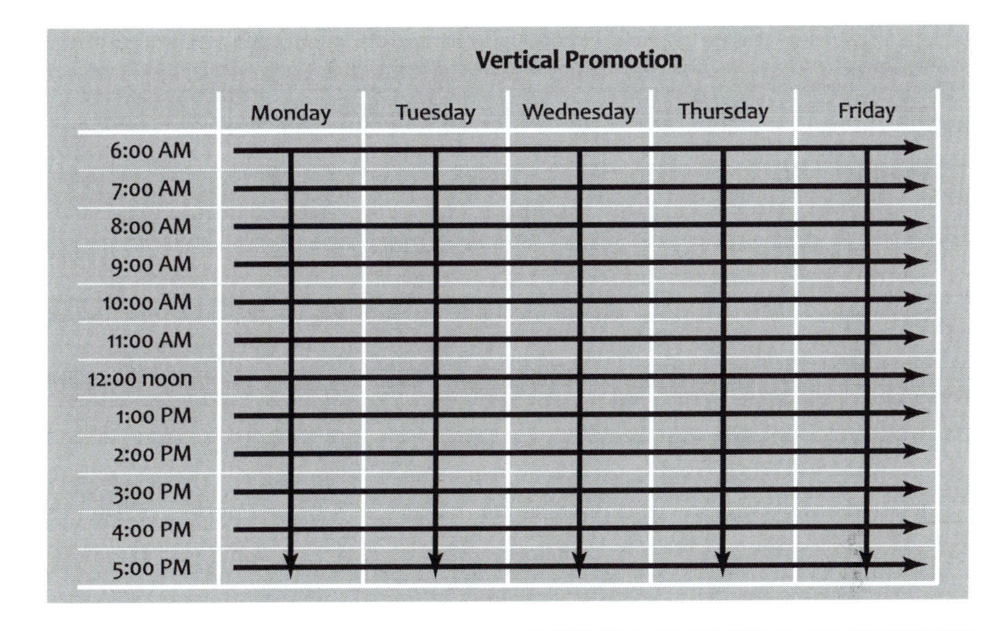

Figure 10.5 Vertical promos are announcements that promote events, contests, and station public service on an hour-to-hour basis. Announcers should always promote the next 20 minutes, next hour, next daypart, or show to create interest and build listenership. Combined with horizontal promos, a powerful promotional grid is formed.

recliner to watch the rest of their show. Worst-case scenario, the commercial or promo fails to sell the product.

However, when a person is in his or her car driving down the road and a commercial or promo comes on the radio, the person has no other activity to fall back on besides listening to the radio. The only choice he or she has is whether or not to change stations. An interesting promo or commercial holds a listener's attention until the music returns, whereas a bad promo or commercial sends the fickle finger of fate to press the next preset on the radio and change stations. On a radio station, a sloppy commercial or promo not only fails to sell the product, it costs the station listeners and Nielsen Audio ratings. Bad commercial and promo production is just as responsible for a drop in station ratings as are bad jocks and poor music selection.

Station Imaging

The Purpose of Station Imaging

Station imaging consists of the live and recorded elements that are used between segments of a station's programming to tie everything together, creating a station's signature sound. Through a singular, consistent voice, imaging creates a mood, or tone, and establishes a texture for a radio station. Imaging can be a powerful audio marketing tool that constantly reminds listeners what station they are listening to and why they should continue to listen. Overall, imaging creates the public face of the radio station and its entertainment products.

The Promotional Voice

If imaging creates the face of the radio station, then the promotional voice constitutes the features of that face. It is how listeners recognize that station. Station imaging requires a distinctive voice, a real powerhouse, to stand out and set your station apart from the competition (see audio demo, cut 10-4). The voice can be male or female—which is determined by the kind of image the station wants to create.

The promotional voice should not be someone on your station staff, and it is often outsourced to a professional voice talent (see figure 10.6). In many cases, stations within a regional or national group swap voices among the stations. The promotional voice should appear only on the station's imaging. It should never be heard as a jock on the station, it should never be heard on commercials on the station, and it should never be heard on another station in the same market. The station's promotional voice is the personality of the station and has a very special relationship with the listener.

Figure 10.6 You have heard this man's voice! A professional voice from outside a market can make a big difference in the sound of a radio station. Because of their national exposure, professional voices often have great creative ideas for station imaging and branding. This particular voice is heard on more than 100 radio stations and has been used on thousands of national and international television commercials.

A well-established promotional voice attracts attention because the listeners, whether they are aware of it or not, relate to the voice as a friend. They know that when this voice speaks, it is something special and deserves their attention. In effect, the voice is the radio station speaking directly to them. Even when listeners are not listening closely, the promotional voice catches their attention, either consciously or subconsciously.

There are two types of station imaging, liners and sweepers. With either of these types, the four keys to success are keeping the message simple and repetition, repetition, repetition.

Liners

A liner is a dry voice (without music or sound effects) that promotes the talent, the music, a station promotional event, or the station's social media (see audio demo, cut 10-5). Liners keep listeners informed about what is happening and promote the station's branding and positioning statement. Liners help to create a sense of urgency; if the listener doesn't stay with the station a few minutes longer, something is going to be missed. Liners can either be recorded by the station's promotional voice or done live by the on-air talent. Some stations use a mix of live and recorded liners.

Sweepers

Sweepers are liners that have been produced to create the station's signature sound. Sweepers are designed to reinforce the music format and the station's attitude. A well-produced sweeper triggers an immediate listener response. Listeners instantly recognize the signature sound and their radio station (see audio demo, cut 10-6).

Some of the various elements often included in a sweeper are the promotional voice, sound effects, music elements that match the type and energy of the station format, and various effects created through audio processing. Depending on the station format, a sweeper can be as short as 7 seconds or as long as 20 seconds. Although short in length, sweepers are full-blown, multitrack productions and require excellent production skills to make an effective impression (see audio demo, cut 10-7).

Sweepers are often designed and produced to make music transitions smoother between two dissimilar cuts of music. For example, sweepers are produced that transition from a fast-paced song to a slow-paced song, and vice versa. There are many variations on transition sweepers, including sweepers to take the station from music into traffic, weather, or news (see audio demo, cut 10-8).

Station Identification: The Legal ID

Station identification is a legal requirement of the FCC. Radio stations must make station identification announcements when they sign on and off for the day and hourly, as close to the top of the hour as possible, at a natural programming stop-set. Legal station identification consists of the station call letters, immediately followed by the station's city of license, for example, KOKO, Warrensburg or WDBO, Orlando.

Once the legal obligation is met, the station is free to enhance the ID by including other cities the station serves or by adding a station positioning statement. For example, "Serving Altmonte, Mt. Dora, and Zellwood, we're WDBO, Orlando." In this case, the station has to be careful that it does not make a false claim to serve a city that it does not. Also, be sure to check the station license for the correct call letters and city of license. Contrary to popular belief, most stations do not have AM or FM included as a part of their assigned call sign. Starting in 1923, stations east of the Mississippi River were required to use call letters starting with a W; those west of the Mississippi with a K. Stations that went on the air prior to this ruling, such as KDKA in Pittsburgh, Pennsylvania, or WHB in Kansas City, Missouri, were allowed to keep their original call sign. Other exceptions have been made for cities covering areas just east and west of the Mississippi River.

Something as simple as a station ID can also negatively affect station branding and positioning if it gets in the way or sends a mixed message to the listener. Station IDs can be used to position the station, or they can be buried. The city of license will affect the station's choice here. For example, in the Orlando, Florida, market there are a number of radio stations that are licensed to smaller towns outside the Orlando **metro area**, but their signal coverage over Orlando makes them valuable broadcast properties. If a station boldly states that it is from Cocoa Beach, which is 50 miles to the east of Orlando, people in Orlando are not going to identify the station as an Orlando station.

Some stations prefer to give their call letters and city of license buried between two songs near the top of the hour, with little or no attention drawn to the ID, as is the case for many stations around Orlando and other major cities. Another way to mask an ID is to include it creatively in a produced station sweeper so that it blends in and sounds just like all of the other imaging that is targeted at positioning and branding the station (see audio demo, cut 10-9).

Jingles

Jingles are positioning statements and station branding that are sung and set to music. They were once thought to be absolutely critical to a station's success. For many years, it was thought that a new package of jingles each year was what kept a station fresh and exciting. Today, jingles run hot and cold, depending on the format. A quality, custom jingle package from one of the major studios can easily top the $5,000 mark (see audio demo, cut 10-10).

With jingles the question is not so much whether they work or not, but whether or not they fit, musically, with the format, positioning, and branding of the station. Although jingles can certainly be written to carry out positioning and branding, it is critical that they fit the format and sound design of the station. Some markets are jingle rich; others use very few, if any.

Image Placement

As a production person, you need to remember that listeners are going to hear the imaging you produce over, and over, and over, so it is imperative for you to do quality work. The placement of station imaging is a market- and format-based decision. Every station treats imaging in a different manner. A new station needs to hammer home the station's message and personality to an audience that is likely listening for the first time. In this case, the station needs to overdo it and constantly remind the listeners what station they are listening to and what the station does better than everyone else. Because the station is "burning," or using the imaging a lot, you are going to have to produce a steady stream of creative materials so that the station sound remains fresh.

Established radio stations use imaging to help keep the listeners they have, to hook new listeners, to direct listeners to specific dayparts the station wants to build an audience for, and to the station's website and social media. The overall goal is to increase time spent listening to the station or increase the number of website and social media followers.

Station staff and listeners hear imaging in a completely different way. Jocks and production people hear every liner or sweeper played. Jocks get tired of liners and sweepers very quickly and often complain that the station is running too many. There is an old adage that applies to just about everything in radio. When the staff is sick and tired of a liner, sweeper, promotional announcement, or even a particular commercial, the audience is just beginning to really notice it. The audience does not listen for every liner, sweeper, or promo as the jock and production person does, and so they react differently to these programming elements.

Jocks should not be allowed to go into or out of a stop-set without forward-promoting some aspect of the radio station with a liner or sweeper. Jocks should constantly be selling the station's signature sound and products with your production work. It is important for jocks to promote what is going to happen in the next 20 minutes, the next hour, or on the next jock's show. As previously stated, every jock on the station should also be cross-promoting the next day's morning show and the station's website and social media.

At a very minimum, a station should use a liner or sweeper once every quarter hour. At the opposite end of the spectrum, a station running a tighter rotation of liners and sweepers might choose to run a series of two songs back-to-back, then a sweeper, then two songs, and so on.

Contest Promotion

Station contests are an extension of promotion. The goal of a contest is to cause the station's listeners to notice who they are listening to in the first place and to get them to listen longer, to increase their time spent listening, and to get them to visit the station's website and follow the station's social media. Another purpose of a contest promotion is to focus, or guide, listeners to a specific daypart the station wants to build an audience for. Creative production is what sells the contest or promotion to the listener.

A good contest is one that offers your target audience a prize it can relate to and would like to have. The prize has to be large enough to attract and hold a listener's attention. A six-pack of soda is not a very impressive prize, but a new smart phone with a year's free unlimited service is. Of course, the universal prize every listener likes the best is cash!

A contest has to be presented to listeners in a manner that is simple and easy for them to understand. Complicated contests fail every time because listeners have a hard time grasping all of the intricate rules. Such contests appear as though you are trying not to give away the prize. The contest has to be winnable; the station wants to have a winner! The station's website and social media are powerful tools and are the perfect place to register for contests, conduct instant contests, do contest follow-ups (such as, "Sorry, you are not a winner this time but here is a coupon for your next visit to . . ."). Additionally, websites and social media generate valuable databases of listener information.

One of the simplest contests a station can play is Hi-Lo. The station starts with a mystery cash jackpot. Listeners are invited to call the station at various times of the day and guess the jackpot. If they hit the exact amount in the jackpot, they win it. If they miss the jackpot amount, they are told that their guess was either high or low. Each time a listener misses the jackpot the station adds a fixed amount of cash to the jackpot, usually the station's dial position translated into dollars. For example, if the station is at 106.1 on the dial, it adds $106.10 each time the jackpot is missed.

It will take about four days of play for the audience to figure out exactly how much is in the jackpot through the process of elimination. Listeners get so involved that they keep hour-by-hour tracking charts of the amounts guessed. The last four or five plays of the game are very exciting as listeners get within pennies of the amount and finally hit it. When the jackpot is won, the station starts a new mystery cash jackpot and repeats the contest.

Contests require extensive preplanning due to the prize element and the legal requirements that must be met under both state and federal laws. Contests are detail-heavy events involving published contest rules, prizes that must be awarded, and sometimes insurance policies to cover an event in which listeners are going to be involved. Contests are planned as much as a year in advance to coordinate the timing of the contest with other events occurring in the market at the same time.

A contest promo is just like a commercial for a business—it needs to stick out and grab the listener's attention. Just like a commercial, the focus needs to be more on the listeners and the benefits they receive by listening to the radio station than on the radio station telling everyone how great the station is. The focus of contest promos should be on the question every listener asks: "What's in it for me (WIFM) if I listen to your radio station?" Contest promos have to be written from the listener's viewpoint, in simple, easy-to-understand terms. Production people need to understand that contest promos deserve just as much attention as a station's best client—and maybe even a little more.

The Great Tease: Getting the Listener Interested

Before a station can have a contest, there has to be someone interested in participating. A key element to the success of a contest is the tease. Teases are the well-planned hints that personalities let "slip" on the air, the mysterious promos you created, and the promos that give listeners a clue as to what is going to happen, but still leave them guessing. The tease is planned in advance and carried out with precision between the programming, promotion, production, and sales departments. Sometimes the tease starts as early as two weeks before the contest begins. Often the tease begins on the station's website and social media, causing site visitors to become listeners. The station can use its database of online visitors to send emails and text messages directing people to return to the site for details or listen to a particular show.

Occasionally, for competitive reasons, the tease is carefully planned so that only a select few within the station's promotion department really know what is going to happen. The station wants the element of surprise to thrill listeners and drive the competition absolutely nuts trying to figure out what the station is up to. Some stations go to the length of having the contest tease promos produced by a voice talent outside the market. In this type of situation, the talent is required to sign a nondisclosure agreement, which is a legal, binding agreement stating that the talent cannot tell anyone what is being done, and will face serious legal action if he or she does tell.

Care has to go into the production of a tease. The goal is to get the listener's imagination working, trying to figure out what the station is up to. During the tease phase it is important not to raise the listener's expectations too high or in the wrong direction. The key to a successful contest promo is to tease the audience, generate interest, and get people talking about what it might be. Tease promos should always undersell the event, and the station should deliver more contest than was promised during the tease.

Contest Promos

Once a contest is launched, a series of promos and online activity is required to support it. A contest's generic promos tell the listeners, in terms they can understand, who you are, what you are doing, and the benefits of listening to your station or going to your website. Contest explanation promos explain who can enter the contest, how the contest works, and what a listener has to do to win. If a contest is complicated, and some are, the contest explanation should be broken down into a series of two or three easy-to-understand promos that rotate. The station website and social media are also a good place for "complete contest details."

Contest promos use many of the same elements of regular station promos, with the exception that there is a prize, or prizes, to be won. Contest promos must create a sense of urgency, that the listener must listen to the radio station, and he or she must listen often, or he or she is going to miss out on a great prize. Just like station promos, contest promos can be designed as cross-promotional, vertical, or horizontal in nature. Contest promos encourage the listener to listen for the next 20 minutes, the next hour, the next day, or over the weekend for what is going to happen on Monday. Likewise, the station website and social media should encourage multiple visits and can follow up each visit with a text message or email to entice the site visitor to return or to listen to the station during a specific daypart.

If a contest involves winning phone callers, every call is recorded, and every winning call turned into a promo. When a winner screams and yells on the phone it translates into a promo that generates a lot of excitement about the contest. A winner's promo should be produced quickly so that it can be on the air within an hour of the event and run throughout the rest of the day. Likewise, if contest prizes are awarded outside the station, such as at remotes, every winner should be captured on video. The audio from the video goes on air and the complete video clip goes on the station website and social media. Still photos or video of winners picking their prize up at the station serves to make it more real for the listeners. The video clips and photos help keep the excitement at a high level. Contest promos are the fuel that keeps the contest moving in a forward direction; they require creativity and fast work on your part.

Contest Postpromotion

Once the new car, incredible vacation, or truckload of cash is given away, there are still a lot of promos to be produced. Contest postpromotion is just as important as the contest itself. Contest postpromotion should last at least five days on air after a major contest and should become a part of the station website and social media with photos and videos. Just like the promos that launched the contest, the postpromotion promos should stress the benefits the listener gains from listening and what the station is doing for the listeners. The station wants listeners to know who won the grand prize and how happy and excited the listener was to win. Tell listeners on air; show them online. A great contest postpromo is one in which the winner tells listeners why they should be listening to the station.

Legal Requirements of Contest Promotion

As a production person, you should be aware that a station must comply with the FCC and the Federal Trade Commission rules and regulations, in addition to any applicable local and state laws regarding contests. In most cases, these rules, regulations, and laws are to prevent deceptive contest promotion and fraud in the awarding of the prize. Just the hint of fraud in a station contest can quickly turn into a public relations nightmare, costing the station listeners and revenue from advertisers. For this reason, approved contest promo copy prepared by the promotion director cannot be changed on a whim. He or she has carefully written it to satisfy all the legal requirements as well as the station's promotional needs.

Although you are probably not an attorney, you should be aware of general contest requirements and your state's laws in this regard. General contest rules and regulations are specified by the FCC and include:

> A Licensee that broadcasts or advertises information about a contest it conducts shall fully and accurately disclose the material terms of the contest, and shall conduct the contest substantially as announced or advertised over the air or on the Internet. No contest description shall be false, misleading or deceptive with respect to any material term. . . . Material terms include those factors which define the operation of the contest and which affect participation therein. (47 CFR Part 73, §73.1216)

Further, the FCC states that "the [Federal] Criminal Code prohibits fraud by wire, radio or television, and violation of this provision may lead to Commission sanctions against FCC licensees."

Depending on the contest, certain legal disclaimers, such as the odds of winning the contest, may be required to be included in the promo. It's your job as the production person to weave them creatively into the promo. An innocent error of omission in editing the copy on your part could cost the station a fine or cause a contest to come under scrutiny. The station's website and social media are an excellent place to post complete contest rules and any necessary legal information. Listeners can be directed to the station website for "complete contest rules and odds of winning."

Suggested Activities

1. Select three radio stations in your market and see if you can determine each station's positioning statement. Analyze the positioning statement and see if it accurately describes the station and what it does. Using the criteria found in this chapter, try to develop a better positioning statement for any of the stations that more accurately describes the station.
2. Select three radio stations in your market and determine whether each is actively branding their station. If so, write a description of the station's branding concept and the "promise" the station is making to their listeners. Make a list of at least 10 areas of the station's packaging in which the branding is carried out (billboards, T-shirts, etc.).
3. Select a station in your market and develop a new branding concept for the station, including a new positioning statement. Based on what you develop, write a package of 15 liners for vertical and cross-promotion of the station's programming.
4. Using a calendar, develop a promotion time line for a concert promotion in which there are two station visits by the music group, a contest to win 20 pairs of tickets, and a grand prize of two front-row seats and VIP backstage passes, along with limo service and dinner with the group. Work backward from the last day of postpromotion, to the concert, to the award of the grand prize, the award of the 20 pairs of tickets, the start of the contest promos, the start of the contest tease promos, and the dates by which promos have to be produced.
5. Visit your state attorney general's website, or phone the office, and find out about the contest laws in your state. Are there any types of contests that are prohibited? Can you use alcoholic beverages as a prize in your state? Do you have to register a contest with the state, or post a bond with the state, to ensure the award of the prize?

■ Websites for More Information

For more information about:

- *contest insurance and radio contests,* visit SCA Promotions at www.scapromotions.com
- *custom jingle packages,* visit TM Studios, Inc. at www.tmstudios.com
- *custom station imaging, liner, and sweeper services,* visit Autumn Hill Studios at www.jefflaurence.com
- *Federal Communications Commission rules and regulations regarding contests,* visit the FCC at www.fcc.gov
- *federal rules, regulations, and laws regarding contests,* visit the Federal Trade Commission at www.ftc.gov

■ Pro Speak

You should be able to use these terms in your everyday conversations with other professionals.

barker—A person who attempts to attract people to an event or place for entertainment purposes. He or she often talks about the uniqueness or novelty of the entertainment. A radio station can be a barker for its website.

branding—A station's marketing plan that seeks to provide a clear differentiation from the other stations in the marketplace; based on the station's overall identity and attitude as defined by the listener.

core audience—The primary target demographic of a radio station.

daypart—A broadcast industry term that classifies a specific time period of radio listening. For example, 6:00 AM to 10:00 AM, Monday through Friday is generally accepted as the morning-drive daypart.

metro area—A metropolitan city's area as defined by the Office of Management and Budget; also generally accepted by the major rating services as a designated geographical area.

Nielsen Audio—The primary radio audience measurement service. Ratings are often referred to as "the book," since the ratings were originally published in book form each quarter. Stations now take electronic delivery of the data.

positioning statement—A unique selling proposition, or slogan, used by a station to separate itself from the competition.

sound design—The overall sound quality of a radio station that sets it apart from others.

time spent listening—An audience ratings measurement indicating the number of quarter-hour time segments the average person spends listening to a radio station in a given time period, such as a week.

11

Fieldwork

Taking the Station on Location

HIGHLIGHTS

Location Planning

When listeners think that they are going to get a chance to see the magic of how a radio show is created, they are drawn to a location, much the way Dorothy in the *Wizard of Oz* was drawn to pull back the curtain and see the wizard at work. For a listener, there is always the surprise of seeing what the on-air staff looks like for the first time (never what is imagined) and seeing the equipment that makes radio possible.

Radio routinely goes on location for commercial remotes from sponsors' businesses and civic events. These live broadcasts are an important part of promotion and sales. Stations also spend a lot of time on location producing local sports and news coverage. Regardless of whether it is a commercial remote, a sports event, or news coverage, a successful remote broadcast requires a lot of careful planning.

Promotion and Sales

Unless a breaking news story is involved, most remote broadcasts are a joint effort of the station's promotion and sales departments. The reason a business hosts a remote is the interest, excitement, and increased sales that it can potentially generate. Stations do remotes for the sales revenue a remote generates and the promotional value gained from the radio station staff interacting face-to-face with listeners.

The promotion and sales departments can even turn a public-service event, such as a dog-adoption day by the Humane Society, into a revenue-generating event. For example, a local business hosts a dog-adoption day, paying for the remote broadcast as a community service. As a result, the station gets to be associated with a high-profile community event. (Who doesn't like puppies?) The business gets increased customer traffic, sales, and tremendous public-relations value from the event. (Everyone loves puppies.) The Humane Society gets a major promotion, free. (They love puppy adoption!) This is an example of a win, win, win situation.

A lot of preplanning is required on the part of everyone in the station for a successful remote. The planning usually starts in the sales department and is fine tuned by the promotion department, or vice versa. The promotion department also works in conjunction with the engineering department to make sure the remote is technically feasible. At some point, production is called in to produce ads for the client, station promotional announcements, and create social media content for the event.

Depending on how a station is structured, the person in charge of producing the remote varies. At some stations, a production person is charged with overseeing the technical, and sometimes the talent, side of a remote. Other stations send an engineer and a remote producer from the promotion department. In smaller markets, the talent may be his or her own producer and technician.

Planning for a remote, from the sales department's perspective, takes place between the salesperson and the client to determine what the goals are for the remote. In some cases, the goal is not so much retail sales as it is making listeners aware of a new business's location, launching a new product, or promoting a grand opening. During the planning phase, salespeople tend to look at remotes from the business side: What can the station do for the client? The client looks at the remote and asks, "What's in it for me?" Will the station's listeners be the kind of people that are interested in what the client has to offer? Is the station capable of generating increased customer traffic and sales and creating a positive impression on customers? During the negotiation process between the client and the salesperson, the salesperson will likely turn to the promotion department for ideas and input on how to make the remote a success.

Before any in-depth remote planning can take place, someone from either the sales or the promotion department needs to conduct a site survey of the client's business. The purpose of the survey is to check for things such as:

- Where can the station vehicle be parked?
- Is there adequate parking for listeners?
- Is there adequate electrical power?
- Where will the on-air talent be located?
- Where can station banners and other promotional materials be positioned?
- How early can the station get on the property to set up before the event?

Depending on the station, the engineering department may do its own site survey to make sure it's technically possible to get a signal back to the station from the site. All of these little details are summarized on a site-survey form to assist in planning the event (see figure 11.1).

Regardless of a remote's specific purpose, the promotion department looks at a remote from the standpoint of creating an event, or a **draw**, for the client. The

WRPC Remote Event Specifications

104.1 FM *Classic Hits*

DATE: 5/29
TIME: 10 am – 12 Noon
CLIENT: Wild Eddie's Performance Used Cars

Name of Event: Wild Eddie's Performance Used Cars
Promo Start Date: Promo 3576 STARTS Mon 5/24
Account Executive: Jeremy Powell
Station Promotional Event: No
Booked Thru Programming: No
Logged Thru Traffic: Yes
Event Type: Live broadcast
Business Name: Wild Eddie's Performance Used Cars
Customer #: 88-9679
Business Contact: Tyler Sellers
Business Phone: 555-5011
Address: 3418 Hendersonville Rd.
Directions: I-26 south to exit #42, left 2.8 miles, on the right
Copy Talking Points: Prepared by client/will give to staff on arrival
Talent Requested: Aaron Michaels
Talent Assigned: Aaron Michaels
Remote Talent Fee: $300
Attire: Standard station attire (logo golf shirt/slacks)
Drinks: Yes
Drink Brand: Any
Vendor Drink Trailer: Coca Cola ___ Pepsi _x_ Other ___ None ___
Inflatable Station Logo: Yes
Contest Registration Box: Yes
Registration Box Info: Win a Car Contest
Food: Yes, Subway party subs
Remote Prizes: T-shirts _x_ Lisc Plates _x_ Keychain _x_ Ball Caps ___
Vehicle: Station van
On-Site AC Power: Yes _x_ No ___
Overhead Power Lines: Yes ___ No _x_ If yes, note conditions below
Engineer: Tim Neese
Frequency: VHF ___ UHF ___ Wireless _x_ ISP ___
Special: client paying $175 extra for station mascot to appear.
NOTES:
Park station van next to street and use generator power if needed.
Station banner to go below client's street sign.

Figure 11.1 A site survey is a good way for everyone associated with a station event to know what is going to occur and what his or her role is in the event. Site surveys are usually electronic forms so that all of the departments of the radio station have easy access 24/7.

event aspect of the remote often involves contests, free station promotional items (T-shirts, etc.), product samples, free food, free soft drinks, **vendor involvement**, and products and prizes offered by the client.

The promotion department is in charge of making sure the station's positioning statement and branding are carried out at the remote through the use of the station vehicle, banners, staff attire, station mascot, and so forth. Remotes are unique in that the promotion department is charged with taking an ordinary day and turning it into something special. Some of the things involved in promotion planning for an event include:

- Station vehicle to be at the business, parked near the street to act as a billboard.
- Real-estate-sized yard signs directing listeners to the remote, if necessary.
- Station banners on the business or in front of the business.
- Presence of the station's salesperson who sold the remote to act as the liaison between the client and the station staff during the event.
- Station promotion staff to coordinate the remote so that it comes off as planned.
- Station-supplied promotional items (T-shirts, stickers, ball caps, free stuff, etc.).
- Vendor-supplied promotional items (manufacturer or vendor **swag**).
- Business-supplied promotional items (T-shirts, ball caps, gift certificates, etc.).
- Any necessary station, business, or vendor contest materials (contest entry boxes, point of purchase signs, etc.).
- Food vendors.
- Soft drink vendors.
- Station mascot, if applicable.

Depending on the size of the event, the preceding list includes just a few of the physical items and considerations that might be necessary. A final but crucial consideration is the business's regular customers. Have you physically laid out the remote to allow space for the client to do business? The worst of all possible scenarios is a business so jam-packed with listeners that sales come to a standstill.

All of these elements have to be coordinated in advance so that both the station staff and the business owner are aware of how the remote is going to be staged. In addition to the live broadcast, most stations will require a station staff member to oversee, and coordinate with the client, the live social media coverage of the remote and any follow-up social media coverage after the event.

News and Sports Location Planning

Planning a remote with the news and sports departments presents an entirely different set of circumstances. The difference is that rather than the station creating the event, someone else creates the event and the station's role is to cover it.

Typically, news remotes involve a breaking story or disaster coverage from a major event. In any case, preparations for such events are often made far in advance, even though the specifics cannot be known or planned for. As an example, many news departments have what are called go bags or jump-and-run bags, intended for a reporter to pick up and go with. The bags may contain a smart phone, recorder, mic, **mic flag**, cables, spare recording media (flash memory cards), and a tablet or small laptop computer with a wireless device and audio recording and editing software. Depending on the location, the "bag" may contain a press parking permit and a credit card for location expenses such as parking. It is not unusual for a couple of meal bars to be stashed in the bag. Similarly, networks and contacts are often set up in advance to share emergency information among stations. When a natural disaster occurs the contacts can be called at other stations around the state or region to get reports and live updates.

Another form of news remote occurs when a station takes one of its talk shows "on the road." To some degree, the talk show becomes a mini-newsmaker. Such remotes involve both sales and promotion and are often treated as a regular commercial remote.

Although the rights to broadcast a sporting event may belong to someone else, the station's sales and promotion department will likely be involved in selling sponsorships and promoting the station's involvement. Sports remotes require a lot of planning from the standpoint of what the station would like to do and what the rights holder of the sporting event permits the station to do. This ranges from high school events to major national teams. For example, your station may not be permitted to broadcast a major sporting event because a national network has secured the broadcast rights. However, the rights holder to the event may permit the station to host a tailgate party in the parking lot prior to and during the event. With the support of the station's sales and promotion departments, a station can host a large remote and get a lot of mileage out of being associated with a major sporting event.

Even when a station is granted rights to broadcast a sporting event, there are generally strict guidelines as to what is permitted with regard to banners and station promotional items at the event. The station and the rights holder must agree on exactly how the station is to promote and broadcast the event and what promotional activities are permitted during the event.

Another form of sports remote is taking a sports talk show on location prior to a sporting event or featuring a major sports figure in a public appearance. Some sports talk shows are produced entirely on location as a copromotion with a business that serves as a cohost for the show; for example, the show is produced at a local sports-themed restaurant or bar on a regular basis. The business receives the benefit of people coming to see the show, and the station receives the visibility that helps to build audiences.

Depending on the station, the role of the production person is to prepare the promotional announcements and commercials, and possibly to serve as location producer or advisor to the promotion and sales departments on what is required technically to broadcast the event.

As mentioned earlier, regardless of the type of remote, most stations will require someone on the station staff to be in charge of the social media coverage of the remote as it is occurring and any follow-up social media coverage after the event.

Remote Broadcast Transmission Methods

From a technical standpoint, the first rule of remotes is KISS: keep it simple, stupid. The key element to the success of any remote broadcast is using a simple, reliable, remote transmission system to return the program's audio to the station and to have a backup for that system. Redundancy is your friend. Unlike a studio broadcast, if there is a problem on location, you cannot dash down the hall to another studio and go on the air. In this section of the chapter we are going to examine remote systems, moving from the simplest to the most complex. Keep in mind, though, no matter how sophisticated the equipment, the goal is to put the talent's voice on the air. Always have a backup.

Plain Old Telephone Service (POTS)

In the beginning there was copper wire and the telephone, and only the telephone. To this day, a hard-wired telephone is the simplest and most reliable way to get someone on the air in a pinch. The talent dials the station. The station uses a device called a **phone hybrid** to interface the telephone line to the station's broadcast equipment and puts the call on the air: simple, but effective. Remember that the original goal is to put the talent on the air and, in a worst-case scenario, POTS can do just that. Many a sporting or news event has been covered using a cell phone held like a microphone because the primary remote equipment failed at the last minute before a broadcast. It's not pretty, but it works.

POTS audio quality is limited to what you would normally hear over a phone (see audio demo, cut 11-1). However, a smart phone with the addition of a wireless interview tool interfaces a professional microphone, headphones, and auxiliary inputs and outputs to any cell phone with Bluetooth wireless technology (see figure 11.2). The audio quality using this kind of a setup is considered HD voice quality and what you would expect from a digital connection.

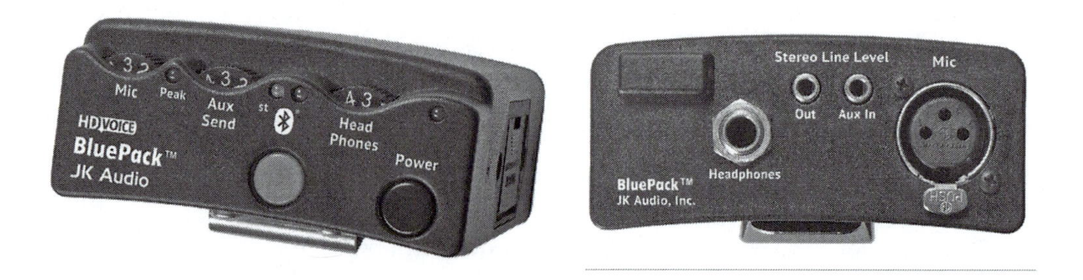

Figure 11.2 A phone hybrid is just what its name implies, a combination of devices. This phone hybrid pairs with your smart phone like a Bluetooth headset. It is designed to clip onto a belt or clothing and provides an audio mixer, headphone amplifier, and stereo line input and output, as seen here.

Codecs

Digital audio technology allows the use of everything from POTS lines; 4G LTE wireless devices or smart phones; Audio over IP (AoIP); and audio via satellite using a coder/decoder (**codec**) at each end of the connection. Codecs are **duplex devices**, meaning that studio-quality audio can be sent and received simultaneously on a single connection.

A codec is a hybrid device combining a phone, **A/D** and **D/A** converters, a small audio mixer, and the necessary software for modems to match whatever means of transmission it is connected to (see figure 11.3). The codec typically has at least one telephone line jack (POTS), a modular connector for Ethernet or IP, a USB connector, and various audio input and output connectors. Additionally, the codec may have a module for 4G LTE wireless connectivity or a module for satellite audio connectivity. What it really comes down to is you have a remote truck in one small box that can go virtually anywhere, connect to just about anything, and get a signal back to the radio station using various mediums, from POTS to satellite.

Figure 11.3 A codec is a hybrid device combining a phone, A/D and D/A converters, a small audio mixer, and a modem. The codec physically connects to a POTS line, the Internet, or a wireless module. Codecs are extremely popular since they are virtually a remote setup in a box. Above the codec is a wireless microphone receiver for the sideline reporter.

At the remote, once the codec is connected to a hard-wired phone line, IP port, a smart phone, or wireless connection, the technician either dials the station codec's phone number or enters the web address of the station codec. In some cases, the codec is preset by the station engineer and all the technician at the remote site does is press a connect button. The remote and station codecs link together and complete the digital connection. Depending on the quality of the phone line, FM–quality audio can be achieved over POTS and studio quality is the norm with Voice over IP, a smart phone, or wireless connection (see audio demo, cut 11-2).

A codec is a duplex device, meaning that it is capable of sending and receiving audio simultaneously on a single line, as already described (see figure 11.4). The feature is often used to provide a remote announcer with an interruptible foldback (IFB) loop. While the audio from the remote is going to the station, the radio station's program signal (IFB feed) is being sent back down the line from the station to the announcer's headphones. The reason it is called an interruptible foldback loop is that the producer back at the radio station can interrupt the station's return feed and talk to the announcer in his or her headphones. Typically, the program monitor is in one ear of the announcer's headphones and the IFB is in the other. Needless to say, it takes some practice to be able to talk on the air and simultaneously listen to cues in headphones.

One thing to be aware of is that codecs can introduce a delay of up to an eighth of a second in the audio during the analog-to-digital encoding/decoding process. A remote announcer monitoring the station in headphones on the IFB could hear his or her voice an eighth to a quarter of a second after speaking. To prevent the announcer from hearing the delay, the station sends the remote announcer a station

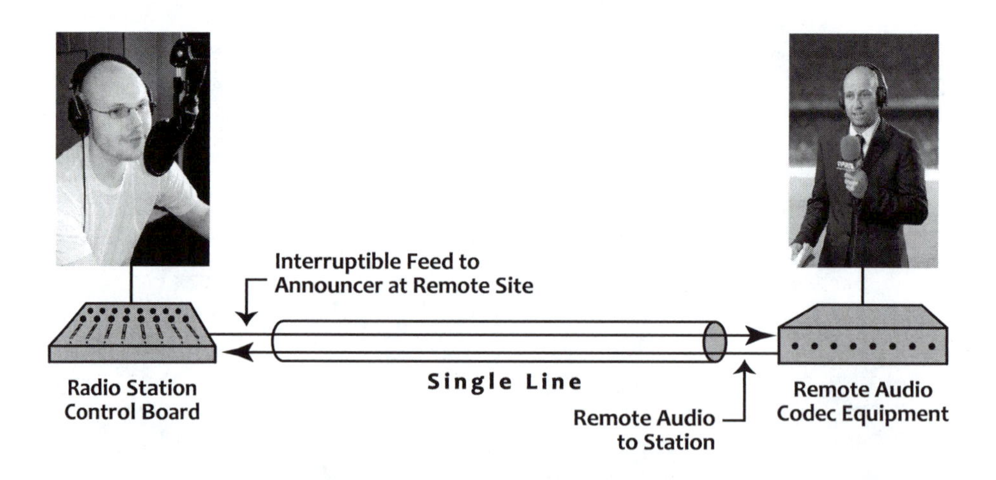

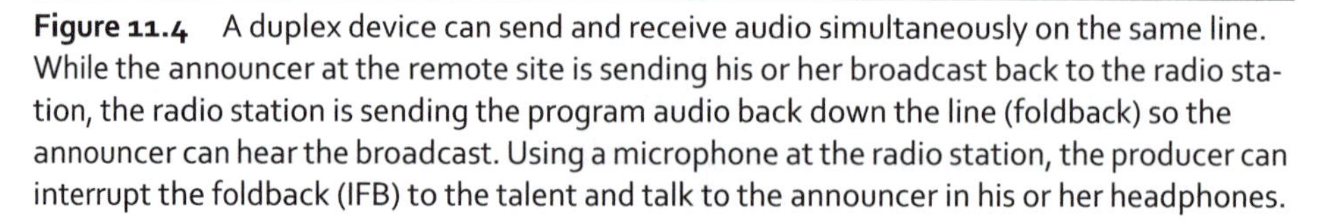
Figure 11.4 A duplex device can send and receive audio simultaneously on the same line. While the announcer at the remote site is sending his or her broadcast back to the radio station, the radio station is sending the program audio back down the line (foldback) so the announcer can hear the broadcast. Using a microphone at the radio station, the producer can interrupt the foldback (IFB) to the talent and talk to the announcer in his or her headphones.

program mix, minus the announcer's voice. This is called a mix/minus feed, which is the program mix of the station minus one or more audio sources. Most broadcast consoles are equipped to easily create a mix/minus feed for the announcer. A simple way to accomplish this is to route every channel on the control board *except* the remote announcer's channel to the program and audition buses. The announcer's channel is routed only to the program bus so it can go on the air. The audition bus is used for the announcer's monitor feed because it includes every channel on the control board except the channel with the announcer's own voice. The announcer hears everything on the station except his or her voice (see figure 11.5).

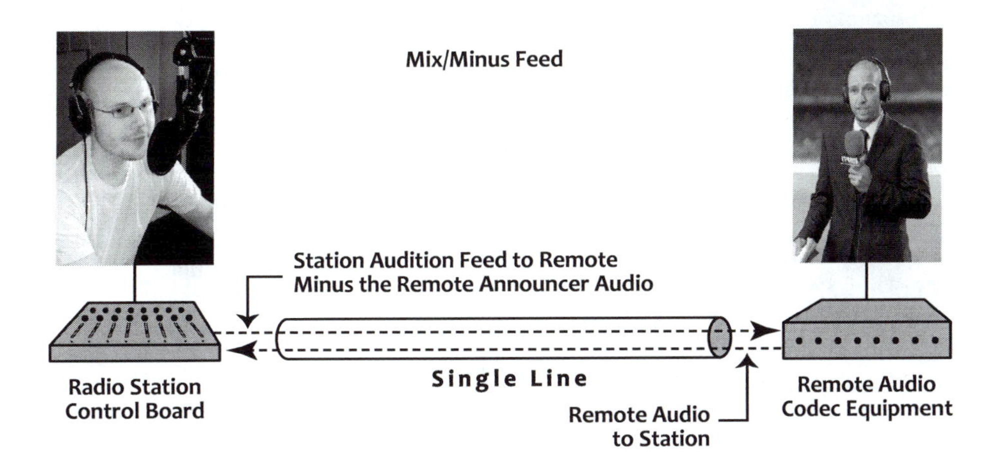

Figure 11.5 A mix/minus feed to an announcer at a remote broadcast is typically created by putting everything on the station's control board into both program and audition except the announcer's audio feed to the station, which is only put in program. Audition then has the radio station's entire signal (mix) minus the announcer's audio feed to the station, which is then sent back to the announcer. Therefore, the announcer hears everything but his or her own voice.

Smart Phone Codecs. Smart phones are valid broadcast tools. Thanks to 4G LTE connectivity, smart phones are capable of hosting codec apps that can deliver FM-quality audio via 4G LTE wireless or WiFi. The codec app can also send studio-quality audio files back to the station. The mobility of the smart phone makes it ideal for sports and news reporting. Such apps often include recording and play-back features sophisticated enough to allow the reporter to be on the air live and playback recorded audio cuts at the same time (see figure 11.6 on the next page).

ISDN Codecs. ISDN codec technology is outdated and has been replaced with newer, and less expensive, AoIP technologies. However, you may still find a legacy ISDN codec being used within a group of stations simply because it works and the stations have not updated the technology (see audio demo, cut 11-3).

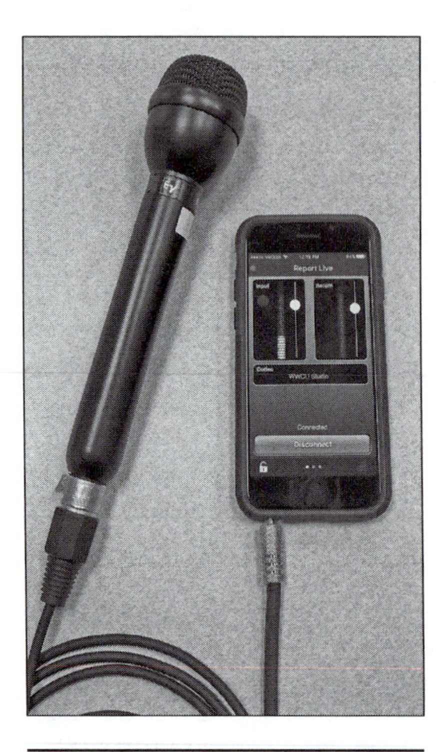

Figure 11.6 A smart phone equipped with a codec app becomes a powerful handheld remote broadcast tool. Such units are ideal for news and sports reporting on the go.

Remote Pickup Units

A remote pickup unit (RPU), or transmitter, is exactly what its name implies. It's a small, low-power transmitter used to send a radio signal back to the radio station.

RPUs are physically small and often have a built-in mixer; they operate on AC line current, batteries, or 12-volt power from a vehicle. Rather than being called a remote pickup transmitter or RPU, these units are most often referred to by the brand name Marti, even when they aren't of that brand. Marti (named after George Marti, who founded the company) is known for building what are universally termed "bulletproof" transmitters and receivers. It has been estimated that at least 80 percent of the radio stations worldwide use at least one of Marti's products. Don't be surprised the first time you see a 30-year-old Marti RPU in service at a radio station, just like the one in figure 11.7. The units operate in the VHF (very high frequency) and UHF (ultrahigh frequency) radio bands and deliver good-quality audio (see audio demo, cut 11-4).

At a remote, the transmitter is set up with the antenna aimed at the radio station's receiv-

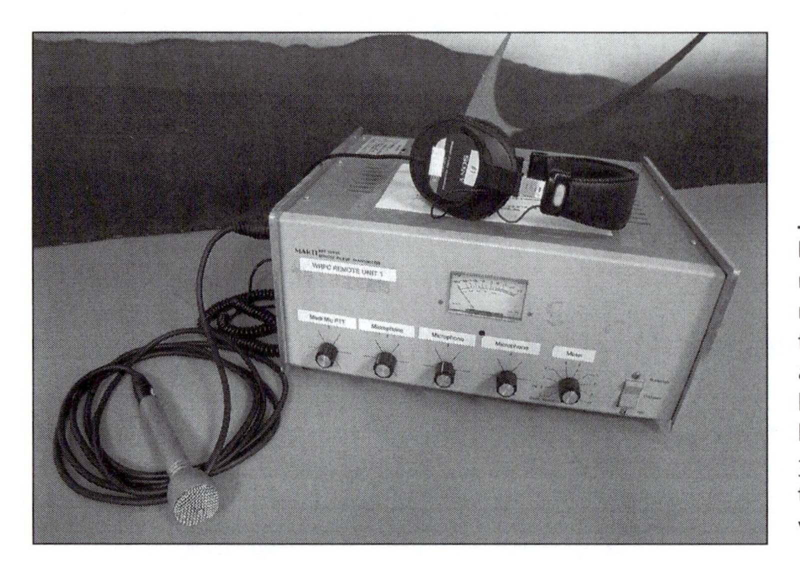

Figure 11.7 Marti remote pickup transmitters are legendary for their toughness and ability to just keep working. Don't be surprised to see a 30-year-old unit, like this one, still in service at a radio station.

ing antenna. Various antenna configurations and a direct line of sight to the station's tower can greatly enhance the range of the transmitter (see figure 11.8). Many radio stations outfit a station van with a telescoping mast and twin directional transmitting antennas. Under ideal conditions, such systems can transmit 45 miles with good-quality audio. A Marti receiver at the station picks up the signal and feeds it to the control board. Unlike duplex codecs, a remote pickup transmitter is a one-way communication system, from the remote to the station.

A disadvantage when using a Marti RPU is that several pieces of equipment are necessary to go on the air, including a transmitter, antenna cable, antenna mast, and antenna. Care is required in the setup to make sure the antenna is pointing in the correct direction (use a map and a compass if you need to). Sometimes buildings, structures, and the surrounding terrain can reflect, or block, the signal from

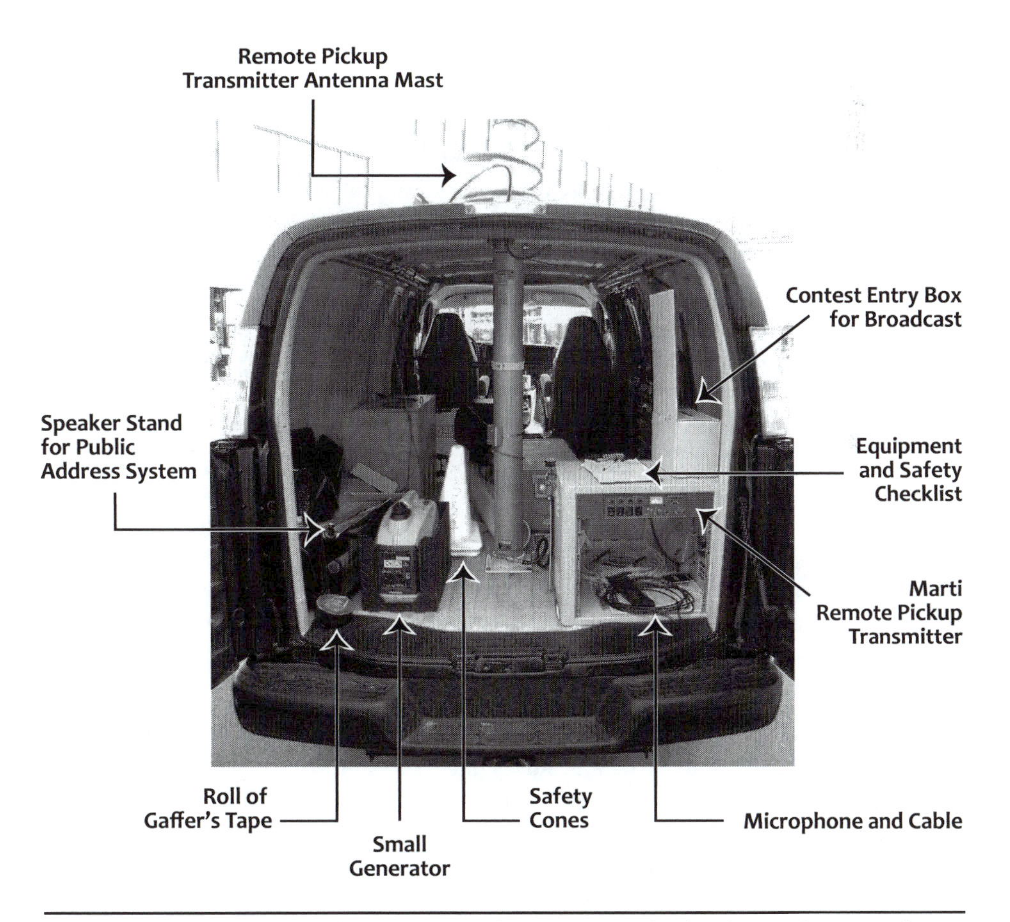

Figure 11.8 A remote broadcast van is a radio station on wheels. This van has everything to do a remote broadcast safely, including a portable generator to run the Marti remote pickup transmitter.

reaching the station, in which case you have to move. It is always advisable to do a site survey and system test the day before a remote.

The advantage of a Marti is the ability it gives you to transmit good-quality audio from just about anywhere in your community on short notice. Another advantage is that the station does not have to depend on others, such as the phone company or the web, to return the signal to the station. Marti transmitters are still used by some stations for remotes, news, and sports coverage. The transmitters can be easily mounted in a station news vehicle for on-the-spot news coverage. The Marti's functionality can be expanded beyond its built-in audio mixer by plugging a production mixer into either a line or mic input.

Basic RPU technology has been around a long time and most stations have, and continue to use from time to time, RPU transmitters because of the reliability, good-quality audio, and the independence that they give a station to do a remote when and where it wants to without having to depend on anyone else.

Remote Pickup Safety. Each radio station's RPU system requires specific safety training. Safety practices must be followed when raising any kind of antenna mast near power lines. A station van should be equipped with an automatic mast safety device to keep it from coming in contact with a power line. Parking a station van under power lines and inadvertently raising the telescoping antenna mast into the power lines has killed several people. An automatic mast safety device will prevent this from happening. The author once watched a competing radio station finish a remote broadcast at a music concert and attempt to drive off with the antenna mast still up. They were fortunate. The power line that they hit just sheared off the antenna mast, and no one was injured. You do not even have to have a remote truck in order to be in a dangerous situation. Imagine two excited radio sportscasters sitting on open aluminum bleachers, holding golf umbrellas, broadcasting a high school football playoff game in the rain. At their feet was an unprotected Marti transmitter plugged into 110-volt electrical power. In the excitement of the muddy, rough-and-tumble game, it never occurred to the broadcasters that either lightning or the 110-volt power feeding the Marti could kill them—and that's why safety training and safety consciousness is essential.

Satellite Telephone Service

If you are headed into the boonies, there are several satellite telephone technologies available to return a radio signal to a station. The only restriction is that you have to be able to see the sky to use a satellite (sat) phone. A number of different classes of satellite phones are available, and they are not as expensive as you might think. Sizes range from what appears to be an ordinary cell phone to a satellite uplink the size of a laptop computer.

A handheld sat phone uses either standard, land-based cell phone service or satellite phone service, depending on what is available (see audio demo, cut 11-5). The phone operates like a regular cell phone except that a whip-type antenna is raised from the side of the sat phone for satellite service (see figure 11.9). Satellite phone rates are competitive and, depending on the service plan, sometimes are less than $100 per month for unlimited talk time. Just like a cell phone, a sat phone can be adapted to connect to phone hybrids and codecs, allowing microphones and headphones to be interfaced with the phone.

At the opposite end of the spectrum are portable satellite uplinks, which appear similar to a laptop computer. The uplink cover is raised at an angle and folding panels open to the left and right, forming the transmit/receive antenna (see figure 11.10 on the following page). A built-in **GPS** locator tells the operator in what direction to aim the antenna to reach the nearest satellite. The uplink can be aimed at more than 45 **low-earth-orbit** (**LEO**) satellites. The uplink is not only capable of transmitting a regular satellite phone call but also transmitting an audio codec signal, allowing broadcasters to send studio-quality audio back to the station. Typically, a mixer is plugged into the codec to create multiple microphone inputs.

Although sat phones may initially appear beyond the reach of the average station, a station does not have to purchase a satellite phone to be able to use one. Many companies rent sat phones on a weekly or monthly basis, making the cost reasonable, even for a local station.

Figure 11.9 Satellite phones provide the opportunity for a radio station to go anywhere to do a remote, such as on lakes, the ocean, and in the mountains. This phone has two modes of operation, land-based cell phone and satellite. The small antenna on the right is used in cell-phone mode. When the angled antenna on the left is folded out and extended, the phone switches to satellite mode.

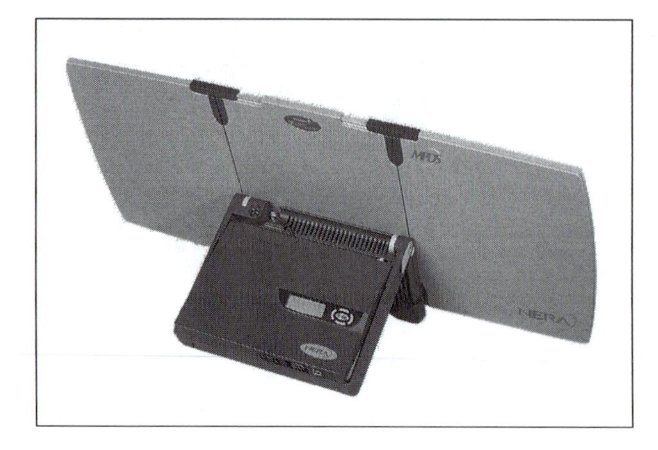

Figure 11.10 A portable satellite uplink is a powerful tool for news, sports, or other live remotes. The uplink shown here is about the size of a laptop computer and is not only capable of transmitting a regular satellite phone call, but also transmitting a digital audio codec signal, allowing broadcasters to send studio-quality digital audio back to the station. Typically, an audio mixer is plugged into the uplink to create multiple microphone or line inputs.

Producing Audio for Remotes

Remember, the first rule of producing is always KISS: keep it simple, stupid. There is not a single book, or list, in the world to tell you how to produce every remote you might possibly be assigned. There is, however, a simple system you can use to help you prepare for virtually any remote broadcast. There are three basic types of radio remotes: commercial, sports, and news, but you can use the same simple system to prepare for any of them. It's a low-tech checklist that you create for any given broadcast, starting with the talent and following the signal path backwards to the radio station. Your list should have three elements: transmission equipment, monitoring equipment, and tools.

Commercial Remotes

Commercial remotes are, just as their name implies, remotes produced at a business with the sole purpose being to increase customer traffic and sales. Below is a checklist for a basic commercial remote from a client's business using a codec/mixer. Note that some items have backups, just in case Murphy's Law comes into play. Follow the signal path: the talent speaks into a microphone, the microphone needs a windscreen, the microphone sits in a mic clip, the mic clip attaches to a mic stand, the microphone needs a cable, and so forth.

Transmission Equipment

1. Morning show talent—two (2) announcers.
2. Two (2) dynamic microphones—one for each announcer. (Should you choose condenser mics, be sure to take additional batteries or a backup power supply.)
3. One (1) *backup* microphone—in case one stops working.
4. Two (2) microphone windscreens—one for each mic.

5. Two (2) microphone stands with mic clips—one for each announcer.
6. Four (4) microphone cables (length to be determined from site survey)—one main and one *backup* for each announcer.
7. One (1) codec/mixer for the mic cables to plug into.
8. Two (2) six-outlet AC power strips—one main and one *backup* for the codec/mixer and other gear (e.g., monitor) to plug into.
9. Two (2) power cords to feed the power strips (length to be determined by site survey)—one main and one *backup*.
10. Two (2) LAN or modular connector cables (length to be determined by site survey)—one main and one *backup*.
11. One (1) cell phone—for studio-to-site communications and as *backup* should the codec/mixer fail.

Monitoring Equipment

1. Three (3) sets of headphones—one for each announcer and one *backup*.
2. Two (2) monitors (either boom box or tuner, depending on site survey)—one for the announcers' headphones to plug into and one *backup* (the backup is always a boom box, regardless of what type of monitor the first one is).
3. PA system for guests (listeners' air monitor).
4. Two (2) patch cables from the announcers' monitor to feed a PA system—one main and one *backup*.
5. Speakers (left and right) for the PA system.
6. Speaker stands.

Tools

1. Two (2) rolls of 2-inch wide gaffer's tape—never use duct tape.
2. Bag of assorted-length plastic wire ties.
3. Bag of patch cords and adapters, including headphone splitters, audio adapters (mini to one-quarter inch phone and similar), headphone extension cords, etc.
4. One (1) Leatherman tool or well-equipped Swiss Army knife.

All of the equipment on the preceding list is required for a well set up remote. Of course, your station's remote might be as simple as a smart phone with a codec, a microphone, and a set of ear buds. How the gear is transported varies from station to station. A station van serves two functions. It is a place to put your stuff, and it is a rolling billboard. With a built-in power source (portable generator), a vehicle also becomes a mobile operations center that serves as a base of operations for a remote. For example, instead of playing with lots of mic cables, mount two wireless microphone receivers in the vehicle and a receiving antenna on the vehicle's roof. With a handheld wireless mic, the announcer then has the freedom to roam the store with the business owner, push price and product, and mingle with customers. Since everything is mounted in the vehicle, there are fewer wires and a much easier setup. The merchant, playing the station on his or her storewide system, provides the station monitor. Depending on the location and line of sight, professional wireless microphones have a reach of up to 1,000 feet.

Prior to doing a remote with wireless microphones, it is imperative that you coordinate the frequency of the wireless microphones with any that might already be present at, or near, the location you are broadcasting from. This is particularly true of sports venues or other public facilities. Just like everything else, don't forget backup microphones and mic cables, should the wireless fail.

There is a downside to using such a clean and simple set up. Listeners don't get to see any of the magic. No knobs, buttons, flashing lights, or bouncing meters. This is where you depend on the promotion department to pick up the slack with banners, decorations, and to create an air of excitement so that the focus is on the event and not on the station hardware.

Regardless of whether your microphone is wireless or hard wired, what pickup pattern makes the most sense for location use? If you want to give listeners a sense of being there with you, try an omnidirectional microphone. An omni picks up sound from all directions and works best in a controlled environment. However, if there is going to be significant background noise at an event, the omni becomes a problem. The omni picks up just as much noise as it does announcer, with the voice fighting just to be heard over the noise. When you know background noise is going to be a factor, switch to a super-cardioid microphone and use the directional pattern to isolate the announcer and block the background noise. Any ambience you desire can be added with a crowd mic adjusted to an appropriate level. For an in-depth review of microphone patterns, see chapter 3.

Sports Remotes

The process of making a checklist for sports remotes is the same as for a commercial remote. For sports remotes, start by first substituting broadcast headsets with attached microphones in place of the mics for the announcers.

A headset mic is ideal for sportscasting, allowing the announcer the freedom to move about and have his or her hands free to keep stats or take notes. A headset mic also helps keep the desk in front of the announcer free of wires and visual obstructions. A headset mic is usually highly directional, blocking out all but the announcer's voice. The headphones provide sound isolation so the announcers can hear each other clearly on the program or air monitor. The headphones also provide a convenient way to deliver an IFB signal to the announcers. Broadcast headsets are often wired so that one ear is the program feed and the other is the IFB feed, and they are usually wired with a volume control for the headset and a cough or kill switch so that the announcer can momentarily turn the mic off.

The use of headset mics requires the addition of a crowd microphone for ambience to make the listeners feel as if they are there. A crowd mic should be placed either in the press area, aimed down at the crowd, or on the sidelines, aimed up at the crowd. A crowd mic gives a producer the ability to control the exact ratio of the announcer's voice to the crowd sounds. A crowd mic is normally mixed at a low background level to provide presence and follow the action. Crowd mics are also handy to turn up when the national anthem is played or "the crowd goes wild."

The method of transmission for sports remotes varies depending on the location. Prior to the remote, it is necessary to determine what telco (telephone company) and IP services are available at the sports venue, and whether there are **courtesy lines** available for use. Also, check to see if there are any passwords necessary to connect to the web or to use WiFi, if it is available. And finally, if using IP at a university or major sports venue, make sure there is not a firewall in place that could block your signal from getting out of the press box.

Radio sports remotes are typically produced in monaural (one channel). However, if you are using a stereo codec you can add sonic width and depth to the game by replacing the crowd mic with a stereo mic (see chapter 3). Ideally, a stereo crowd mic should be positioned near midfield or midcourt.

In addition to the headset and crowd mics, sportscasters often plug a laptop or portable recorder into a line input on the mixer to play back pregame interviews. Some sportscasters prefer to use a laptop computer to record, edit, and play back interviews (see figure 11.11). Such devices can also be used to record the game for instant replays of the play-by-play action. To handle this many inputs, a small audio mixer is required to feed the codec's input.

Figure 11.11 Some broadcasters use a laptop computer to record, edit, and play back interviews. Such devices can also be used to record the game for instant replays of the play-by-play action.

Another popular addition to a sports broadcast is a wireless microphone(s) for courtside or sideline interviews. Here again, create a checklist for the wireless system, including frequency coordination, the microphone, receiver, batteries, and so on, and make it a part of your overall checklist. As a remote setup becomes more technical in nature, a field producer becomes necessary to accompany the play-by-play and color announcers.

Sportscasters often listen to an **off-air monitor** when they are within range of the station. On the road, the talent can be fed a mix/minus monitor signal from the station on the IFB, interrupting the feed when the producer at the station needs to talk to the announcers in their headsets.

Just as with commercial remotes, there is usually someone at the station to act as a control-board operator to put the game on the air, play back commercials, and so forth. However, it's not necessary if the station is equipped with the appropriate computer server and software. Many stations use automation and live-assist systems that permit remote access via the web. With a laptop computer at the remote site, the producer logs into the station server and remotely takes over operation of the radio station, producing the entire broadcast from the remote site.

A sports remote can be as simple or as complicated as the broadcast team wants to make it. Keep in mind that a low-tech written checklist, starting with the talent and working backwards to the station, makes sure you take what you need for the remote. You can even keep the list on your smart phone, retaining individualized lists and notes for future reference should you ever return to that site. Physically observe each item that is packed for the remote and check it off the list, making sure not to forget the backups. Likewise, when it's time to leave a site, go over the checklist again to make sure everything you took to the remote comes back to the station.

News and Special-Event Coverage

Back in the dark ages of pay phones and cassette machines, a reporter's best friends were a roll of quarters and an official telephone company out-of-order sign to tape on the closest phone to the event. When the event was over, you rushed to your out-of-order phone, peeled off the sign, and started feeding the phone quarters. Pay phones are a thing of the past. Most news remotes, or **electronic news gathering (ENG)**, are simple, one-person jobs, and each reporter generally does his or her own work as both talent and technician.

News reporters have a number of effective, portable digital recorders to select from in order to capture sound, including a software option that turns your smart phone into a recorder/playback unit (as discussed earlier in the chapter). Also, smaller, lighter laptop computers and tablets are ideal ENG solutions featuring the ability to provide support for an entire location news operation (see chapter 5). If your laptop can connect via a wireless or 4G LTE connection you can essentially take your entire newsroom on location.

Often election coverage is a wireless operation, with reporters using laptops, smart phones, and other digital recorders equipped with wireless technology. In some instances, candidates' speeches are made available as digital audio files and can be downloaded instantly to laptops for editing and transmission to a station's newsroom. Audio cuts from speeches can be converted to MP3 files and then transferred over the web or attached to emails and sent to news editors. Another advantage of using a laptop for news gathering is that a reporter can archive in their entirety a number of the candidate's speeches, along with all of their edited audio cuts as a reference. Using wireless technology, reporters have found themselves filing stories from rental cars and as they traveled on campaign buses and aircraft equipped with WiFi. Combined with news-writing software and a microphone, a laptop is an anytime, anywhere newsroom.

For the most part, the challenge in news recording comes in capturing the sound at its source. Usually, this involves interviewing the newsmaker. Occasionally, newspeople are asked to capture background sounds such as sirens, chanting protestors, or some other easily identifiable sound that can help support a story. Most small digital recorders have built-in microphones; however, it is best to use a professional microphone plugged into a digital recorder. Save the built-in mic for a last-resort backup.

A newsperson can select either a nondirectional or a super-cardioid microphone for location recording. Remember that the omnidirectional mic captures everything, including the interview subject and any background noise present, which is fine in a controlled environment. On the other hand, a super-cardioid mic provides a great deal of control over what the microphone picks up and is especially useful if the news event is in a noisy location (see chapter 3). Some newspeople choose to be very selective and use a handheld hyper-cardioid shotgun microphone to isolate and pick up only the newsmaker. The whole microphone selection process for a newsperson comes down to what the person is comfortable working with and the quality level he or she desires.

For special-event coverage, such as a press conference or a major speech by a political figure, a **pool audio feed** is often provided. Pool feeds are created with a **breakout box** or a **multiple outlet box** (**mult**). A breakout box is a device that takes a single audio feed and, using amplifiers, splits it into multiple feeds at standard broadcast levels so that each member of the media receives a good, clean audio feed (see figure 11.12 on the following page). Breakout boxes are provided by whoever is holding the press conference or special event.

An extension of the breakout box is a mobile wireless Internet terminal. Set up at a special event, the mobile terminal provides rock-solid wireless Internet access to as many as 150 reporters at once.

Occasionally, a reporter encounters a situation in which there has been little preparation on the part of the group holding the press event, and the newsperson is asked to take a feed from the public address system. There are two ways of doing this. The first is simply to place your microphone in a direct line with the public address speaker, a few feet away so you get a good audio level. The second method you can use is to tap directly into the PA system. A word of caution: any time you take a direct tap off a public address system, you need to be aware that PA systems,

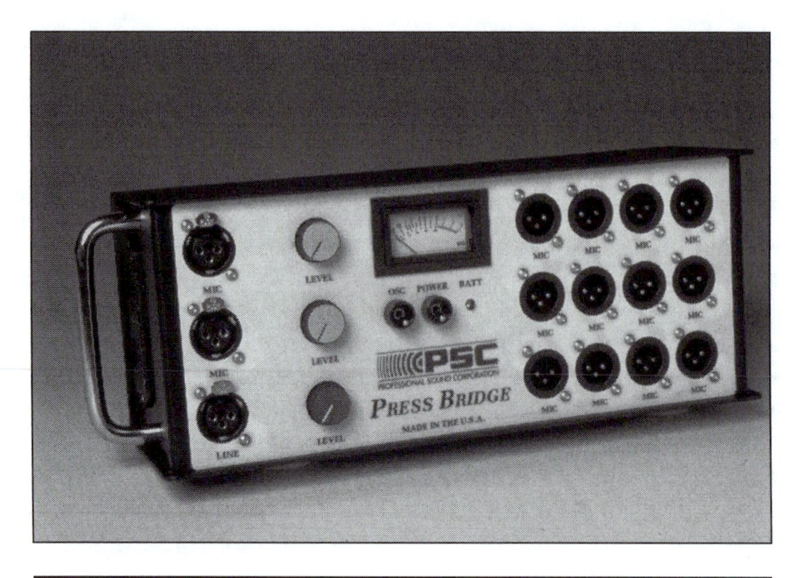

Figure 11.12 Breakout, or pool feed, boxes are provided to ensure the press receives good quality audio at press conferences or special events. Sometimes called a mult box, a breakout box takes a single audio feed and, using amplifiers, splits it into multiple feeds at standard broadcast levels so that each member of the media receives a good, clean audio feed. The example here can provide 12 feeds at mic level to the media.

especially inexpensive ones, are not noted for their outstanding audio quality. Also, you may have to use your spare connectors and cords to rig a patch cord or figure out a way to get the signal from the amplifier to your gear. Look first for a headphone output on the front of the PA amplifier as a place to start, followed by an auxiliary output, which is likely located on the back of the amplifier.

Once you have captured the location sound, there are any number of ways to get it back to the station, such as a Marti, smart phone, codec, or wireless connection. Because a lot of news is done on the fly, the simplest system often works the best.

■ Suggested Activities

1. Select a site for a mock commercial radio remote. Visit the site and conduct a site survey. Decide upon the talent and the transmission and monitoring systems. Create a checklist of the equipment that would be necessary to do the remote.
2. Select a sporting event for a mock radio sports remote. Visit the site and conduct a site survey. Decide upon the talent and the transmission and monitoring systems. What unique or action sounds could you mic to make the broadcast more interesting? Create a checklist of the equipment that would be necessary to do the remote.
3. A city council is meeting to make a major announcement regarding the construction of an industrial park. The six council members will appear at the site and stand as a

group to break ground with a gold shovel. If your station is covering this major event, how would you mic the event so that each council member's comments can be recorded? How many ways could you mic the event? Make a checklist of equipment for the method you choose.

Websites for More Information

For more information about:

- *breakout boxes,* visit Professional Sound Corporation at www.professionalsound.com
- *codecs for POTS, AoIP, and wireless,* visit Tieline Technology at www.tieline.com
- *handheld satellite phones,* visit Globalstar at www.globalstarusa.com
- *radio remote pickup transmitters,* www.bdcast.com/products/audio-data-links/marti
- *remote phone hybrids,* visit JK Audio at www.jkaudio.com

Pro Speak

You should be able to use these terms in your everyday conversations with other professionals.

A/D—Analog-to-digital audio converter.

breakout box—A device designed to allow audio distribution at press conferences through multiple mic or line-level outputs.

codec—Coder/decoder; a digital audio device that codes analog audio into digital audio and decodes digital audio back into analog audio.

courtesy lines—Communication connections that are available in the press area of a sports venue for visiting sportscasters to use without charge. Depending on the facility it could range from a POTS line, a broadband Internet connection for AoIP, to a WiFi connection.

D/A—Digital-to-analog audio converter.

draw—Something to attract people to an event or a location; a gimmick or enticement.

duplex devices—Audio devices that allow bidirectional communications on a single line or radio-frequency transmission.

electronic news gathering (ENG)—Location spot-news coverage.

GPS—Global Positioning System. A satellite-based navigation system made up of a network of satellites that enables pinpoint location identification.

low-earth-orbit (LEO)—A term used to describe satellites orbiting earth at 300 to 500 miles high.

mic flag—A plastic nameplate, or cover, that goes on a microphone with the station's call letters or logo.

multiple outlet box (mult)—A mult is the same as a breakout box, which is designed to allow audio distribution at press conferences through multiple mic or line-level outputs.

off-air monitor—An actual air signal taken from a radio for monitoring purposes.

phone hybrid—An electronic device used to connect broadcast equipment with a telephone line so that audio can be sent and received on a phone line or cell phone.

pool audio feed—A single-point audio feed for the press, usually accomplished with a mult or breakout box.

swag—Stuff we all get. Promotional items given away by a manufacturer, business, or radio station.

vendor involvement—When the vendor of a product or service provides support to a local business in the form of product to give away or expert staff to help demonstrate the product. This kind of involvement with a local business can be quite varied from manufacturer to manufacturer.

The Web

Highlights

Background

In the 1950s, television was going to kill radio, in the 1990s independent dot-com radio stations were going to decimate traditional radio, and in 2001 it was satellite radio that was going to put an end to AM and FM radio stations.

Radio is the ultimate chameleon. From its very beginning, radio has been forced by competing media and new technologies to change its colors on a regular basis to remain competitive. Radio has been through a number of challenges in its history, and each time it has emerged renewed and strengthened. To be successful, radio must remain in a constant state of change, moving forward and adapting to new technology and listener trends.

Typically, any new technology initially brings panic. Those that are going to succeed have to overcome the fear of something new and then turn the challenges they face into new opportunities. Early on the radio industry saw the technological potential of the web and the rich opportunities it promised. The industry decided that it was better to embrace the web and master it rather than the other way around.

The early decision to commit to the web and stream on-air signals was just the beginning. Web audiences grew fast in the early 2000s. According to Edison Research, the online radio audience has more than doubled since 2010 and will likely continue to grow. Even better, online listening is not a substitute for on-air listening. Nearly 90 percent of online listeners report they continue to listen to AM and FM radio. In fact, 90 percent of the US population age 12 plus continues to listen to traditional on-air radio each week. The web is a valuable station resource to extend its positioning and branding concepts and to impact listeners' lives, making radio an interactive medium.

Radio is also undergoing a transition from a mass medium to an individual medium thanks to smart phones and broadband Internet access. By signing up on a station website a listener can create an individualized web experience, including localized weather, news, and traffic alerts sent via text message along with your children's school lunch menu! The options are limitless.

Home radio ownership is declining as people transition from fixed-in-place hardware to mobile hardware to listen to and interact with radio. Over 200 million smart phones are now in use, providing the connectivity to listen anywhere, any time. Radio's "traditional" online listener is also changing; more people are now listening on smart phones rather than on their lap or desktop computers. A much larger piece of personal mobile hardware is your car. AM/FM radio continues to dominate as the number one audio source in cars at over four times the usage of the CD player and over seven times the usage of online radio.

So why do production people need to understand the web, and their station's involvement with the hardware and software that supports it? The industry has come to require that each person in the station take a holistic approach, understanding how the different departments work together to form the whole. Radio employees are not isolated in their own tiny little bubbles within the station. Now more than ever, it is important for you to consider the entire station and understand your role in its operation. Although you might not be involved in every aspect of the station's web presence, you should understand how the web and the various social media platforms contribute to the overall success of the radio station.

Competing with Another Medium

The web is a dual-edged sword with respect to radio and other media. The web is a competitive threat to radio, with the ability to steal listeners and station advertising revenue. On the other hand, the web presents an incredible opportunity to expand a radio station's reach, to provide a tangible presence for listener interaction, and to provide expanded revenue streams for stations.

Trends in media convergence, and the web's ability to present traditional media in a nontraditional way, make it abundantly clear that the web is a natural for radio. A radio listener's options include listening on air, online, on your phone, and on demand. Essentially, the radio industry doesn't care what device you listen on, the important thing is you're listening! These delivery platforms, combined with audio and video streaming, contests, social networking, podcasts, video games, local business and service directories, text messaging and message casting, interactive advertising, etc., make radio a powerful tool for advertisers to reach customers. Further, as podcasts of station shows have become more popular, the term binge listening is now starting to be used to describe listeners that listen to multiple podcasts.

The clear advantage that radio has is that few, if any, mediums have radio's skill and ability to drive on-air listeners to their web content. As mentioned in chapter 10, radio is the best barker in the world. When thinking about the radio industry and the web this author is reminded of something baseball pitching legend Satchel Paige (according to ballplayer Dizzy Dean, the greatest pitcher of all time) once said: "Don't look back. Something may be gaining on you."

To remain competitive radio must be constantly moving forward.

Radio and the Web

The first involvement with the web for most stations is a website and the streaming of its on-air signal. However, streaming the on-air signal is a microscopic aspect of the opportunities available to a station on the web. The web offers the opportunity for a station to become not only something to listen to, but a place to go. A station's website is the focal point for a number of opportunities to interact with the station and with other listeners who like the station. Well-designed sites give the feeling of a social setting, that listeners are part of a group or family. Station websites bring listeners and advertisers together.

With all of the digital advertising platforms on the market, choosing the right one can be confusing for an advertiser. Progressive radio stations are merging the various digital platforms in one location and providing clarity for the advertiser in an advertising world filled with an almost unlimited number of choices. A station's website and social media platforms can provide many of the solutions to an advertiser's digital marketing concerns. This positions the station as a valuable resource for multiplatform advertising.

It's easy for a radio station to make the case to a local business that not only should the business advertise on the radio, but on the station's website and social media platforms as well. Consider this: the explosion of broadband access at home, work, and on smart phones has enabled 90 percent of radio listeners to have access to the web. The average home has six devices that can connect to the web. Over 70 percent of homes have their own WiFi network to connect all of those devices. A radio station can make a good case to a local business to increase revenues by targeting radio listeners and directing them to the station website and social media platforms.

The most compelling reason for a station's involvement with the web is advertising revenue. Total interactive (web) advertising revenue has reached the $60 billion mark annually. Of that money, 40 percent is being generated in the local markets, offering stations an excellent opportunity to increase revenue.

How Radio Uses the Web

One of the primary reasons advertisers use radio is that the medium is local. Radio allows advertisers to target specific demographic and psychographic groups so that their advertising is more effective in reaching potential customers. An advertiser buys commercials on a station because the advertiser values the audience and the business the local audience can bring. The web provides a station the opportunity to more deeply penetrate and engage a local audience.

Local businesses are challenged by online commerce. People like the ease of not having to leave their home to shop; people like making a car service appointment from their smart phone; people like ordering dinner and having it ready to pick up on the way home from work. A local business has to be competitive to succeed.

Typically, when local advertisers think of the web, they think on a global scale. Radio, whose specialty is targeting local customers, has transformed this worldwide medium into a powerful tool for local businesses. Although the web and radio are two distinctly different mediums, when combined they form an incredibly powerful marketing tool.

Web Words: A New Vocabulary

To understand what everyone in the station is talking about, it is necessary to learn a few industry terms that describe important aspects of the station's website. There is one term that you may be familiar with that you have to get out of your vocabulary—hits. Some people equate the number of hits on a website with the number of visitors to a website. When someone uses the term hits in relation to the number of visitors to a website he or she has just told everyone in the room they have no idea what they are talking about! A hit occurs when one file from a website is downloaded from a web page. A web page is made up of a number of individual files. If a web page had 37 different files that made up the page and someone viewed the web page, 37 hits would be recorded. Advertisers do not want to know the number of hits; they want to know the number of page views or **page displays**. How many times was the full page displayed to a viewer?

Probably the most often-used term is **broadband**. The term refers to the type of web connection or access in use. To a consumer, broadband typically refers to a high-speed connection like a cable modem, a **digital subscriber line** (**DSL**), or a wireless connection. For a radio station, a broadband connection might be a T-1 leased line that can transmit data at 1.544 Mbps (megabits per second). A T-1 line is often referred to as a "big pipe" in the industry and is necessary to serve a large number of listeners simultaneously. In chapter 13 you will get an in-depth explanation of Internet connection types and data speeds.

Interactive is a broad term that describes a radio station website(s) and everything associated with the website. Even the department in the radio station that oversees the station website is often called "interactive."

A website is hosted on a **domain**. Think of a domain just as you would the radio station's dial position. The domain is the location of the website. Typical domains include .com for commercial; .gov for government; .edu for education; and .org for a public or noncommercial organization. Some domains are geographic and tell you what country the website is located in, such as .uk for the United Kingdom or .fr for France.

Advertisers also want to know the number of unique users that have visited a page, and how long those users spent viewing the page. The Nielsen Company conducts media research and estimates that the average site visitor will spend about one minute per page viewed; this is called the **session length**. The unique data for each website, such as number of visits, number of page views, number of unique visitors, where those visitors came from, etc., is known as **metrics**. Think of metrics as your website's "ratings."

Using metrics, a radio station can calculate advertising rates. One method is to charge advertisers based on **cost per thousand (CPM)** (M is the Roman numeral for 1,000). For example, an advertiser buys 50,000 page views. The advertiser's message will be delivered 50,000 times to individual viewers and then the ad will be removed from the website. Some advertisers demand that viewers do more than just see their ad, the advertiser wants the viewer to click on the ad and go to their website. This is referred to as a click-through, and the rate is called **cost per click (CPC)**. The advertiser only pays for those viewers that click through to their website.

The universal unit of measurement for an image or an ad seen on a computer screen is the pixel. A pixel is one tiny dot that, when combined with others, creates an image. Advertising and website elements are measured in pixels. A typical horizontal banner or display ad measures 728 pixels wide by 90 pixels high. The standard size for a vertical display ad measures 160 pixels wide by 600 pixels high. Regardless of how large or small the computer screen is the ad will always be proportionately the same size.

Pop-ups are those annoying ads that place themselves on top of or in front of whatever content you are viewing. A less offensive ad is the pop-under ad. A pop-under hides under the web page you are viewing and when you close the page there it is. Most people dislike pop-ups so much they use software to block them. **Gateway ads** are those ads that play just before an audio or video stream begins (rich content). The best gateway ads are 10 to 12 seconds long.

Web Management

In chapter 1 you learned about how a radio station is organized. Radio stations also have organizational structures for their interactive departments or divisions. Groups or clusters of stations typically have a website for each station. It is not unusual for a group to have four to seven websites to create and maintain.

Heading up the station's interactive operations is an Internet manager who is responsible for all aspects of the station's website(s), including sales, coordination with traffic and billing, content, and their technical operation. Essentially, the Internet manager is responsible for everything related to the station's interactive presence on the web. Typically working directly under the Internet manager is the Internet content manager.

The Internet content manager is the program director for the station's website(s) and is in charge of everything that appears on them. Content includes various things like the station's social networking sites, news and weather content, contest registrations, text messages to listeners letting them know when their favorite artist is coming on the radio, and the viral videos used to draw people to the station website. Content managers also "repurpose" content that others create.

The content manager works closely with the programming and promotion departments and the station's webmaster.

The webmaster is the person responsible for maintaining the station's website(s). Key in the webmaster's job responsibilities is ensuring that the station servers and software are operational and are maintained for maximum reliability. Webmasters are also responsible for tracking the volume of traffic through the site and watching site metrics to determine visitor loads and meet their demand. The webmaster is the person responsible for uploading ads and content to the website(s) and making sure the site(s) looks good and is 100 percent functional. How many times have you been irritated at a link that didn't work? Finally, the webmaster is skilled at web design, layout, scripting languages or code, and website functionality.

The fourth person on the interactive team is the Internet sales manager, who is responsible for the sale of advertising, sponsorships, and all of the unique features that can be sold as services that tie into the radio station. Often the Internet sales manager works for the general sales manager of the radio station. In some stations there is a separate interactive sales team; in most stations though the regular salespeople are expected to sell both radio and interactive.

What Web Consumers Want

There are two types of radio station website consumers: listeners and advertisers. Both want something very specific from the station's website. Listeners are after four universal things from every website.

1 *Fast downloads.* Speed is king. Listeners will not wait for a slow download.
2. *Rich content.* Rich content includes interactive content. Listeners want sound, video, and things that involve Flash, Java, and JavaScript programming languages (which cause the site to come to life and to be interactive).
3. *Easy navigation.* Listeners do not want to hunt for content; getting to it has to be intuitive and user friendly.
4. *Up-to-the-minute information.* The web is a "got to have it now" place; a website has to be ahead of the listener.

Unlike listeners, advertisers are after something totally different. Advertisers want three things.

1. *Large ads.* Advertisers want to be seen. A typical horizontal ad is 728 pixels by 90 pixels; vertical ads are 160 pixels by 600 pixels.
2. *Integrated ad content.* Tie both the web and the radio ads together and link them to the advertiser's website. Give someone a cause to click through on the ad.
3. *Advertisers want placement.* The best position is at the top of the website. This is only natural because some monitors cannot display an entire site top to bottom and the advertiser wants to make sure they are seen. We read left to right, so where do you think is the best place to put a vertical ad? On the left side as high as it can go on the site.

While the consumer wants listed here appear simple on their face, they are often forgotten as a key to website success. A station has to keep its consumers in mind in layout and design and touch on each one of these points to be successful.

Programming a Website

Although it may sound like the programming department's job to program the station website, the fact is a production person is likely going to be involved in some aspect of the project. Just as it is important for you to understand how the radio station is programmed and how production plays a vital role in the execution of the programming, it is important for you to understand the overall impact a website has on a station and your role in its success.

A radio station must give listeners a reason to visit its website other than simply saying, "Visit our website!" For a moment, forget about streaming the station's on-air signal on the website and concentrate on visual and interactive content. How can you make the site work without the music? Make it interactive. Remember, a listener is constantly asking, "What's in it for me?" A station must offer something on its website that listeners value, or something that solves a problem listeners have, to get them to visit the website (see figure 12.1). You need to think of programming for your station's website as radio with graphics, pictures, and video.

The creation of the station's website requires the work of several people from programming, promotion, and production. The website must be artistically and graphically pleasing to the station's target demographic. The graphic design of a

Figure 12.1 Listener service features are a reason for listeners to visit a station's website. Look carefully at this example—it is filled with interactive links to involve the listeners in the station and its activities.

country station's website is going to be a lot different from the design for a hard rock station. Regardless of the style, the site must have fast downloads, interactive rich content, must be easy to navigate, and must have up-to-the-minute information. Those who sign up and become members of a station website are entitled to even more content, much of it individualized just for them.

Additionally, many stations have a password-protected website for advertising clients that features the sales department. The sales website can serve as a production showcase featuring demos of the station's best production work. The sales department's website can offer materials created from the Radio Advertising Bureau's research on the benefits of radio advertising. The station can also offer sales and marketing information about the specific demographic and psychographic groups that compose the station's audience.

As a further service to clients, many stations offer access to a secure area of the sales website dedicated to their account, allowing advertisers to check their account information and current advertising schedule, generate an online invoice, and pay for their ads with a credit card. Clients can also download or play a copy of their current ad.

A really great station website requires a lot of people working together to keep it up-to-date and fresh, all working under the watchful eyes of the content manager.

Website Content

With everyone in the station potentially contributing ideas to the website and posting on social media, usually the content manager has his or her hands full staying focused on the site's programming goals. Chief among the goals is maintaining the station's positioning statement and branding concepts. Below is a very limited list of the things a station could use its website for. Notice that streaming the station's on-air signal is not even on the list; it's the easiest thing to do (see figure 12.2).

- client advertising and links to clients' websites
- printable client coupons with links to clients' websites
- restaurant and club listings with links to their websites
- concert promotion and advertising with a link to the promoter's website
- concert ticket advertising and links to ticket vendors
- station contests, contest rules, contest entry forms
- station clubs or fan groups associated with shows
- purchase-and-print tickets to station-sponsored events
- live studio webcams
- instant music requests directed to the jock in the studio via a social networking site or text message
- photos of staff and station events
- interviews with station staff
- on-air staff schedules

- remote schedules or station event schedules
- music lists, even the hourly playlists with a click-to-play feature for top songs
- advertising links to music companies or fan websites
- station music charts
- music news
- music message boards for event notification or music discussion
- music videos for visitors to watch or download
- music or station surveys
- cross-promotion of other stations in your market group
- weather
- news headlines
- news text messaging for school closings, weather emergencies, sports scores, etc.
- community public-service links
- access to free video games
- links to topics that have a wide range of appeal (such as health, beauty, and pets)
- audio on demand, podcasts of music shows and news and public affairs programs
- programs created specifically for the station website that are not on the air

Many items on the preceding list involve audio and video production in some form. Every one of them is not only a reason to visit a site, but also has commercial

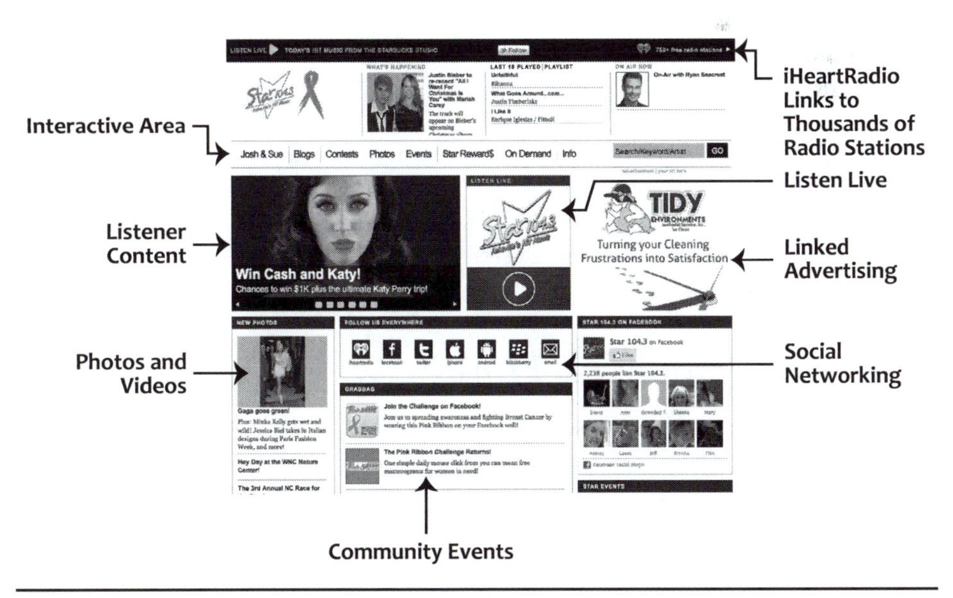

Figure 12.2 This website is an example of a site that takes full advantage of just about every aspect of potential listener involvement, from links to advertisers to joining a station club.

possibilities for generating revenue. For example, a station can partner with retail websites that sell things its listeners are interested in. The radio station receives a per click-through fee or a percentage of each sale made from a click-through. The technology is very robust and allows a station to do just about anything it can dream up, including placing webcams at remotes.

Technical and Economic Considerations

Just as technology allows a station to be incredibly creative with its website, it also limits what a station can do with a website. One of the unique qualities of the web is that the creator of a website cannot control the quality that the end user experiences. For example, a webmaster creates a site using a 32-inch monitor and the latest software and hardware. The webmaster has unlimited broadband access to view the site with, and on his or her computer the site loads fast and looks incredible.

However, what the end user experiences is another matter altogether. The end user may be using a much smaller monitor and a computer that's a couple of years old. The end user could be using a small laptop. The site probably looks good, but do the pictures and video clips load quickly? Does the audio stream function well and sound good? Ultimately, is the end user experiencing the same functionality and seeing what the webmaster is? Key to the webmaster's job is taking into account all the different possible site visitors and available end-user platforms when designing the site.

Another aspect of website design is how does the website look and function on a smart phone or tablet? There are two dominant smart phone operating systems. The station website has to be designed to look and sound good on both.

Economically, **streaming** on the web is quite different from broadcasting an on-air signal. Traditional radio broadcasting involves a signal that listeners are free to listen to whenever they like. No matter how many listeners tune in, the broadcaster's operational costs remain the same. However, a station streaming audio on the web has to purchase enough bandwidth from an Internet service provider for a given number of listeners to hear the station. Every new listener potentially adds to the station's costs. The other economic consideration is the cost of music copyright fees to stream the radio station's on-air signal. This will be discussed in detail later in the chapter.

Audio Production for the Web

Putting an audio signal on the web would appear to be a pretty simple task at first. Tap into the on-air audio someplace in the station and feed it to a computer hooked up to the web. Sending good-sounding audio to the web is not that simple, though. Sending an audio file, like a commercial, on the web is a one-time data transfer; streaming audio is a continuous, real-time event. Trying to send a full-frequency, high-quality audio signal takes an incredible amount of bandwidth, both on the part of the sender and on the part of the receiver. Since it's already assumed that the consumer is an average site visitor, then something has to be done to make the audio fit on the web so that listeners can hear it and the station can afford to send it.

The first problem that broadcasters encounter is the fact that, unlike for the radio station, there are no FCC-mandated technical standards for transmitting the signal to a listener via the web. Because broadcasters don't know what kind of computer, smart phone, tablet, or web connection their listeners are using, they typically prepare audio based on the average site visitor so that as many people as possible will be able to hear the station.

The audio processing required to originate a good-sounding webcast is as technically involved as the on-air broadcast processor. Web audio processors are different from their broadcast counterparts. Many of the companies that produce the sophisticated on-air broadcast audio processors also produce complementary audio processors for the web. Web audio processors use **perceptual audio coding** or data compression (this is similar to the process used in creating MP3 audio files). Perceptual audio coding is a method of compressing, encoding, and finally decoding audio that takes advantage of how humans perceive sound to achieve a significant reduction in the size of the data stream. Properly done, it entails little or no perceptible loss of audio quality. Basically, the audio processor discards everything the listener can't hear, such as the very low and high frequencies. Based on how your ears perceive sound, other unnecessary audio data may be eliminated from the stream or may be enhanced.

Most radio stations use a third-party service called a **content delivery network** (CDN) that specializes in audio streaming that can easily handle thousands of streaming connections at the same time. Likewise, most stations use a similar service to host the station website. These third-party services provide a great deal of analytical data about listeners, including the time of day they are listening, the length of time they are listening, and the overall audience size. The service can also provide the listener's location and the device they are using to listen on. Some services also provide the ability for the station to "target" streaming listeners with different commercials from those playing on the air. In fact, it is possible to send targeted ads to the streams of individual listeners!

The quality of the audio sent to the codec for streaming is important. The audio should be the same quality that is going on the air. Additional compression, limiting, and equalization should be avoided, as the most commonly used streaming audio codecs do not work well with audio that has been heavily processed.

Streaming

Respecting the Music

Whether or not a station chooses to stream an on-air signal is the station's decision; however, it is a decision that must be made in consultation with legal counsel. Technologically, it is easy to put an audio stream on the web. However, to do so within the applicable copyright laws requires some basic knowledge and legal assistance to ensure that all of the music's license conditions are met prior to putting copyrighted works on the web.

Performance Rights. The two aspects of copyright that broadcasters are most concerned with are performance rights and sound-recording rights. Performance rights are the rights to publicly perform a music work. In the case of a radio station, it is the right to broadcast the music. Usually, the artist, composer, author, or publisher holds the performance rights to the music, depending on who created the work and contractually owns the rights. Because it is impossible for musical artists to keep track of the thousands of broadcasters in the country using their work, artists sign contracts with performance-rights organizations to manage their works.

Radio stations pay fees to the three performance-rights organizations that represent songwriters and musicians: the American Society of Composers, Authors, and Publishers (ASCAP), Broadcast Music, Incorporated (BMI), and the Society of European Stage Authors and Composers (SESAC). The fees each station pays are based upon station music-use reports (music logs) and station financial statements. Based on the estimated airplay of each artist's work, the organizations divide the station fees received among the various artists they represent.

Sound-Recording Rights. The sound-recording rights to a song are the rights associated with the physical recording itself, that is, how the musical notes were recorded, edited, mastered, and released to the public—the creative work of the recording process. The company that recorded, produced, manufactured, and distributed the recording usually holds the sound-recording rights, although some artists retain the sound-recording rights to their work.

Radio stations do not pay sound-recording rights fees to broadcast music works on the air. Years ago, when such laws were originally proposed, the broadcasting industry prevailed before Congress and laws were passed in the broadcast industry's favor. This is not the case for streamed music on the web. The **Digital Millennium Copyright Act of 1998 (DMCA)** established, for the first time, that owners of a sound recording had the right to authorize digital public performances and to be paid for such performances.

The DMCA covers webcasting, cable and satellite digital audio services, and future forms of digital transmission yet to be determined. Traditional on-air radio broadcasts remain exempt from the DMCA.

To stream audio on the web radio stations pay sound-recording rights fees to SoundExchange, a nonprofit organization designated by the Copyright Royalty Board of the Library of Congress to collect and distribute the fees to artists. The fee structure for sound-recording rights on the web is based upon detailed computerized records that are required, by law, to be maintained by the webcaster. Unlike with radio broadcast performance rights, the operator of a website pays a per song, per listener fee. In addition to the performance/listener fee, the website operator pays an annual license fee.

As a production person, you must abide by your company's policies on webcasting and placing audio clips on the web. When it comes to questions and decisions regarding copyright, the station's legal counsel makes the final determinations for the station.

Audio Transmission Formats

When audio is transmitted using the AM or FM band, a radio station engineer makes sure the station's signal complies with the technical rules and regulations of the FCC. This ensures a uniform signal that all AM and FM radios can receive. Likewise, manufacturers certify that the radios they build comply with certain FCC rules and regulations. The end result is that the audio quality the listener receives is very uniform.

There are a large number of variables and costs, including the original audio source, the streaming server and software, the content distribution network, and the required bandwidth, that affect the number of simultaneous listeners a station can serve. To try and cover each possible situation in detail is beyond the scope of this text.

The final decision is going to come down to which choices make the best business sense for your station to adopt and which ones the station's engineering staff is most comfortable working with.

Suggested Activities

1. Visit at least three major-market radio station websites. Make a list of the different categories of links that are on each site, such as those for news, weather, staff photos, and so on. Do you see any trends as to what these radio stations are featuring on their websites?
2. Using three major-market radio station websites make a list of the commercial or income-producing links on the site. Do you see any trends as to the types of commercial links the stations are promoting, such as restaurants, online music services, and so forth?
3. Look at three major-market radio station websites. Are the stations doing any web-based contests? If so, what kind or style are they? Can you enter the contest on the web, or do you have to listen to the station first?
4. Visit the website of your favorite radio station. How well has the radio station carried its on-air positioning and branding concepts to the web?

Websites for More Information

For more information about:

- *copyright laws*, visit the United States Copyright Office at www.copyright.gov
- *music performance rights*, visit the American Society of Composers, Authors, and Publishers at www.ascap.com; Broadcast Music, Incorporated at www.bmi.com; and the Society of European Stage Authors and Composers at www.sesac.com
- *sound-recording copyrights and webcasting*, visit the Recording Industry Association of America at www.riaa.com and SoundExchange at www.soundexchange.com

■ Pro Speak

You should be able to use these terms in your everyday conversations with other professionals.

broadband—An Internet connection capable of transmitting data at a speed over 200 Kbps. Typical broadband connections include cable TV modems, DSL lines, and wireless connections.

content delivery network (**CDN**)—A system of computer servers that distributes web content, audio, and video streams simultaneously to thousands of users regardless of their geographic location.

cost per click (**CPC**)—A method of paying for an ad on a website. The advertiser pays a fixed rate each time someone clicks on the ad and is directed to the advertiser's website.

cost per thousand (**CPM**)—A universal media term for the cost to make 1,000 impressions. M is the Roman numeral for 1,000.

Digital Millennium Copyright Act of 1998 (**DMCA**)—This act established that owners of a sound recording had the right to authorize digital public performances and to be paid for such performances.

digital subscriber line (**DSL**)—An Internet connection made over a regular, twisted-pair telephone line capable of carrying data and a telephone conversation at the same time. Popular for home and small business use.

domain—A unique name for a location on the web. Domains can be either generic (such as .com for commercial websites or .org for nonprofit organizations) or geographic (such as .uk for the United Kingdom or .fr for France).

gateway ads—A brief ad that appears before the delivery of rich content on a website. For example, before you get to see a video clip you see a 10 to 12 second sponsor ad.

interactive—A generic term referring to anything involving a radio station website, such as advertising, website sponsorships, streaming audio, and video.

metrics—Data collected about the performance of a website, including such things as number of pages displayed or page views, session length, number of unique visitors, etc. Often referred to as website ratings.

page displays—The number of times a web page is successfully downloaded and displayed on a user's computer.

perceptual audio coding—A method of compressing, encoding, and decoding audio data files that takes advantage of the properties of the human perception of sound to achieve a size reduction of the data file with little or no perceptible loss of quality.

pop-ups—Ads that open in front of a website as it opens and block the view of the website. A pop-under ad is the opposite. The ad loads behind the web page and displays when the page is changed or closed. Both are the most disliked type of advertising on the web.

session length—The time a website viewer spends on a particular website or viewing a particular web page.

streaming—A term used to describe transmitting audio or video on the web. Usually used to describe a continuous, real-time audio feed, such as an on-air signal.

From Here to There

Radio and Audio Transmission in a Digital World

HIGHLIGHTS

A few thousand years before radio was invented people were broadcasting. Broadcasting is an old farming term for sowing seeds, or casting them over a broad area. The term was applied in the early twentieth century to radio since radio messages were cast out to listeners over a broad area. In this chapter, you're going to learn about the methods of producing and transmitting digital audio. You will learn the basics behind **AM** and **FM** broadcasting, high-definition radio, audio transfer on the web, as well as satellite radio services.

Audio Transmission on the Web

The days of shipping a commercial or program on CD are long gone. Thanks to the web and data-compression techniques high-quality audio is easily sent via the web. The key word in the last sentence is quality. What makes this possible is our constant need for speed—the demand for lightning fast data transmission and a "got to have it now" mentality so as to push a deadline right until the last moment. The web gives us the tools to do that.

For competitive reasons, advertising agencies often wait until the last moment to send a commercial to a station, knowing the commercial is only going to take a few seconds to get there. The commercial gets transferred to the station's main server and is ready to go on the air in minutes. Likewise, production work can be exchanged with other stations in local, regional, or national broadcast groups over local or wide area networks.

There are many creative ways audio can be moved about on the web. The technology is a moving target that constantly changes as new software and hardware are developed. In this chapter you're going to learn some of the more widely used methods to transfer audio, keeping in mind that it could all change next week with someone's new data-compression software or networking concept. Streaming the station's on-air signal on the web is discussed in detail in chapter 12.

Connection Speeds

The quantity of data, or audio, that a radio station can send and receive on the web is directly related to the available bandwidth, or connection speed. Ranked by speed, from slowest to fastest, some of the viable connection options include digital subscriber lines (DSL), cable modems, wireless connections and 4G LTE services, or a dedicated T-1 line. Not all services are available in all areas.

A DSL line uses a standard copper phone line to transmit both phone calls and data simultaneously via a special modem. A typical DSL line can provide data at speeds of up to 6 to 15 Mbps, depending on the distance to the nearest telephone company switching center. The maximum distance a subscriber can be located from a switch center is typically 3 miles. Subscribers pay for the service based on where they are located and the speed of service they select. DSL is not available in all areas. A DSL line is a private connection and the connection speed remains fairly constant. Although DSL service fees vary from area to area, one thing remains constant with all web service providers: the faster the data speed, the higher the monthly fee.

Cable TV web connections are probably the most popular high-speed web connections due to their packaged cost and availability. A web connection from a cable TV provider is, in theory, capable of speeds of up to 300 Mbps. Just as with DSL, subscribers pay for service based on where they are located and the speed of the service they select. Although monthly fees are relatively low for the slower speed packages, cable TV web access is a shared service. If your neighbors all have cable TV web access and log on at the same time you do, the connection speed slows down proportionally.

Wireless Internet connections are based on using a service typically offered by a cellular service provider. The current wireless technology is called 4G. It is the fourth generation of wireless cellular development as specified by the International Telecommunication Union (ITU). The 4G standard specifies that a wireless device have a high mobility (driving down the road in a car) connection speed of 100 Mbps and a stationary (sitting in a coffee shop) connection speed of 1 Gbps. However, at this time cellular providers do not have the infrastructure or equipment to meet these high rates of data transfer. Various cellular providers now brand their primary 4G offering as 4G LTE. The LTE stands for long-term evolution.

Typical 4G device connection speeds are anywhere between 2 to 50 Mbps depending on a number of variables. That is considerably less than the ITU standard of 100 Mbps. What has to be remembered is that not all smart phones or connection devices meet the ITU standards and most run at different data transfer speeds. Also, connection speeds can vary on a square foot by square foot basis. In other words, your physical location will affect wireless connection speeds, just like it does when you're making a phone call. And finally, your phone carrier will regulate connection speeds. For example, one carrier tells customers to plan on a consistent 5 to 12 Mbps of usable data.

WiFi "hotspots" in a business or other location are entirely dependent on the type of connection the provider has purchased for you to access over their WiFi service.

Another option is a **T-1** line. A T-1 line is a dedicated private line that simultaneously transmits and receives data at a constant speed of 1.544 Mbps. Typically, a group of radio stations in a market can share a T-1 line, thanks to the line's large capacity to transmit and receive data. If just being used for general web browsing and sending emails a T-1 line can accommodate hundreds of users. However, it is not unusual for a group of stations to have more than one T-1 connection, using one for a local area network (LAN), another for a wide area network (WAN), and a third for all of the on-air studios. A shared arrangement like this allows the group to put the business-oriented LAN and WAN behind a secure firewall (either a physical electronic device or software that prevents hackers from breaking into the networks), while using the third T-1 connection to allow the studio talent the freedom to surf for material and interact with listeners via social networking sites. Depending on location and availability, a T-1 connection typically costs a few hundred dollars a month depending on services desired.

The very fastest practical leased line is a **T-3**, which can transmit and receive data simultaneously at the blistering speed of 44.736 Mbps. A T-3 line is only used for a large facility requiring a huge amount of bandwidth, comparable to an Internet service provider. Depending on location and availability, a T-3 connection typically costs a few thousand dollars a month depending on services desired.

Data Compression and Transfer

Ethernet and Local Area Networks

Ethernet is a computer networking technology most often associated with a LAN. A LAN is just what the name implies: the computers and digital audio equipment within a radio station or station group are all linked together so that they can share audio and data. The primary purpose of a LAN is to eliminate the miles of audio cable necessary to wire a radio station or station group. The audio cables are replaced with a much smaller number of Ethernet cables. Ethernet cables can carry multiple audio sources simultaneously, handling as much as 1 Gbps of data. In some stations, fiber-optic lines are also used to transmit audio and data.

One of the major advantages of a DAW is networking with other workstations in the same facility through a LAN. For example, a radio station might have a control room, two production rooms, and a newsroom, each with a workstation. All four workstations can be networked together to share files. When a commercial or news story is produced, it is instantly accessible anywhere in the radio station.

Ethernet versus IP Audio. The major difference between Audio over Ethernet (AoE) and Audio over IP (AoIP) is that AoE is uncompressed broadcast-quality audio. AoIP, while broadcast quality, requires some type of audio compression to be sent in real time on the web.

Wide Area Networks

A WAN is the next step up from a LAN. Broadcasters in a regional or national group use WANs so that stations can share files over large geographic areas. Stations share files with other stations for any number of reasons, from voice tracking radio shows to delivering custom production or commercial dubs to sharing news stories. A WAN uses AoIP protocols.

Internet File Transfer Protocol

Before the World Wide Web, with its graphics interface and point-and-click icons, there was the Internet. Technically speaking, the World Wide Web operates on the Internet. The Internet is a vast network of networks over which large data files are easily transferred using file transfer protocol (FTP).

To send and receive files using FTP, all that is required is an Internet connection and FTP server-client software on your computer. The FTP server software is at one end of the connection and the client software is at the other. The FTP server turns your computer into a very simple server, minus all of the pictures and graphics. The FTP server only delivers files.

To retrieve a file from a server, you need FTP client software that is the equivalent of a very simple web browser. Again, no flashy pictures or icons. When the FTP client is launched, it asks for the web address of the computer you want to connect to and your client login name and password. The client connects to the server and two side-by-side windows are displayed. One window lists the files on your computer, and the other lists the files on the FTP server that are available for you to download (see figure 13.1). When you have identified the file you want, double click on the file and the transfer begins to your computer. You can also highlight the file and click the directional arrows between the two windows to start the transfer.

One of the reasons that FTP works so well is that it has a very powerful error-checking feature. Each data packet, or segment of data that is received, is compared to the original file on the server. If the data packet is identical to the original, the FTP client accepts it and then the next packet is transmitted. If there is an error, the client asks the server to resend the packet. The FTP software reports a successful file transfer only when every bit of the data has been successfully

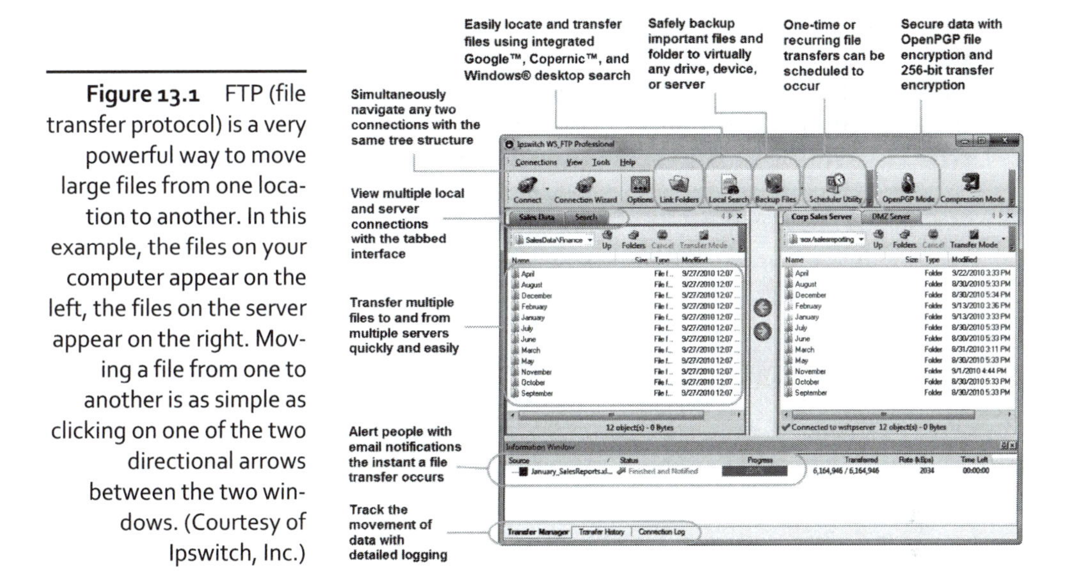

Figure 13.1 FTP (file transfer protocol) is a very powerful way to move large files from one location to another. In this example, the files on your computer appear on the left, the files on the server appear on the right. Moving a file from one to another is as simple as clicking on one of the two directional arrows between the two windows. (Courtesy of Ipswitch, Inc.)

received. Due to the accurate error-checking ability of the FTP client and server software, quality in equals quality out.

Quality is one of the big things that FTP has going for it. At broadband connection speeds, it is not necessary to use any type of data compression or encoding. Consequently, it is important to understand that any production you put up on your server for someone to retrieve had better be good, clean, quality audio.

Millions of data files are transferred every day using FTP on the Internet. Although FTP is not as fancy as a WAN, it certainly is just as functional. FTP gives smaller regional broadcast groups and individual program producers the ability to transfer audio files without the expense of leased WAN lines. Independent production and recording studios use FTP to transfer audio files across the country from studio to studio or to advertising agencies and back.

The final aspect of FTP that makes it so interesting is that there are a number of free versions of the server and client software available, such as WS_FTP Professional, by Ipswitch.

MP3 and MP2 File Transfer

There is one thing that potentially stands in the way of transferring quality audio on the web, the bandwidth and speed needed to send large audio files. As stated earlier, AoIP has to be compressed.

In 1987, work began at the Fraunhofer Institute in Germany to develop data compression and perceptual audio and video coding to reduce the size of audio and video files and still retain the perceived quality of each. The International Organization for Standardization foresaw the coming need for video and audio data compression and formed the Moving Picture Experts Group (MPEG) to standardize these methods. Two popular data compression technologies were developed as a result of their research: MPEG Audio Layer-3 (or MP3, as it is popularly known) and MPEG Audio Layer-2. MPEG-2 is primarily a professional format used in the television and motion picture industry. Some broadcast software companies also use MPEG-2 as their preferred file-storage format and Video over IP transmission method.

An MP3 codec (coder/decoder) compresses, or shrinks, a CD-quality audio file by a factor of 12, maintaining the original perceived CD sound quality. This dramatic reduction in file size makes the transfer of quality audio files relatively easy for just about anyone. Thanks to the small file size, huge amounts of music or audio can be stored on hard or flash drives.

A key to successfully using MP3 audio compression is the quality of the codec. There are many free MP3 codecs on the web, however, although there are some good ones, many leave a lot to be desired when it comes to quality audio. Most professional MP3 codecs are based directly on the original Fraunhofer research and research done by Thomson Multimedia.

One of the features of most MP3 codecs is an adjustable bit rate. The bit rate determines the audio quality and the storage space necessary for the data. Typically, a 128 Kbps bit rate produces 15 kHz, CD-quality audio using a 12:1 compression ratio. An audio file for a 1-minute commercial is approximately 10 megabytes and compresses to about an 850-kilobyte (less than 1 megabyte) MP3 file. Although it results in a slightly larger file size, many producers use higher bit rates (of 192 or 256 Kbps) for a more accurate reproduction of the original audio.

Web Hosting and Emailing Audio Files. Two of the most popular ways for professionals to transfer MP3 files easily on the web are posting the MP3 file to a web page and attaching the file to an email. Websites work well for file retrieval. There are several ad-posting services that specialize in hosting commercial and program files for stations to download. Likewise, there are services like Dropbox that offer secure cloud storage, personal clouds, and easy file drop off and retrieval for individuals as well as businesses. Web distribution of a commercial provides a substantial cost savings over sending flash drives to individual stations. If necessary, the web page can be secured with a password to prevent unauthorized downloads.

An MP3 audio file can be attached to an email just like any other file. Emails can be addressed to one person or to a number of people for mass distribution.

Integrated Services Digital Network

ISDN technology is outdated and has been replaced with newer, and less expensive, AoIP technologies. However, you may still find a legacy ISDN codec being used within a group of stations simply because it works and the stations have not updated the technology.

An ISDN codec is capable of sending and receiving CD-quality audio with a frequency response of up to 20 kHz over a single ISDN telephone line. Telephone companies no longer install ISDN lines.

Analog Radio Transmission

HD Radio has taken the broadcasting industry in a new direction. Although the basics of radio transmission might not be interesting to everyone, they are important from the standpoint of how a traditional analog radio signal gets from point A to point B. It is the analog radio signal onto which the digital HD signal is piggy-backed. The following is a rather nontechnical discussion of the AM and FM transmission methods upon which HD Radio is based.

AM Transmission

Transmitting a radio signal involves attaching, or superimposing, an audio signal onto a much-higher frequency electromagnetic wave, called a radio-frequency **carrier wave**. Although very few people realize it, the carrier wave is the frequency or dial position of a radio station. The process of superimposing audio onto the carrier wave is called **modulation**. There are two common forms of modulation or attaching an audio signal onto a radio wave: AM and FM.

AM radio was the first method of transmitting voice signals, and was developed over 100 years ago. AM is the abbreviation for amplitude modulation. If you remember from chapter 2, the term amplitude refers to the height of a sine wave. The term modulation means change. Modulating, or changing, the amplitude of a radio-frequency carrier wave creates an AM radio signal.

The "standard" AM radio band in the United States is between 535 kHz and 1,605 kHz (see figure 13.2). The FCC assigned 540 kHz as the first radio station in the band; the last station is at 1,600 kHz. Stations are separated by 10 kHz intervals across the band. For example, the first station is at 540 kHz, the next is at 550 kHz,

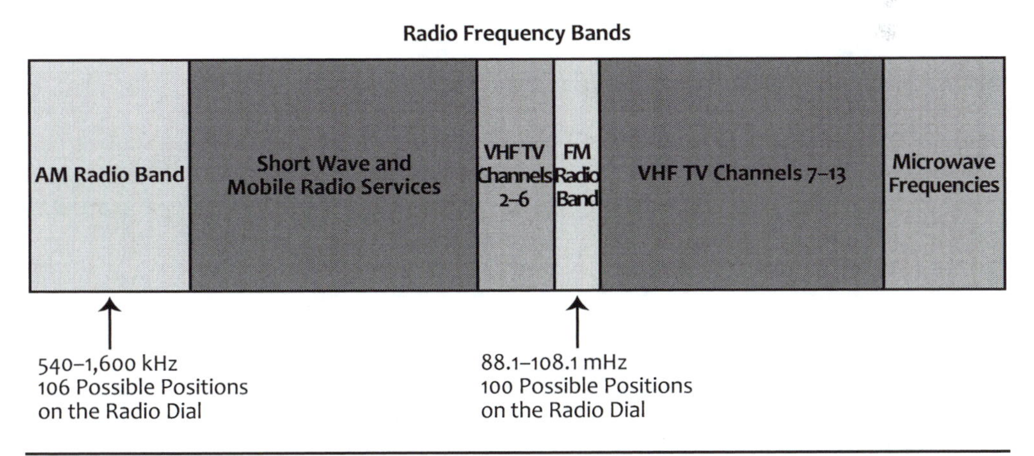

Figure 13.2 This chart shows how the AM and FM radio bands relate to other frequencies in the radio-frequency spectrum. A complete chart of all available radio frequencies is quite extensive and requires a large wall chart to display properly. Such charts are available for purchase from the FCC.

then 560 kHz, and so forth. The FCC coordinates and assigns all of the frequencies to avoid interference.

In addition to the standard AM broadcast band, the FCC has assigned the extreme upper and lower areas of the AM band for specific purposes. On the low end of the band, 530 kHz is assigned to travelers' information services. On the high end of the band, 1,590 through 1,700 kHz are assigned to a class of regional stations sometimes used for traffic or emergency information.

An AM transmitter generates a high-power, radio-frequency carrier wave on the channel, or frequency, assigned by the FCC, such as 740 kHz. The carrier wave is so named because it is used to carry the audio to the radio receiver. A silent radio carrier is unmodulated, or unchanged, from its original state.

To broadcast audio on an AM station, the audio signal from the station's control board is sent to the modulator in the transmitter. The modulator uses the rapid changes of the audio signal's amplitude to modulate, or change, the amplitude of the high-power carrier wave. The modulator superimposes the audio onto the powerful carrier wave that is sent to the station's transmitting antenna. The amplitude of the carrier wave becomes identical to, and tracks with, the amplitude of the original audio signal (see figure 13.3).

When the AM radio signal arrives at the receiver, the radio's circuitry detects the carrier wave and demodulates the signal. The demodulation process separates the audio from the carrier and sends the audio to the radio's amplifier and speaker.

AM radio broadcasts with a limited audio-frequency response because of the narrow **bandwidth** assigned to each station by the FCC. AM radio is limited to

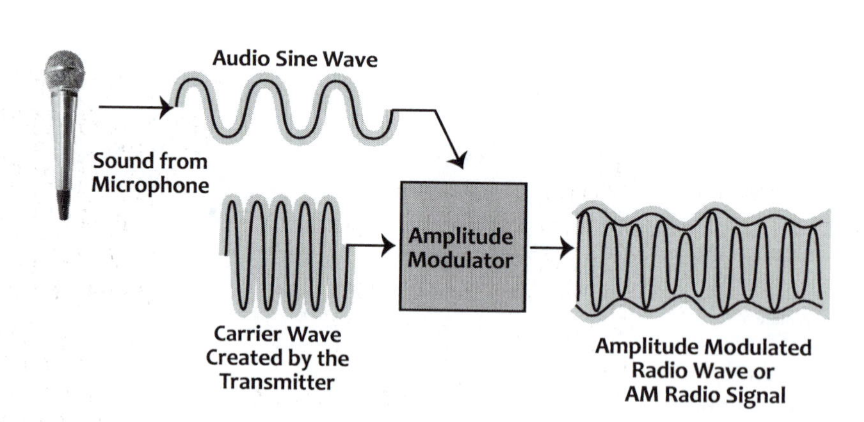

Figure 13.3 AM radio uses the image of the audio sine wave to modulate, or vary, the amplitude (height) of the AM carrier wave. When the AM radio signal arrives at the receiver, the radio's circuitry detects the carrier wave and demodulates the signal. The demodulation process separates the audio from the constantly changing carrier wave and sends the audio to the radio's amplifier and speaker.

transmitting the audio frequencies between 50 Hz and 7.5 kHz, or about half of our hearing range. Because of this limited audio-frequency response, AM does not have the bright, open sound of FM. Another drawback to the radio frequencies in the AM band is that they are subject to atmospheric noise and static from a number of sources, especially lightning.

Typically, an AM radio station broadcasts a single channel of audio (monaural). Although a number of methods of broadcasting AM stereo were developed in the early 1990s, broadcasters and receiver manufacturers never agreed upon an industry standard.

An AM radio station antenna is the radio tower, or a set of vertical cables mounted just off the face of the tower and running parallel to it. AM radio station towers are energized with high voltage and are dangerous. The tower is insulated at its base and where the guy wires connect to the tower. *NEVER* go inside the safety fence at the base of an AM tower when it is energized, unless the station engineer escorts you. Under some conditions, high levels of radio-frequency power can be encountered near the tower and can be dangerous to your health. The technical term for radio-frequency energy is nonionizing radiation. In simplest terms, it's like walking into a microwave oven. Observe all safety precautions and warning signs (see figure 13.4).

Classes of AM Stations. AM radio stations are classified by the FCC into four power and frequency classifications, labeled Class A through D, on local, regional, and clear channels. Such classes allow a maximum number of stations to serve the country without interfering with one another.

Local channels are assigned by the FCC to stations that are intended to serve their community of license and a limited area around it. Local channels are limited to a maximum power of 1,000 watts. Local stations may remain on at full power during the nighttime hours.

Figure 13.4 Although you cannot see or feel radio waves, under some conditions, high levels of radio-frequency power can be encountered near a radio tower. Such high levels of power can be hazardous to your health and you should obey all safety signs.

Regional channels are assigned by the FCC to stations that are intended to serve larger geographic areas. A regional channel operates with higher transmitter power both during the day and at night, provided the station uses a directional antenna to aim the signal in a specific direction to prevent interference with other stations.

Clear channels are assigned to only a few, select radio stations that the FCC classes as dominant. A typical clear-channel station has a maximum power of 50 kilowatts (50,000 watts) and, using an omnidirectional antenna, provides nighttime service to a 750-mile radius from the station. The maximum number of clear-channel stations assigned to a single frequency is two. In such cases, the two stations are located great distances from one another to prevent interference. Examples of clear-channel stations include WCBS (New York), WGN (Chicago), and KFI (Los Angeles).

Groundwave Propagation. An AM radio station signal can travel much farther than an FM signal because of the radio frequencies and antenna designs involved. An AM radio signal travels over the earth's surface primarily by conduction. The technical term for the signal's ground-hugging ability is **groundwave propagation**. Daytime AM signal coverage is usually limited by the FCC to 100 miles for the most powerful stations.

Skywave Propagation. As the sun starts to go down, AM radio signals begin to travel great distances, thanks to a phenomenon called **skywave propagation**. This occurs as the long wavelengths of AM signals interact with the radically shifting layers of the ionosphere several miles above the earth's surface. The phenomenon starts at sunset and continues through the night. AM signals often travel hundreds, or thousands, of miles as the signals travel up to the ionosphere and are reflected back down to earth. Because of this phenomenon, some stations have to reduce power, change the direction they transmit in, or go off the air to prevent interference with other stations.

FM Transmission

Edwin Armstrong developed FM radio in the early 1930s. Due to a series of patent fights with RCA (which Armstrong eventually won), the implementation of FM broadcasting was delayed at least 20 years.

FM is the abbreviation for frequency modulation. Rather than modulate the amplitude of the radio carrier wave, as in AM broadcasting, FM modulates, or changes, the frequency of the carrier wave. An FM transmitter generates a high power, radio-frequency carrier wave (the carrier-wave frequency is the channel assigned by the FCC, such as 95.3 mHz). When looked at on an oscilloscope, the unmodulated sine waves of the radio carrier appear identical in amplitude and frequency (see figure 13.5).

The FM radio band in the United States includes the frequencies from 88 mHz to 108 mHz and is located between channels 6 and 7 in the television broadcast band (again, see figure 13.2). The FCC assigned 88.1 mHz as the first radio station in the band; the last station is at 107.9 mHz. Stations are separated by 200 kHz

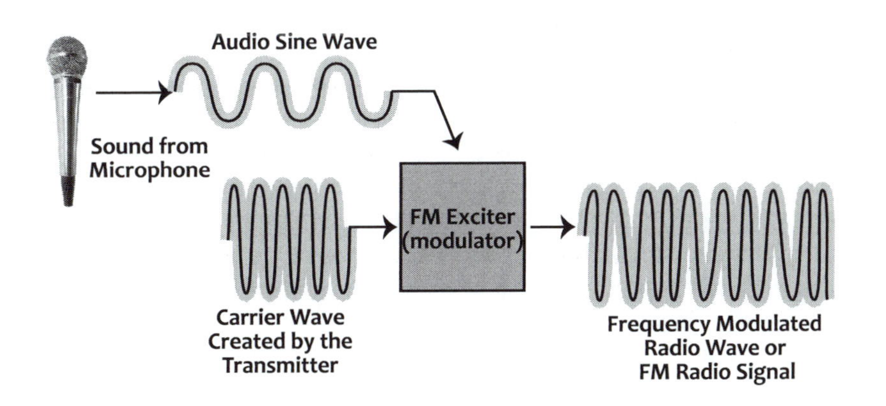

Figure 13.5 The FM exciter uses the image of the audio sine wave to modulate, or vary, the frequency of the FM carrier wave. When the FM radio signal arrives at the receiver, the radio's circuitry detects the carrier wave and demodulates the signal. The demodulation process separates the audio from the constantly changing carrier wave and sends the audio to the radio's amplifier and speakers.

intervals across the band. For example, the first station is at 88.1 mHz, the next is at 88.3 mHz, then 88.5 mHz, and so forth. There is room for 100 channels in the FM band. The FCC assigns the channels to radio stations based on location and power to prevent interference with other stations.

To broadcast audio on an FM station, the audio signal from the station's control board is sent to the modulator or, as it is often referred to, the FM exciter. The modulator uses the variations in the amplitude of the audio signal to modulate the frequency of the high-power carrier wave being sent to the station's transmitting antenna. As the audio's amplitude varies up and down, the modulator tracks it and changes the frequency of the carrier wave slightly above or below the station's carrier frequency. The maximum change permitted by the FCC in the carrier frequency is 75 kHz above or below the carrier.

When the FM radio signal arrives at the receiver, the radio's circuitry detects the carrier wave and demodulates the signal. The demodulation process separates the audio from the carrier and sends the audio to the radio's amplifier and speakers.

Unlike AM radio stations that are spaced closely together on the band, FM radio stations are spaced much farther apart (200 kHz). The wider separation allows FM signals to have a wider bandwidth and thus transmit a wider audio frequency range. FM radio can transmit audio frequencies from 50 Hz to 15 kHz, or very close to the average person's complete hearing range. To deliver the additional audio information to a radio receiver for stereo FM (i.e., transmitting two audio signals, left and right channel), a sub carrier is attached to the main radio carrier wave.

Stereo FM broadcasts are not subject to the electrical interference and static experienced by AM broadcasts. FM **multipath distortion** is a problem that plagues FM broadcasts in mountainous regions or large cities with tall buildings. Unlike the AM signal, the FM signal does not have a groundwave that follows the contours of the earth. Instead, an FM signal travels in a straight line until it hits something, like a tall building or the side of a mountain, and is reflected in another direction. Multipath distortion typically occurs in a car radio as the original signal and reflected signals arrive at the car radio at slightly different times. The radio has difficulty receiving the various reflected signals and the result is an undesirable distortion of the audio's high frequencies. If the car is moving, multipath distortion can cause the radio signal to fade in and out rapidly.

An FM signal is transmitted from an antenna mounted at, or near, the top of a tower (see figure 13.6). The tower is simply a supporting structure for the antenna. However, you should *NEVER* climb an FM tower while the antenna is energized. High levels of radio-frequency power (nonionizing radiation) can be encountered that are dangerous to your health. Due to the higher frequencies involved in FM, it can be even more dangerous than the nonionizing radiation given off by an AM tower. Tower climbing is dangerous. Observe all safety precautions and warning signs when near any FM tower.

Classes of FM Stations. In contrast to an AM signal, which follows the curvature of the earth, an FM signal simply travels in a straight line from the antenna off into space. The earth's surface curves out of sight about every 90 miles, and so

Figure 13.6 An FM antenna is mounted to the side of a tower with the tower serving only as a supporting structure. The two antenna elements pictured here are part of an antenna that can have several elements, depending on the power of the station.

the line-of-sight FM signal path effectively limits the range of a full-power FM station to approximately 60 to 90 miles, depending on terrain.

An FM station operates at its FCC-assigned power 24 hours a day. The FCC has established eight different classes of FM stations, based on antenna **height above average terrain** (**HAAT**) and the **effective radiated power** (**ERP**) needed to cover a specific geographic or service area. The specific classification of a station is based on a set of complex calculations designed to deliver a specific power level of radio signal at a specific distance from the station's antenna. Because FM signals travel by line of sight, the higher the antenna, the further the signal will travel. The combination of antenna height and power determines how far an FM signal will travel.

As an example, a Class A FM station operates with 6,000 watts of power and an antenna height of up to 100 meters (330 feet). At the other end of the spectrum, a Class C FM station operates with 100,000 watts of power and an antenna height of up to 600 meters (1,980 feet). Sometimes geography works in a station's favor. For example, a radio station placed its FM transmitter on top of a 5,000-foot-tall mountain. Due to the extreme height, the station was required by the FCC to reduce its power to only 200 watts. Although 200 watts of power is not at all impressive, the height of the antenna more than compensated for the low power. Since there is nothing to block the view of the antenna for many miles, the line-of-sight coverage for the station is excellent.

AM and FM HD Radio Technology

The analog technology used in AM and FM broadcasting remained essentially unchanged for more than 50 years. With the rapid development of digital audio it was only a matter of time before a digital transmission system was developed to deliver high-quality digital AM and FM broadcast signals. In late 2002, iBiquity Digital Corporation received approval from the FCC for the implementation of their system, called HD Radio.

An HD Radio broadcast signal uses an AM or FM radio station's existing FCC-assigned frequency and bandwidth. The system offers a number of advantages to broadcasters and consumers alike, the first of these being greatly improved audio quality. HD AM sounds like present day FM and HD FM sounds like a CD.

A second major advantage of HD Radio digital transmission is backward and forward compatibility using present AM and FM radio station frequencies. A radio station upgrading to HD Radio technology continues to broadcast its existing analog signal for as long as necessary. Consumers can continue to listen to their favorite stations until they choose to buy an HD radio. The analog to digital changeover is similar to when television made the transition from black and white to color.

HD Radio transmission opened the door not only to high-quality audio, but also to a whole new range of digital information and entertainment services that stations can offer. Sophisticated data-compression techniques further enhance this ability, allowing more data to be transmitted within the original bandwidth of an AM or FM station.

Such services include full color information displays on the radio, providing listeners station information, song titles, album photos, etc. Further, HD radio has the capability to allow listeners to use iTunes tagging for artists and songs for later purchase and to "live pause" the program and resume listening. A station can provide real-time updates for weather, traffic, sports, news, and emergency information. Along with HD radio broadcasts, all of the additional features are free and there's no data usage fees associated with HD radio.

The HD Radio Transmission System

AM and FM HD Radio transmission systems are very much alike. Both systems simultaneously transmit an analog and digital audio signal as well as data services. HD radio is also content rich. In addition to the main channel, an HD station has three other channels to transmit programming or data on.

To create an HD Radio signal, a station's analog transmitter is equipped with a device to combine and piggyback the analog audio, digital audio, and data signal onto the radio station's frequency. In the simplest of terms, a "hybrid" analog/digital signal is created. At some point in the future, when most people have converted to digital radios, a station may choose to discontinue transmitting the analog signal. To go deeper into the complex technical aspects of the HD Radio signal transmission is beyond the scope of this text.

The average cost to upgrade a radio station to an HD Radio hybrid analog/digital signal is approximately $75,000 to $300,000. Older transmitters and equipment must be replaced, since they are not capable of transmitting the sophisticated analog/digital hybrid signal.

There is no mandatory implementation date for HD Radio technology; the changeover is occurring gradually as each station sees the benefits to be gained from the conversion to HD Radio broadcasts. HD radios can now be purchased in electronics stores, or online, and all 33 US automakers now offer HD radios as either standard or optional equipment. However, the technology used to transmit the signal is like a moving target—it is likely going to change in the future as software and hardware advance with regard to digital transmission and desired feature sets.

HD Radio System Benefits

There are a number of major benefits to be gained by stations converting to the HD Radio digital-transmission system; chief among them is audio quality. Traditional analog AM and FM audio quality is dependent on the bandwidth of the radio signal. Digital signals do not require as much bandwidth as do present analog signals, so it is easier to transmit a high-quality digital audio signal.

HD Radio technology provides FM-like audio quality for AM, giving the once nearly forgotten band a new life. An additional benefit for AM, thanks to new sophisticated radios, is that the static and noise that were once common on the AM band are eliminated. Likewise, FM HD Radio signals provide improved, CD-like quality.

HD radios use digital processors and powerful algorithms to provide improved reception by electronically sorting through all the incoming signals. By intelligently selecting the most powerful signals, noise, static, and distortion can be eliminated under most conditions.

Competitive Benefits of HD Radio Technology. There are at least five major competitive benefits to be gained when a station switches to HD Radio technology. First, the station gains the technical improvements in the station's audio quality and the marketing strategy that labels the station as high-definition. Second, the ability to transmit data services offers opportunities to increase radio station income from new, nontraditional revenue sources. Third, intelligent receivers improve reception and eliminate the common problems of static, noise, and distortion, possibly improving and increasing audience ratings. Fourth, HD Radio technology offers growth opportunities for broadcasters, retailers, automobile dealers, and aftermarket stereo shops to partner and sell new radio products using radio advertising. Fifth, the improved digital audio quality allows land-based radio stations to compete effectively for listeners against satellite radio companies whose big selling point is digital quality.

Software-Defined Radios

Software-defined radios (**SDR**), although related to HD Radio, are standalone products designed to receive either analog or digital radio signals.

In a traditional radio design, the received radio signal is routed through a number of different electrical circuits before you ever hear it. The design of the radio's internal components is extremely critical, since the components process the high-frequency radio waves. On the other hand, software-defined radios, rather than being filled with a number of different circuits and components as traditional radios are, have a tiny microchip. Essentially, an SDR is a small computer designed to decode radio signals with incredible accuracy. An SDR receives a radio signal and immediately converts the signal into digital data. The chip's software processes the data and extracts the usable radio signals, eliminating noise, static, and interference from other radio stations.

Because the radio is designed around software, it can be included in many computer-based devices that you would not normally think of as having a radio, such as your smart phone. **NextRadio®** is an app that takes advantage of the FM receiver chip that is in all current smart phones. Just like a regular radio, the chip receives the FM broadcast signals and the NextRadio® app integrates them into your smart phone. Because the FM chip is a radio and is receiving broadcast signals there are no data use fees and no Internet access costs to worry about. An additional benefit to listening to the radio on a smart phone over listening to streamed music is that the radio uses three times *less* of the smart phone's battery power than streaming music does.

A listener can choose to add a number of features such as album graphics, the ability to instantly purchase a song, add things to a favorites list, enter radio station contests, and buy concert tickets. The additional feature-rich content uses the phone's data plan to make radio interactive. The additional features use about 20 times *less* data than streaming music. Most major Android cellular devices have activated the FM feature in their smart phones, and most carriers support the service.

HD Radio Production, a New Dimension

HD Radio production presents a whole new set of challenges, particularly to AM radio. For the first time, an FM listener has CD-like quality and an AM listener has near FM-like quality. With new HD radios and high-quality, factory auto sound systems, the listener is able to hear with clarity what he or she has been missing up until now. For the first time, the listener can hear what you hear in the studio. Suddenly, it is more important than ever to be aware of audio quality.

The Challenges of HD Radio Production

For FM broadcasters, the leap to HD Radio technology is not that severe if they have a modern digital facility. However, if ever there were a case to apply the old computer adage "Garbage in equals garbage out," this is it. Heavy audio processing, such as compression, limiting, and sloppy equalization, is not only going to be noticed, but also perceived as annoying. Poor audio quality becomes a valid reason for a listener to tune out.

For AM radio, the change to HD Radio technology is much more dramatic. In some cases, stations have to be completely redesigned and reequipped for digital audio. Keep in mind that for over 80 years, AM radio was limited to broadcasting only those audio frequencies between 50 Hz and 7,500 kHz, or about half the human hearing range. For years, AM broadcasters didn't really worry too much about full-frequency audio quality. It wasn't necessary. Often, the mantra regarding AM audio quality was "Don't worry, it's just AM." HD Radio technology provides AM with a dramatic change in audio quality, an entirely new sound, and one that gives AM stations a chance to be more competitive with FM stations.

HD Radio technology is a moving target. As the technology continues to develop, stations will have to improve their audio quality to keep up with listener demand.

Satellite Radio Services

Digital satellite radio services are an alternative to **terrestrial-based** AM and FM radio stations. There are several major differences between satellite radio services and terrestrial-based radio stations, the primary one being that satellite services charge a monthly subscription fee. There is one major satellite radio service provider: SiriusXM Radio.

The Satellite Radio Transmission System

The satellite transmission system starts with a typical on-air studio. Instead of feeding an AM or FM transmitter, though, the studio feeds a satellite modulator. The modulator generates the ultrahigh-frequency signal (gigahertz) that transmits the audio to a series of satellites in **geosynchronous orbit** (the satellites remain in a stationary position with respect to the earth) 22,000 miles above the United States. The satellites retransmit the digital signals back down to earth in a broad **footprint**, or coverage area.

To receive a typical television satellite signal, a reflective dish is required to pick up and focus the weak signal. Satellite radio services use very high-power transmitters so that only a small antenna, which can easily be mounted on a car or in a portable radio, is required. Radios capable of receiving satellite radio services can also receive terrestrial AM and FM signals. Although we typically think of satellite-based radio as being in an automobile, there are also portable satellite radios and units designed for the home or business.

A question that often brings up concerns is what happens when a satellite radio-equipped car goes under a bridge or into a tunnel: Can the car's antenna "see" the satellite (receive the signal)? This is actually not a problem, because satellite radios receive the digital signal and store several seconds of it in a storage or buffer area of the radio before playing it back. When a car passes under a bridge, the radio continues to play the signal stored in the buffer and then catches up to the satellite once the car emerges from under the bridge. This is done seamlessly, and the listener is not aware of what is happening. In long tunnels or parking garages, loss of signal is possible, in which case the listener can switch to terrestrial AM/FM signals. However, in a tunnel or under a large mass of concrete and steel, such as a bridge, even terrestrial-based signals may not come in well.

Competitive Benefits of Satellite Radio

Satellite radio features over 175 channels of programming, including over 70 music channels mixed with numerous channels of news, talk, and information programming depending on which subscription package the listener has chosen.

When originally conceived, there was great debate about how fast satellite radio would outdate terrestrial-based AM and FM radio stations. The reality has turned out to be that they are two different broadcast services serving two different needs. Each service has a number of advantages and disadvantages. The primary disadvantage of satellite radio is that listeners must pay for something that traditionally, for over 100 years, has been free.

Satellite-based radio offers the distinct advantage of being able to broadcast to the entire continental United States. From the listeners' perspective, this allows them to enjoy the same channel as they travel, no matter where they go. For people who travel a lot, like truck drivers, the service has value. Because satellite radio services are national radio services they are not capable of offering the detailed local programming (such as local personalities, news, and sports) that listeners often enjoy during morning- and afternoon-drive times. To overcome this disadvantage, satellite providers provide weather and traffic report channels in a number of major metropolitan areas.

Satellite services were the first to broadcast high-fidelity digital signals. This is a big selling point of satellite radio. However, with the advent of HD Radio transmission systems, terrestrial AM and FM stations are now competitive. Prices for satellite radios are comparable to prices for standard AM/FM stereo radios. In addition, car manufacturers include satellite-equipped radios in many models as a standard feature.

As to whether satellite or terrestrial radio is "better," the answer to that question ultimately lies with the listener.

■ Suggested Activities

1. Plan an evening to experience AM skywave propagation. Wait until after the sun goes down and, using either a table-model AM radio with an antenna or your car radio, slowly tune the AM band and see how many stations you can identify. Hold a contest to see who can log a station from the longest distance.
2. Contact the director of engineering for a local radio station and arrange an engineering tour of both an AM and FM radio station. Find out how each is handling the changeover from an analog to digital transmission system in its physical plant.
3. Download a free shareware copy of an FTP server-client program. After loading it on two computers, try exchanging an audio file between the two.
4. Using the MP3 codec that comes with Adobe Audition CC, practice converting audio files to the MP3 format and playing them back again. With your personal web page (or one that you have permission to use), practice posting an audio file on the website and downloading it.
5. Create a one-minute audio file, then compress it using an MP3 codec. Attach the compressed file to an email and send it to a friend.

■ Websites for More Information

For more information on:

- *HD Radio*, visit DTS at www.hdradio.com
- *MPEG Audio Layer-2 and MPEG Audio Layer-3*, visit Fraunhofer IIS at www.iis.fraunhofer.de
- *NextRadio®*, visit www.nextradioapp.com
- *satellite-based radio services*, visit SiriusXM Satellite Radio at www.siriusxm.com
- *skywave and groundwave propagation*, visit the Federal Communications Commission at www.fcc.gov

Pro Speak

You should be able to use these terms in your everyday conversations with other professionals.

AM—Amplitude modulation. The AM radio band extends from 535 kHz to 1,605 kHz.

bandwidth—With reference to a radio carrier wave, bandwidth is the space on the radio band occupied by the radio carrier wave frequency and the audio that has been superimposed upon it. The superimposed audio signal widens the radio carrier by adding sidebands just above and below the assigned carrier frequency.

The carrier wave and the upper and lower sidebands of the signal compose the bandwidth of a signal.

carrier wave—A high-frequency radio wave transmitted by a radio station that has been amplitude or frequency modulated.

effective radiated power (ERP)—When applied to an FM station, the term specifies in watts how much power the station can use to transmit its signal, as measured at the antenna.

FM—Frequency modulation. The FM radio band extends from 88 mHz to 108 mHz.

footprint—A satellite signal's geographic coverage area on the earth.

geosynchronous orbit—An orbit that places a satellite approximately 22,000 miles above the earth so that it rotates at the same speed as the earth, maintaining a fixed position in space.

groundwave propagation—The property of physics that allows amplitude-modulated radio carrier signals to follow the earth's surface through conduction.

height above average terrain (HAAT)—Applied to an FM radio station, the term specifies in meters the height of the station's transmitting antenna above the average terrain that surrounds it.

modulation—The process of changing a silent carrier with either amplitude or frequency modulation.

multipath distortion—Multipath distortion occurs in a car radio as an FM signal and a reflected FM signal of the same frequency arrive at the radio at slightly different times. The radio has difficulty receiving the two signals and the result is an undesirable distortion of the audio's high frequencies. Multipath distortion can cause the radio signal to fade in and out rapidly.

NextRadio®—An app that takes advantage of the FM receiver chip that is in all current smart phones. Just like a regular radio, the chip receives the FM broadcast signals and the NextRadio® app integrates them into the smart phone.

skywave propagation—A phenomenon that occurs as the long wavelengths of AM signals interact with the radically shifting layers of the ionosphere several miles above the earth's surface at night, causing AM signals to travel great distances.

software-defined radio (SDR)—A radio that converts the incoming radio signal into a digital data stream and then processes the data to extract the usable signals.

T-1—The technical designation for a direct-leased Internet connection that is capable of transferring 1.544 Mbps.

T-3—The technical designation for a direct-leased Internet connection that is capable of transferring 44.736 Mbps.

terrestrial-based—Fancy term for earth-based.

14

Programming, Production, and Measuring Success

HIGHLIGHTS

Radio is under attack—it has been from the very start over 100 years ago. As we have discussed repeatedly in the previous chapters, it continues to survive thanks to its adaptability. In case you hadn't noticed, the word digital changed the world. Each new entertainment and leisure technology, and the convergence it brings, is a challenge to radio, an opportunity to steal away some of radio's audience.

Radio has to consider that every 18- to 34-year-old takes extreme pride in being a "millennial." According to numerous studies the two most important things to this age demographic is a smart phone/mobile device and broadband Internet access. This demographic is always connected to something and it frequently multitasks. People in this group often identify themselves and pick friends and social groups based on the common shared belief in a particular technology. For many, the "mobile device" has become a status symbol, fashion statement, and an expression of their own personality. The device is a lifeline to others, a social networking hub, a photo album, etc. As this generation ages their technology will change and move with them and they will be replaced by another younger generation.

For several years the assumption was that traditional radio and television would suffer at the hands of the millennials. However, research indicates that those age 12+ are not picking one medium over another. In fact, they are increasing their daily time spent with radio, television, and the web. Since 2001 the daily use of these mediums has jumped over 20 percent, with the average consumer spending nearly nine hours a day with radio, television, and the web.

Although new technology is impressive, one thing remains clear: it is not digital technology we listen to; what we listen to are the people and the programming that *use* the technology to deliver their messages. To succeed with new generations, radio must embrace the technology and honor its roots of presenting entertaining and informative personalities and delivering real-time interaction with listeners, something an MP3 player or streaming music cannot do. And although satellite-delivered radio signals are a part of the marketplace, they are not local and cannot interact with local listeners.

That is where you come in. As a production person, you are not only going to be involved in production, you are going to be a part of the team of people that has to develop the next generation of radio to appeal to young people, whether it be on air, online, or on demand. The kind of personality-based entertainment radio that is going to dominate the medium in the next few years requires lots of production. As a part of this transformation, you are going to get to help create, develop, and carry out your station's program format.

It is important for you to know that programming a radio station is not based on someone's hunches, guesses, or feelings. Programming is based upon a mix of quantitative and qualitative research, including tools like **psychographic** or lifestyle research. Programming involves talented individuals who are capable of interpreting the data, developing a unique format, and selecting, motivating, and coaching on-air talent. It is not an easy job.

The success or failure of a radio station in a rated market is measured by its Nielsen Audio audience ratings. Businesses, ad agencies, and competitors use these ratings as a benchmark to judge your station's success by. This is not necessarily so in smaller markets where the measure of success comes directly from the experiences of the business community. If Bob's Hardware bought ads on your station and it sold all the widgets it was advertising, you have a great radio station.

It is important that you understand programming and ratings, as you will work closely with the program director, the promotion department, and sales on many projects. Producers get a lot of guidance and cues from these groups because the creative product they are asked to create has to fit within the station's programming, positioning, and branding concepts.

Nielsen Audio Ratings

Two words that strike fear into even the bravest of radio souls are "the book." The book refers to the results of the Nielsen Audio Ratings. The ratings are the one element of the business that no one in the radio station can control, except by doing his or her best possible job. It's not necessary for a production person to be able to explain all of the ratings data and do detailed audience-number breakouts. But, it is important that you understand the basic terms and what the ratings mean, because they will become a part of your daily working vocabulary.

Nielsen Audio conducts audience research several times a year (depending on market size). Typically, if the radio station is in one of the approximately 275 designated market areas, the station is rated 2 to 13 times a year, based on audience research conducted during designated 4- or 12-week survey periods. The survey data is collected either by a written diary that the survey participant keeps or by the portable people meter (PPM). A PPM is a small, cellular-based device that the survey participant wears or carries with them. The device has a very sensitive microphone and is constantly listening for inaudible coding embedded in each radio station's programming that identifies the radio station. The data is retrieved on a regular basis from the PPM. Eventually, the written diary will go away as stations adapt to PPM technology. At the end of each survey period, the ratings data is issued to the subscribing stations in ebook form.

The audience survey is a very detailed piece of research (see figure 14.1 on the following page). In most markets, the data it contains is a major factor in determining how well the radio station will be able to sell advertising. The data is not difficult to comprehend if you understand the three basic types of audience estimates being reported: Average Quarter-Hour, Cume, and Time Spent Listening.

Average Quarter-Hour

The first audience estimate, Average Quarter-Hour (AQH), is comprised of three individual rankings: AQH Persons, AQH Rating, and AQH Shares (again, see figure 14.1). AQH Persons is the estimated number of people listening to a particular radio station for at least five minutes during any quarter-hour period in a specific daypart. The estimate is broken down into Nielsen Audio's standardized demographic age groups and dayparts, including weekdays and weekends. Among the standard age groupings are persons 18–34, 18–49, 25–49, 25–54, and 35–64. The age groupings also are further broken down into men and women. The AQH Persons estimate helps advertisers determine how many potential people can hear their radio ad in any given 15-minute time period.

The AQH Rating is the estimated number of radio station listeners expressed as a percent of the total population of the measured survey area in any quarter-hour period in a specific daypart (again, see figure 14.1). Each percent is referred to as a rating point. To arrive at the AQH Rating, Nielsen Audio uses the following formula:

$$\frac{\text{AQH Persons}}{\text{Survey Area Population}} \times 100 = \text{AQH Rating}$$

This estimate is broken down into the standard demographic age groups and dayparts, including weekdays and weekends. The AQH Rating tells advertisers the esti-

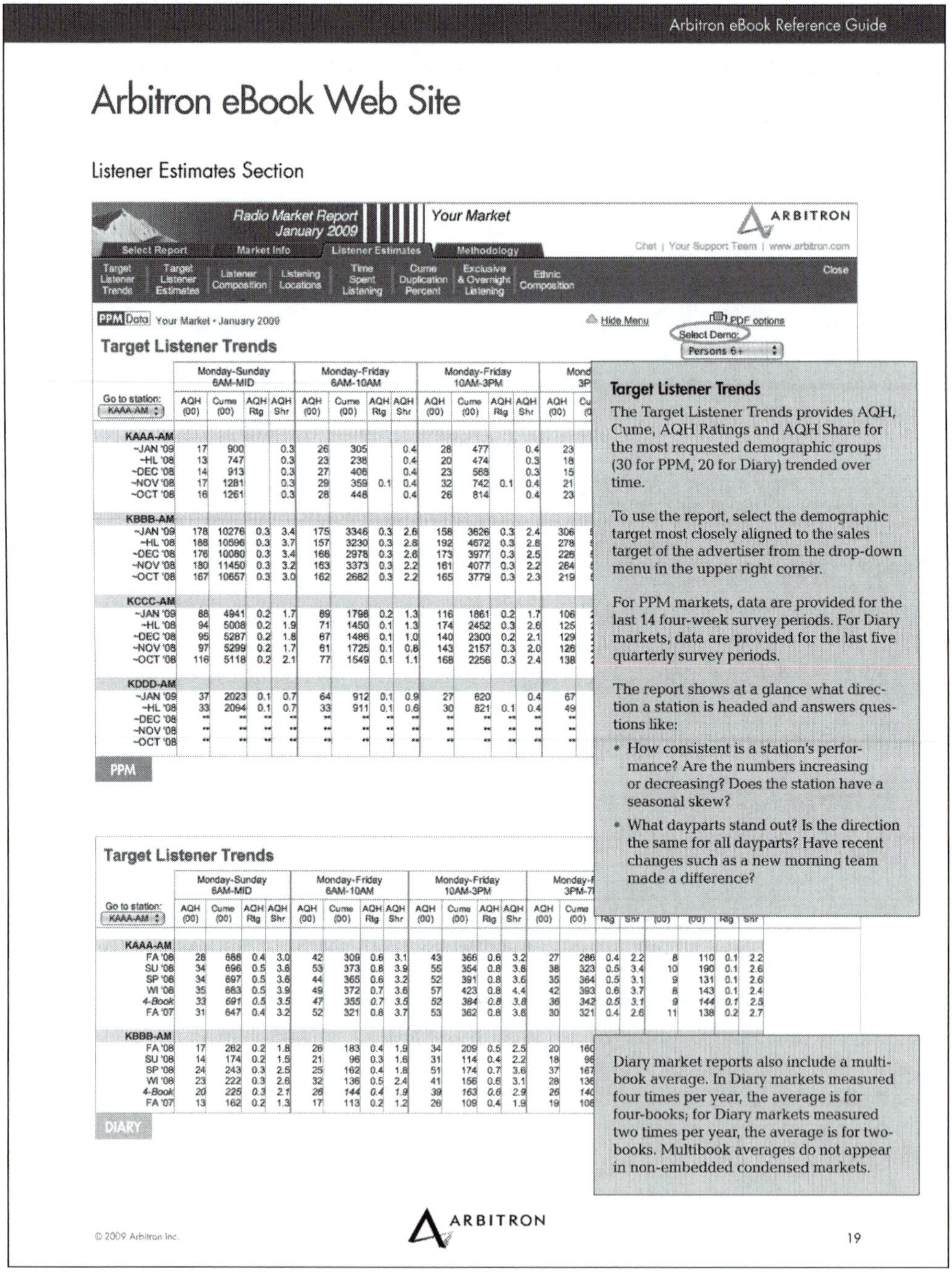

Arbitron eBook Web Site

Listener Estimates Section

Radio Market Report January 2009 | Your Market | ARBITRON

Chat | Your Support Team | www.arbitron.com

Select Report | Market Info | Listener Estimates | Methodology

Target Listener Trends | Target Listener Estimates | Listener Composition | Listening Locations | Time Spent Listening | Cume Duplication Percent | Exclusive & Overnight Listening | Ethnic Composition | Close

PPM Data Your Market • January 2009 — Hide Menu — PDF options — Select Demo: Persons 6+

Target Listener Trends

Go to station: KAAA-AM	Monday-Sunday 6AM-MID				Monday-Friday 6AM-10AM				Monday-Friday 10AM-3PM				Mond 3P
	AQH (00)	Cume (00)	AQH Rtg	AQH Shr	AQH (00)	Cume (00)	AQH Rtg	AQH Shr	AQH (00)	Cume (00)	AQH Rtg	AQH Shr	AQH (00)
KAAA-AM													
~JAN '09	17	900		0.3	26	305		0.4	28	477		0.4	23
~HL '08	13	747		0.3	23	238		0.4	20	474		0.3	18
~DEC '08	14	913		0.3	27	408		0.4	23	568		0.3	15
~NOV '08	17	1281		0.3	29	359	0.1	0.4	32	742	0.1	0.4	21
~OCT '08	16	1261		0.3	28	448		0.4	26	814		0.4	23
KBBB-AM													
~JAN '09	178	10276	0.3	3.4	175	3346	0.3	2.6	158	3626	0.3	2.4	306
~HL '08	188	10596	0.3	3.7	157	3230	0.3	2.8	192	4672	0.3	2.8	278
~DEC '08	176	10080	0.3	3.4	166	2978	0.3	2.6	173	3977	0.3	2.5	226
~NOV '08	180	11450	0.3	3.2	163	3373	0.3	2.2	161	4077	0.3	2.2	264
~OCT '08	167	10657	0.3	3.0	162	2682	0.3	2.2	165	3779	0.3	2.3	219
KCCC-AM													
~JAN '09	88	4941	0.2	1.7	89	1798	0.2	1.3	116	1861	0.2	1.7	106
~HL '08	94	5008	0.2	1.9	71	1450	0.1	1.3	174	2452	0.3	2.6	125
~DEC '08	95	5287	0.2	1.8	67	1486	0.1	1.0	140	2300	0.2	2.1	129
~NOV '08	97	5299	0.2	1.7	61	1725	0.1	0.8	143	2157	0.3	2.0	128
~OCT '08	116	5118	0.2	2.1	77	1549	0.1	1.1	168	2256	0.3	2.4	138
KDDD-AM													
~JAN '09	37	2023	0.1	0.7	64	912	0.1	0.9	27	620		0.4	67
~HL '08	33	2094	0.1	0.7	33	911	0.1	0.6	30	821	0.1	0.4	49
~DEC '08	**	**	**	**	**	**	**	**	**	**	**	**	**
~NOV '08	**	**	**	**	**	**	**	**	**	**	**	**	**
~OCT '08	**	**	**	**	**	**	**	**	**	**	**	**	**

PPM

Target Listener Trends

The Target Listener Trends provides AQH, Cume, AQH Ratings and AQH Share for the most requested demographic groups (30 for PPM, 20 for Diary) trended over time.

To use the report, select the demographic target most closely aligned to the sales target of the advertiser from the drop-down menu in the upper right corner.

For PPM markets, data are provided for the last 14 four-week survey periods. For Diary markets, data are provided for the last five quarterly survey periods.

The report shows at a glance what direction a station is headed and answers questions like:

- How consistent is a station's performance? Are the numbers increasing or decreasing? Does the station have a seasonal skew?
- What dayparts stand out? Is the direction the same for all dayparts? Have recent changes such as a new morning team made a difference?

Target Listener Trends

Go to station: KAAA-AM	Monday-Sunday 6AM-MID				Monday-Friday 6AM-10AM				Monday-Friday 10AM-3PM				Monday-F 3PM-7							
	AQH (00)	Cume (00)	AQH Rtg	AQH Shr	AQH (00)	Cume (00)	AQH Rtg	AQH Shr	AQH (00)	Cume (00)	AQH Rtg	AQH Shr	AQH (00)	Cume (00)	Rtg	Shr	(00)	(00)	Rtg	Shr
KAAA-AM																				
FA '08	28	688	0.4	3.0	42	309	0.6	3.1	43	366	0.6	3.2	27	286	0.4	2.2	8	110	0.1	2.2
SU '08	34	696	0.5	3.6	53	373	0.8	3.9	55	354	0.8	3.8	38	323	0.5	3.4	10	190	0.1	2.6
SP '08	34	697	0.5	3.6	44	365	0.6	3.2	52	391	0.8	3.6	35	364	0.5	3.1	9	131	0.1	2.6
WI '08	35	683	0.5	3.9	49	372	0.7	3.6	57	423	0.8	4.4	42	393	0.6	3.7	8	143	0.1	2.4
4-Book	*33*	*691*	*0.5*	*3.5*	*47*	*355*	*0.7*	*3.5*	*52*	*384*	*0.8*	*3.8*	*36*	*342*	*0.5*	*3.1*	*9*	*144*	*0.1*	*2.5*
FA '07	31	647	0.4	3.2	52	321	0.8	3.7	53	362	0.8	3.8	30	321	0.4	2.6	11	138	0.2	2.7
KBBB-AM																				
FA '08	17	262	0.2	1.8	26	183	0.4	1.9	34	209	0.5	2.5	20	160						
SU '08	14	174	0.2	1.5	21	96	0.3	1.6	31	114	0.4	2.2	18	96						
SP '08	24	243	0.3	2.5	25	162	0.4	1.8	51	174	0.7	3.6	37	167						
WI '08	23	222	0.3	2.6	32	136	0.5	2.4	41	156	0.6	3.1	28	136						
4-Book	*20*	*225*	*0.3*	*2.1*	*26*	*144*	*0.4*	*1.9*	*39*	*163*	*0.6*	*2.9*	*26*	*140*						
FA '07	13	162	0.2	1.3	17	113	0.2	1.2	26	109	0.4	1.9	19	106						

DIARY

Diary market reports also include a multi-book average. In Diary markets measured four times per year, the average is for four-books; for Diary markets measured two times per year, the average is for two-books. Multibook averages do not appear in non-embedded condensed markets.

ARBITRON

Figure 14.1 Although quite detailed in its presentation, the data contained in a Nielsen Audio survey is easy to comprehend providing you understand the three basic types of audience estimates. In many markets, Nielsen Audio ratings are a major factor in determining how well the radio station will be able to sell advertising. (Copyrighted information © 2016, of the Nielsen Company US, LLC licensed for use herein. On September 30, 2013, Arbitron Inc. became part of The Nielsen Company (US), LLC. Arbitron Inc. is now known as "Nielsen Audio, Inc.")

mated percent of the total population in the measured survey area that is listening to your radio station.

The AQH Share is the estimated number of radio station listeners expressed as a percentage of the people who were listening to the radio in any quarter-hour period in a specific daypart (again, see figure 14.1). The estimate is broken down into the standard demographic age groups and dayparts, including weekdays and weekends. The estimate tells advertisers what percent of the survey area radio audience is tuned into the radio station at any given time.

Cume

The second audience estimate, Cume, is comprised of two different rankings: Cume Persons, short for cumulative persons, and Cume Ratings (see figure 14.2 on the next page). The category Cume Persons is the total estimated number of *different* radio listeners who listened to the radio station for at least 5 minutes in any quarter-hour period in a specific daypart. No matter how long the listener stayed tuned to the station, the person is only counted *once.* This estimate is broken down into the standard demographic age groups and dayparts, as well as weekdays and weekends. (In addition, Nielsen Audio provides the Exclusive Cume Persons ranking, which is the estimated number of different listeners in the area who listened to *only one* radio station in a given daypart.)

Cume Persons estimates tell advertisers the total number of different people that listen to the radio station in a given time period, or put a different way, the total number of different people they could reach with their ads. A Cume Rating is also calculated by Nielsen Audio as a percent of the total survey population.

Time Spent Listening

The last category of the Nielsen Audio audience estimate is Time Spent Listening (TSL). This is the number of quarter-hours the *average* person spent listening to the radio station in a given daypart. Obviously, the radio station with the longest TSL has a very loyal audience. This estimate tells advertisers what the chances are of the same listener hearing their ad more than once in a given time period (see figure 14.3 on p. 377).

In addition to the three basic audience estimates, the Nielsen Audio data can be used to develop additional estimates and information valuable to the programming and sales departments. When it comes to audience ratings, there has always been a debate among broadcasters about survey methods and the number of valid survey participants needed to get accurate results. The PPM is a huge step forward toward incredibly accurate survey data collection because the PPM is in essence a neutral third party collecting the data. Although this topic can be endlessly debated from a scientific standpoint, the reality is that advertising agencies, public relations firms, media-buying services, and a radio station's competitors all depend on the Nielsen Audio ratings and use them as a benchmark by which to judge a station's performance.

Arbitron eBook Web Site

Listener Estimates Section *(continued)*

Radio Market Report January 2009 | Your Market | ARBITRON

Chat | Your Support Team | www.arbitron.com

Select Report | Market Info | Listener Estimates | Methodology

Target Listener Trends | Target Listener Estimates | Listener Composition | Listening Locations | Time Spent Listening | Cume Duplication Percent | Exclusive & Overnight Listening | Ethnic Composition | Close

PPM Data Your Market • January 2009 | Hide Menu | PDF options

Listener Cume Composition

Listener AQH Composition ▸ **Listener Cume Composition**

Monday-Sunday 6AM-MID, Cume Person (00)

Go to station: KAAA-AM	Persons 6+	Children 6-11	Persons 12+	Teens 12-17	Men 18+	Men 18-24	Men 25-34	Men 35-44	Men 45-54	Men 55-64	Men 65+
KAAA-AM	900		900		340	15			95	77	153
(%)	100		100		38	2			11	9	17
Rating	1.6		1.8		1.5	0.5			2.1	2.6	6.6
KBBB-AM	10276	770	9506	1341	4443	1175	1245	1106	544	189	183
(%)	100	7	93	13	43	11	12	11	5	2	2
Rating	18.0	13.4	18.6	24.7	19.5	38.6	25.0	22.1	12.2	6.3	7.9
KCCC-AM	4941	570	4371	376	2647	686	900	504	380	76	101
(%)	100	12	88	8	54	14	18	10	8	2	2
Rating	8.7	9.9	8.5	6.9	11.6	22.6	18.1	10.1	8.5	2.5	4.3
KDDD-AM	2023	208	1815	72	811		183	201	134	164	129
(%)	100	10	90	4	40		9	10	7	8	6
Rating	3.6	3.6	3.5	1.3	3.6		3.7	4.0	3.0	5.5	5.5
KEEE-AM	6285	343	5942	531	3632	356	796	1308	973	179	21
(%)	100	5	95	8	58	6	13	21	15	3	
Rating	11.0	6.0	11.6	9.8	15.9	11.7	16.0	26.1	21.8	6.0	0.9
KFFF-AM	9333	538	8795	1005	4756	989	1490	1187	815	177	99
(%)	100	6	94	11	51	11	16	13	9	2	1
Rating	16.4	9.4	17.2	18.5	20.9	32.5	29.9	23.7	18.3	5.9	4.2
KGGG-AM	14284	1014	13270	1461	5541	901	1501	1585	927	387	240
(%)	100	7	93	10	39	6	11	11	6	3	2
Rating	25.1	17.6	25.9	26.9	24.3	29.6	30.1	31.7	20.8	13.0	10.3
KHHH-AM	3157	456	2701	350	1673	245	712	963	170	120	61
(%)	100	14	86	11	53	8	23	12	5	4	2
Rating	5.5	7.9	5.3	6.4	7.3	6.1	14.3	7.3	3.8	4.0	2.6
KIII-AM	7094	506	6586	587	4319	802	1012	1489	789	187	38
(%)	100	7	93	8	61	11	14	21	11	3	1

PPM

Listener Cume Composition

Listener AQH Composition ▸ **Listener Cume C**

Monday-Sunday 6AM-MID, Cume Person (00)

Go to station: KAAA-AM	Persons 12+	Teens 12-17	Men 18+	Men 18-24	Men 25-34	Men 35-44	Men 45-54	Men 55-64	Men 65+	Women 18+
KAAA-AM	688	5	352	29	50	30	84	129	31	331
(%)	100	1	51	4	7	4	12	19	4	48
Rating	9.8	0.7	11.5	7.0	8.4	5.2	14.4	27.9	7.0	10.2
KBBB-AM	262	10	125			20	28	47	31	127
(%)	100	4	48			7	11	18	12	48
Rating	3.7	1.4	4.1			3.5	4.8	10.2	7.0	3.9

DIARY

Listener Composition (AQH and Cume) *(continued)*

Report Features

	PPM	Diary
Demos		
P6+	•	
Children 6-11	•	
Teens 12-17	•	•
P12+	•	•
Persons, Men and Women:		
18+	•	•
18-24	•	•
25-34	•	•
35-44	•	•
45-54	•	•
55-64	•	•
65+	•	•
Estimates		
AQH (00)	•	•
AQH Composition %	•	•
AQH Rating	•	•
AQH Share		
Cume (00)	•	•
Cume Composition %	•	•
Cume Rating	•	•
Daypart		
Mon-Sun 6AM-Mid	•	•

ARBITRON

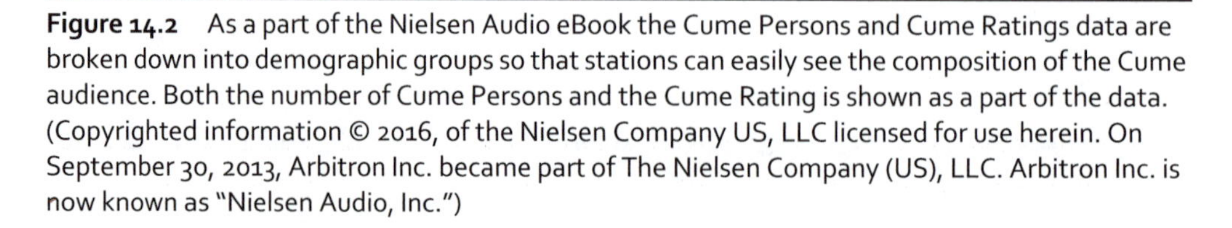

Figure 14.2 As a part of the Nielsen Audio eBook the Cume Persons and Cume Ratings data are broken down into demographic groups so that stations can easily see the composition of the Cume audience. Both the number of Cume Persons and the Cume Rating is shown as a part of the data. (Copyrighted information © 2016, of the Nielsen Company US, LLC licensed for use herein. On September 30, 2013, Arbitron Inc. became part of The Nielsen Company (US), LLC. Arbitron Inc. is now known as "Nielsen Audio, Inc.")

Arbitron eBook Reference Guide

Arbitron eBook Web Site

Listener Estimates Section *(continued)*

Radio Market Report January 2009 | Your Market | ARBITRON

Select Report | Market Info | Listener Estimates | Methodology

Chat | Your Support Team | www.arbitron.com

Target Listener Trends | Target Listener Estimates | Listener Composition | Listening Locations | Time Spent Listening | Cume Duplication Percent | Exclusive & Overnight Listening | Ethnic Composition | Close

PPM Data Your Market • January 2009 | Hide Menu | PDF options

Time Spent Listening

Go to station: KAAA-AM	Monday-Sunday 6AM-MID Hours and Minutes						
	Persons 6+	Persons 12+	Persons 18-34	Persons 25-54	Persons 35-64	Men 18-34	Men 25-54
KAAA-AM							
~JAN '09	2: 15	2: 15	1: 45	1: 15	1: 30	1: 00	1: 30
~HL '08	2: 15	2: 15	0: 30	1: 15	1: 30	0: 15	1: 30
~DEC '08	2: 00	2: 00	0: 15	1: 15	1: 15	0: 00	1: 30
~NOV '08	1: 45	1: 45	0: 15	1: 30	1: 15	0: 15	3: 00
~OCT '08	1: 30	1: 45	0: 30	1: 15	1: 30	0: 30	2: 00
KBBB-AM							
~JAN '09	2: 00	2: 15	2: 30	2: 00	2: 00	2: 45	1: 45
~HL '08	2: 00	2: 00	2: 00	1: 45	2: 00	2: 30	1: 45
~DEC '08	2: 00	2: 00	2: 00	2: 15	2: 30	2: 00	2: 00
~NOV '08	2: 00	2: 00	2: 15	2: 00	2: 00	2: 15	1: 45
~OCT '08	1: 45	1: 45	1: 45	2: 00	2: 15	1: 45	1: 45
KCCC-AM							
~JAN '09	2: 00	2: 15	2: 30	2: 30	2: 15	3: 00	2: 45
~HL '08	2: 15	2: 15	2: 15	2: 15	2: 30	2: 45	2: 30
~DEC '08	2: 15	2: 30	2: 45	2: 45	2: 30	3: 15	3: 00
~NOV '08	2: 15	2: 15	2: 15	2: 15	2: 45	2: 30	2: 00
~OCT '08	3: 00	3: 15	4: 00	2: 45	2: 30	4: 15	2: 30
KDDD-AM							
~JAN '09	2: 30	2: 30	2: 00	2: 45	3: 00	1: 15	1: 30
~HL '08	2: 15	2: 15	1: 00	2: 00	2: 00	1: 15	1: 30
~DEC '08	**	**	**	**	**	**	**
~NOV '08	**	**	**	**	**	**	**
~OCT '08	**	**	**	**	**	**	**
KEEE-AM							
~JAN '09	1: 15	1: 15	0: 45	1: 30	1: 45	0: 45	1: 30
~HL '08	1: 00	1: 00	1: 00	1: 00	1: 00	1: 15	1: 15
~DEC '08	1: 15	1: 15	1: 00	1: 30	1: 30	1: 15	1: 45
~NOV '08	1: 15	1: 15	1: 00	1: 30	1: 30	1: 00	1: 30
~OCT '08	1: 30	1: 30	0: 45	1: 45	2: 00	0: 45	2: 15
KFFF-AM							
~JAN '09	1: 45	1: 45	2: 00	1: 45	1: 45	2: 15	1: 45
~HL '08	1: 45	2: 00	2: 30	2: 00	1: 30	2: 15	1: 45

PPM

Time Spent Listening

The Time Spent Listening (TSL) report is how long listeners spend with a radio station in a week. In a typical scenario, the demographic with the highest time spent listening for a station should match the station's stated target audience.

For advertisers, TSL provides valuable insight on the connection listeners make with a station. TSL is also an essential tool for making programming decisions.

For PPM markets, TSL is reported for 11 demographic targets; for Diary markets, TSL is reported for 10 demographic targets. Diary markets also include a two- or four-book average.

Report Features

	PPM	Diary
Demo		
P6+	•	
P12+	•	•
Persons, Men and Women:		
18-34	•	•
25-54	•	•
35-64	•	•
Estimate		
Time Spent Listening	•	•
Daypart		
Mon-Sun 6AM-Mid	•	•

Time Spent Listening

Go to station: KAAA-AM	Monday-Sunday 6AM-MID Hours and Minutes					
	Persons 12+	Persons 18-34	Persons 25-54	Persons 35-64	Men 18-34	Men 25-54
KAAA-AM						
FA '08	5: 00	2: 00	5: 30	5: 30	1: 30	5: 30
SU '08	6: 15	2: 45	7: 30	7: 15	2: 30	7: 15
SP '08	6: 00	6: 30	6: 30	6: 00	2: 30	5: 00
WI '08	6: 30	1: 45	6: 30	6: 30	1: 45	7: 00
4-Book	*6: 00*	*3: 15*	*6: 30*	*6: 15*	*2: 00*	*6: 15*
FA '07	6: 00	2: 45	4: 45	7: 00	3: 30	4: 30
KBBB-AM						

DIARY

ARBITRON

27

Figure 14.3 Time spent listening to the station is important to the station and advertisers as well. This Nielsen Audio ebook page shows each demographic group and how long they listened to a particular station during a given time period per month or quarter. (Copyrighted information © 2016, of the Nielsen Company US, LLC licensed for use herein. On September 30, 2013, Arbitron Inc. became part of The Nielsen Company (US), LLC. Arbitron Inc. is now known as "Nielsen Audio, Inc.")

Music and Format Selection

One of the first programming tools a program director (PD) learns about is music. Prior to becoming a PD, a person will have personal favorites with regard to artists and styles of music. When it comes to the business of radio, though, personal favorites get set to the side. The first thing a PD learns about music is that music is a commodity. As a commodity, music has a shelf life that is dependent on the wants and desires of the customer, or listener. The shelf life of a piece of music is based on what is in style and popular. The average song has a life of 6 to 8 weeks. Just like any other commodity, music is intended to be bought and sold by the people who produce it. That is how they make their living.

The second thing that a PD learns about music is that music is just one of the many tools that a radio station uses to attract listeners. Just as a carpenter selects a specific type of saw to cut a specific type of wood, a PD selects a specific style of music to reach a specific **target audience** (see figure 14.4).

Sometimes the tool used to attract listeners is not music. There might be a strong demand in a market for a news- or sports-talk format. Other alternative formats that a station might consider include foreign-language (most notably Spanish) programming, religious programming, or ethnic programming, if any of those potential target audiences is not being served. Remember, the PD is seeking to reach a specific target audience, and sometimes that requires some creative approaches.

Program directors who know their craft can work with just about any format, because they understand the tools they have to work with. They understand that it is their job to research and to become familiar with the target audience and then determine how best to reach that audience and attract them as regular listeners.

Researching a Market

Before the competitive advantages of one format over another in a market can be considered, a PD needs to answer three seemingly simple questions about the market: (1) What do I know about the market? (2) What don't I know about the market? and (3) What do I need or want to know about the market? Each question logically leads to the next.

In the process of answering these questions, a PD discovers that there are three elements in a radio market over which he or she does not have control: the station's coverage area (or footprint), the area's population, and the number of competitors the station has. These three elements will impact every decision PDs make. Although they cannot control these elements, their best offense against them is an in-depth knowledge of the radio marketplace.

Footprint. The first element a PD has no control over is the radio station's coverage area, or footprint. The FCC determined the geographic area that the station's signal could reach when the station's license was granted, based on a number of technical factors. The station's engineer can provide the PD with detailed maps and drawings showing where the station's signal is strongest and weakest (see figure 14.5 on p. 380). By examining the geographic area with the station's engineer,

Radio Format Analysis

Individual preferences in radio listening are wide and varied—and radio operators respond by offering a broad range of programming to serve every taste! Here's a list of the current popularity of formats aired on over 11,000 commercial radio stations licensed in the US today. An additional 275 commercial stations are licensed under construction permits for future broadcast; 199 commercial stations are currently dark/off the air. Over 2,300 HD Radio stations are using HD Radio technology, and there are over 1,550 HD2/HD3/HD4 multicast channels on the air. Number of commercial streamed AM/FM radio stations: 7,898; 3,775 noncommercials report streaming. NextRadio reports nearly 3.2 million app downloads of FM radio on smart phones.

Rank	Format	Number of Commercial Stations
1	Country	2,126
2	News/Talk	1,343
3	Classic Hits	920
4	Spanish	866
5	Sports	768
6	Adult Contemporary	601
7	Contemporary (CHR Top 40)	584
8	Classic Rock	498
9	Hot AC	465
10	Religion (Teaching, Variety)	318
11	Oldies	316
12	Rock	299
13	Black Gospel	212
14	Adult Standards	191
15	Contemporary Christian	176
16	Ethnic	167
17	Urban AC	163
18	R&B	150
19	Southern Gospel	138
20	Alternative Rock	114
21	Soft Adult Contemporary	109
22	Modern Rock	107
23	R&B/Adult/Oldies	72
24	Variety	56
25	Jazz	24
26	Rhythmic AC	21
27	Easy Listening	19
28	Gospel	15
29	Classical	13
30	Modern AC	9
31	Pre-Teen	1
32	Other/Format Not Available	53

Figure 14.4 The Radio Advertising Bureau (RAB) has the world's largest library of radio industry research. Typically, the top format for radio stations is country followed by news/talk. (From "Why Radio Fact Sheet," Radio Advertising Bureau [http://www.rab.com/public/marketingguide/DataSheet.cfm?id=6]. Sources: Inside Radio/M Street Corporation, December 2016 [no Canadian or Mexican stations were included]. HD Radio statistics from iBiquity/HD Digital, December 2016; TagStation/NextRadio, November 2016.)

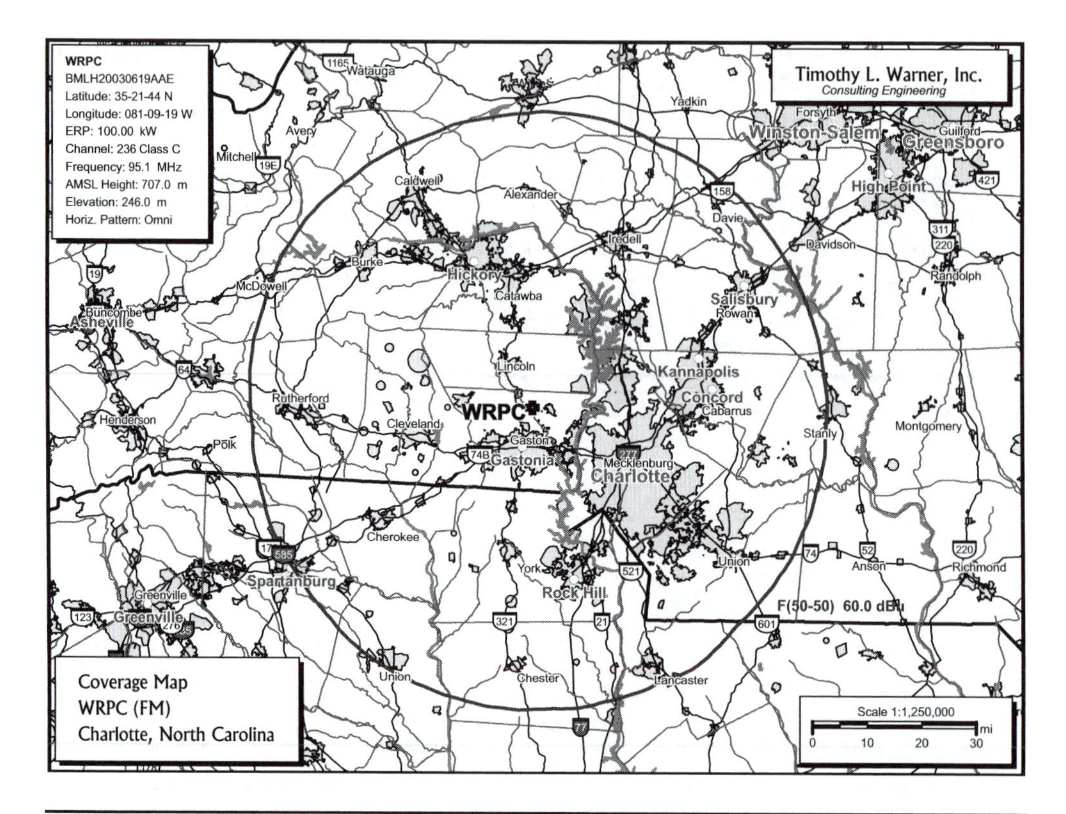

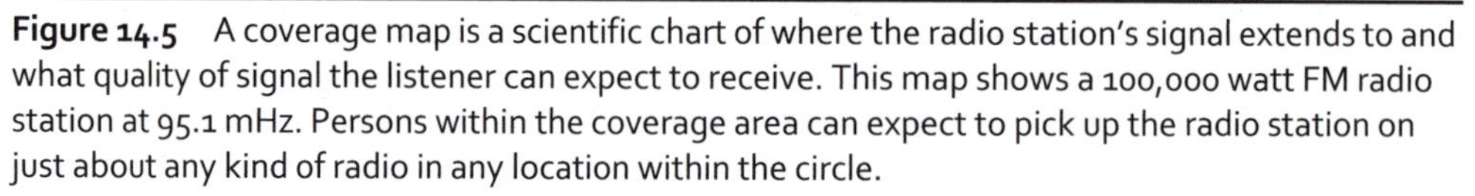
Figure 14.5 A coverage map is a scientific chart of where the radio station's signal extends to and what quality of signal the listener can expect to receive. This map shows a 100,000 watt FM radio station at 95.1 mHz. Persons within the coverage area can expect to pick up the radio station on just about any kind of radio in any location within the circle.

the PD can better define the market and focus program research only on the areas that can receive the station's signal.

The station engineer can also show the PD the competing radio stations' coverage areas, including signal strengths and weaknesses. These detailed engineering drawings are a matter of public record and can be obtained from the FCC engineering database for each radio station. This is truly a case where it pays to know your competition.

Population. The second element that a PD has no control over is the population that lives within the radio station's coverage area. The PD has no control over who moves in and out of the area, or the unique economic, social, or demographic profile of the people who live in the coverage area.

Although PDs can't control the population, they certainly can identify who is in the population through demographic research. This knowledge will be used as a competitive advantage.

Competitors. The third element the PD cannot control is the number of radio stations in the market he or she is competing against. The FCC decides what radio stations go where. Unlike any other type of business, the number of radio stations in a market is controlled and limited. In this respect, the competing stations are limited in size and number. When conducting format research, most PDs carefully examine every radio station in the market. Depending on the size of the market, it is entirely possible that there is more than one radio station attempting to be successful with the same format. The fact is, if a radio station is extremely successful, someone will launch an assault on the station's market position, hoping to steal some of the station's rating points.

As a PD examines these uncontrollable elements, it is important to keep in mind the PD's three original questions: What do I know about the market? What don't I know about the market? What do I need or want to know about the market? Market knowledge and ongoing competitive intelligence are some of the most serious aspects of programming a radio station.

In-Depth Format Analysis

After all of the data has been gathered and analyzed for a market, it is time to start narrowing the format possibilities. Format analysis begins by brainstorming and making a list of all the possible format options. Format analysis is based around a business model, just as with any other business looking for new customers.

One of the interesting things about radio programming is that what other businesses call customers, the radio industry calls listeners. In reality, though, listeners are customers of what the radio station has to offer. Do listeners and customers get treated differently? Although radio's customers are not going to spend any money with the radio station, they do spend something that is just as valuable—their time. If enough of these customers give the station their time, ratings go up, advertising rates increase, and advertising clients buy more time on the radio station, providing the cash the station needs to succeed in business.

Any business looks for two types of customers. The first is a customer who is not being served by any other business. The second is a group of customers that is not being served well. Following this basic plan, a PD first determines whether there is any group of people within the marketplace that is not being served. The PD is looking for a hole in the marketplace, an audience that the other radio stations have overlooked. The group of people the PD would like to find is one that fits into one of the standardized radio demographic groups, such as adults 25 to 54 or young adults 18 to 34. These are called target demographics, or demos. By selecting a demo that fits an accepted industry standard, the target audience is much easier to market and sell to advertisers. The PD also examines the other radio stations in the market for weaknesses to see whether there are any audiences that are being underserved by another station.

Selling the Target Demographic

With a target demo identified, the PD consults with the station's general manager and sales manager on the next step of the format-selection process, which is based upon a question: If the station can achieve market dominance or a strong ratings position with the target demo, are there advertisers willing to pay to reach the audience? In other words, what is the monetary value of the selected target demo? The radio station may be able to deliver a huge audience in a particular demographic segment of the population, but are they consumers? What types of businesses want to talk to the radio station's target demo? Does the target demo buy things, and what do they buy (see figure 14.6)? These are critical questions that must be answered.

For example, a radio station develops a format appealing to 12- to 17-year-olds and has the highest rating in this target demo in a major metropolitan city. This is fabulous—or is it? Are 12- to 17-year-olds good consumers? What kinds of products and services do they buy? Do 12- to 17-year-olds buy new cars, open bank accounts, use credit cards, consume beer, join health clubs, purchase furniture,

Urban Adult Contemporary (Urban AC)

SCARBOROUGH RESEARCH

Purchased Apparel/ Accessories Past 12 Months	Index
Men's Business Clothing	99
Men's Casual Clothing	87
Men's Shoes	91
Women's Business Clothing	134
Women's Casual Clothing	104
Women's Shoes	112
Children's Clothing	116
Infants' Clothing	108
Costume Jewelry	119
Fine Jewelry	115
Cosmetics, Perfumes, Skin Care Items	106

Alcoholic Beverage Consumption Past 30 Days (Among Adults 21+)	Index
Wine	102
Domestic Light Beer	85
Domestic Regular Beer	92
Imported Beer	101
Microbrew	32
Malt Alternative/Malt Liquor Beverages	168

Snack Foods Past 30 Days (Household)	Index
Ice Cream/Frozen Treats	103
Salty Snacks	92
Candy	100
Energy Bars/Nutrition Bars	82

Coupon Usage (Household)	Pct.
Use Grocery Coupons	75%
Use Coupons for Other Goods/Services	44%
Compared With All Coupon Users, Households of Listeners Obtain Coupons From:	**Index**
Newspapers/Magazines/Mail	98
In-Store Circulars/Coupons	106
Loyalty Cards	100
Electronic Sources (E-Mail/Text/ Online)	103
Product Packages	93

Health and Wellness Indicators	Index
Have Health Insurance	96
Have Life Insurance	106
Belong to Health or Exercise Club	98
Regularly Buy Organic Food	93
Past 12 Months:	
Dental Check-Ups or Procedures	85
Vision Check-Ups or Procedures	91
Treated by Dermatologist	77
Past 30 Days:	
Bought Prescription Drugs	98
Looked for Medical Services or Info Online	98
Shopped for Medicine Online	101

Recreation/Hobbies Past 12 Months	Index
Jogging/Running	114
Bowling	116
Free Weights/Circuit Training	102
Basketball	146

The Urban AC audience leans more heavily female and that was reflected in the purchasing patterns of apparel and accessories. Listeners were above the national averages for the purchase of women's clothing and shoes, especially women's business clothing.

Just over half of the audience aged over 21 had wine, beer or other malt beverages during the past 30 days (54%). They were slightly above average for drinking wine and imported beer but were well beyond the norm for choosing malt alternative and malt liquor beverages.

Most Urban AC listeners had health insurance (81%) although they were slightly below the national norm for having this coverage. Two-thirds also had life insurance.

Listeners had more limited participation in recreational activities than most other formats studied for this report but they were above average for jogging, bowling and playing basketball.

continued ▶

Note: An index of 100 is average. Median dollar amounts shown represent the mid-point, meaning that half the group are below the amount shown and half are above the amount shown.
Source: Scarborough Research, Scarborough USA+, Release 2, 2010

48 RADIO TODAY • 2011 EDITION

Figure 14.6 As a group, listeners of a particular music format exhibit specific trends as consumers. This report carefully examined those who listen to Urban Adult Contemporary (Urban AC) music. A couple of interesting facts about Urban AC listeners is that they are nearly 50 percent more likely to engage in basketball as a sport or hobby than the rest of the general population. Urban AC music tends to have a more female audience than male as evidenced by the purchases of women's business clothing. (Courtesy of Arbitron, Inc. © 2011, from *Radio Today*.)

take vacations, dine out at nice restaurants, or buy major appliances? All of these business categories are major radio advertisers and are *not* looking for 12- to 17-year-old customers. Typically, this young demographic group tends to purchase consumables rather than durables. They spend their money on things like fast food, cosmetics, and clothing.

The revenue-producing potential of the target audience is the bottom line with respect to a format. With the chosen audience, will the sales department be able to generate the revenue required to meet the station's income projections and produce the needed return on investment? Is there another target demo that has more revenue potential?

There are no easy answers to these critical questions. Most radio stations do formal market research to find the answers. Stations want to know everything they can about their geographic service area and their target demo. On the quantitative side, stations want to know the number of households, the average number of people living in those households, how long it takes those people to commute to work every morning, the average household income, and the average spendable income, to name just a few data categories. Qualitatively, the radio station wants to know what brands and models of cars these people drive, what kind of beer they like, how many times they eat out each week and what kinds of restaurants they eat in, how many trips they make to the grocery store each week, and even whether they do their laundry at home or in a Laundromat.

Many radio stations hire outside firms to do quantitative and qualitative research, feeling that because the firm is an outsider it will see and find things the station staff may overlook or just not be familiar with. In larger markets, a station or regional group may have an in-house audience-research specialist.

Once the target demo that can produce the most revenue potential for the radio station is clearly identified, the PD takes the next step in the programming process: find out what the target demo wants to listen to, and then give it to them!

At this point in the process, audience surveys, **focus groups**, and music testing are conducted with people in the target demo who reside in the coverage area. In some respects, audience research helps to validate the market research the station has already conducted to identify the target demo. What the radio station wants to know from this research is what the target demo wants to listen to. Although it is easy to make a guess, sometimes local and regional factors come into play and what you thought might work turns out to be just the opposite. Another round of surveys is often conducted with businesspeople to make sure the target demo reflects the businesses' potential customers.

Strategy and Tactics

Once a program format is decided upon, the PD develops the strategy and tactics for the presentation of the format to the listener. Just as in the military, the strategy is the overall plan, and the tactics are the specific elements needed to carry out the plan. For example, market research shows that there is a large population of 18- to 34-year-old males in the station's coverage area that is not being targeted by any other media. The PD decides the best strategy for reaching this very desirable demographic is with Alternative rock music and the promotion of the lifestyle that surrounds it.

The tactics employed to reach and hold this audience could include a high rotation of the top 30 hits, fewer commercial stop-sets, announcers with in-your-face attitudes and edgy language, and lots of concert promotions, special events, and contests. The strategy and tactics developed by the PD determine how the radio station ultimately communicates with listeners.

The Program Director's Duties

Program directors focus on four goals during the format design. First, the format presentation must be appealing and interesting enough to cause the target demographic to listen to the radio station in the first place. Second, the program flow of the format must maintain the listeners' interest to a point that causes them to listen longer. Third, the on-air staff must be trained by the PD in how to communicate with the target listeners so that they return and listen again. And fourth, and this is critical, the format must be designed in a way so that listeners clearly know which station they are listening to, as well as being able to recall the station when speaking with friends and family. In all of the decisions a PD must make about the format, the fulfillment of these four goals must be kept in mind.

Music Mix and Rotation

Program directors are responsible for determining a station's music mix and rotation. They develop schedules that regulate how the music format is going to flow each hour, as well as the type of songs that are played and how often. These detailed hourly schedules are called play clocks (see figure 14.7). The PD creates as many play clocks as necessary to accomplish the programming goals. Play clocks often rotate or change from daypart to daypart. The morning-drive daypart is typically 6:00 AM to 10:00 AM, midday is 10:00 AM to 2:00 PM, the afternoon drive runs from 2:00 PM to 6:00 PM, evenings from 6:00 PM to 10:00 PM, and overnights from 10:00 PM to 6:00 AM. The music played during the morning-drive show is arranged and rotated differently than that in the midday or afternoon-drive dayparts.

Play clocks are arranged in a way that promotes the smoothest music flow and the longest time spent listening. PDs carefully figure out how often the music needs to rotate and how often a song is played. When the number-one hit in the station's format is played, how long is it until it can be played again? When a two-year-old song in the format is played, how soon does it come up again in the rotation? A day? Three days? PDs make these determinations prior to the format ever going on the air, based on format research and what has worked well in similar markets.

Most PDs use a music-scheduling program for this tedious aspect of their job. The approved music list is loaded into a software program, with each song having a number of identifying qualities assigned to it that the program looks for when making a music list for a particular daypart (see figure 14.8 on p. 386). A music

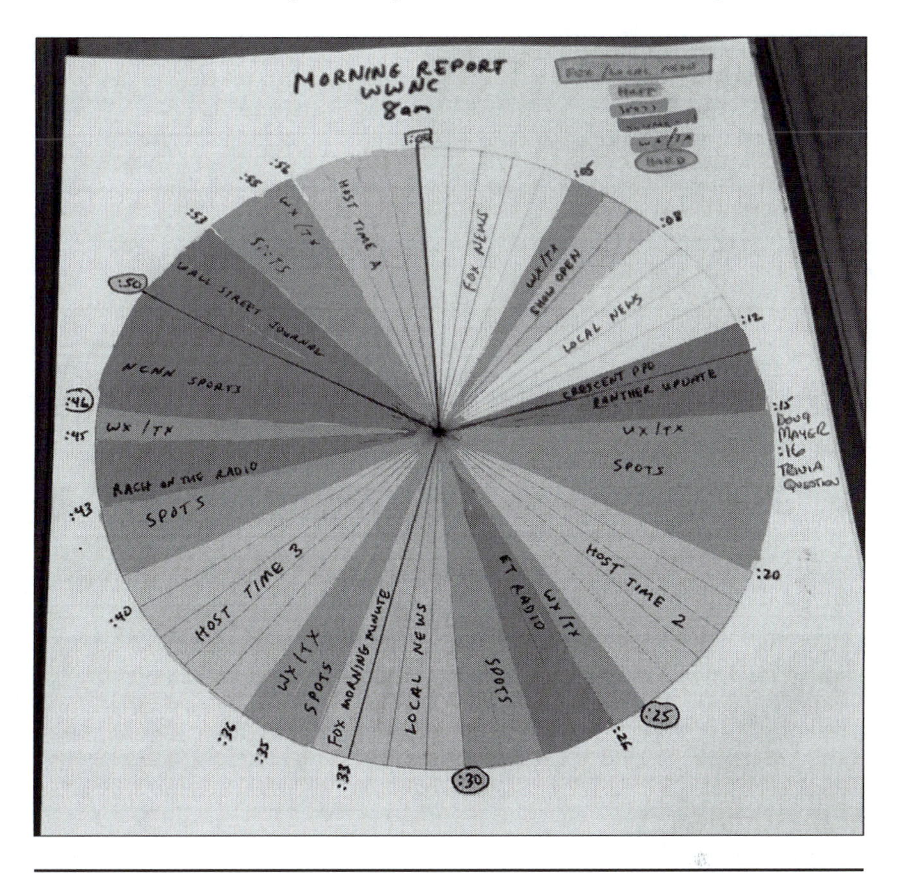

Figure 14.7 A play clock is used by the program director to control the program flow. Sometimes the clock will change throughout the day. This news/talk play clock is for the 8:00 AM hour. Notice that every minute of the hour is accounted for!

software program makes the PD's job much more manageable and frees the PD up to spend time on other aspects of his or her job, such as working with talent.

Program Elements

In addition to programming the music, the PD selects all of the program elements that occur each hour. How many traffic reports will there be? Will there be news, weather, and sports? How do air personalities say the time and temperature? Every element is examined for every minute of every hour. Specific guidelines are established for air personalities in giving the time, promoting music and programming, promoting social media, giving the weather, and knowing whether or not they are able to talk over music, among other things. PDs create and enforce programming rules associated with each element. PDs do not hesitate one second to tell the air staff when something has gone beyond the format limits or violates the station's programming guidelines.

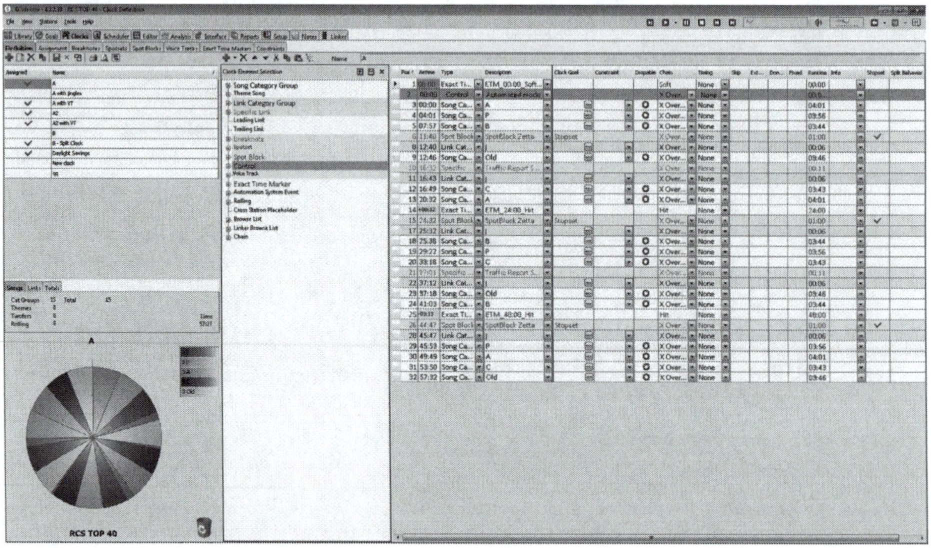

Figure 14.8 Music is critically important. Using music selection software, each song can have a number of identifying qualities assigned to it by the music or program director. The computer looks for these qualities when creating a music list for a particular daypart. (Courtesy of RCS, Inc.)

Selecting a Promotional Voice

Another important aspect of the program director's job is to select the radio station's promotional voice for station imaging. This is the voice that you hear promoting the station with liners and sweepers like, "Seattle's home for classic rock hits, Y-100!" The imaging can be dry, with no music or sound effects, or fully produced, with music and a signature sound effect the station chooses to use as an identifier (see audio demo, cut 14-1).

As a general rule, the PD selects this distinctive voice from outside the market to create a personality and image for the radio station that is unique, standing out from all of the other radio stations. The PD has several options when purchasing imaging. Imaging may be purchased fully produced, with music sweeps and effects backing up the voice, or the PD may decide to let the production department produce everything and just order dry voice tracks.

The promotional voice is sacred. It is always the promotional voice, and it should never be used to voice commercials, since the PD does not want to tie the station imaging and slogans directly to commercials with the same voice.

If a station elects to use singing jingles, these are produced by an outside source to the PD's specifications, so that they complement the music format of the station. Just as with every other element of each hour, the PD sets the rules on how imaging and jingles are used in each daypart.

Formatics and Inventory Control

Just as in a regular business with products on shelves, the blocks of time the radio station has for sale are referred to as its inventory. PDs are particularly sensitive to how the station's inventory intrudes into the format. Each particular format, based on the target demo, tolerates only so many commercials each hour. Once that point is reached, the audience reacts negatively, and the commercials become known as a tune-out factor. Ratings suffer and revenues drop. The PD, in conjunction with the general manager and sales manager, determines the maximum number of commercials that can be sold on the station each hour.

The PD also determines the best commercial placement within each hour, based on the desired flow of the format. Generally, commercials are grouped together into stop-sets, or breaks. The PD determines the commercial load, or how many commercials are played in the stop-set, and in most cases, the order in which the different styles of commercials are played. The PD also sets the rules as to how the commercials are placed with regard to music. For example, can a commercial be butted up directly to music? Likely not, since that means you would be tying the music directly to the commercial—a negative image most PDs would rather avoid.

PDs, along with the general and sales managers, also help develop the station's commercial-acceptance policy. What kinds of commercials are allowed on the air? What kinds of products are acceptable, and what kinds of products do not fit the format? Some products are banned outright and are not allowed on the air. For example, some stations do not accept ads for hard liquor, condoms, or feminine-hygiene products; whereas other stations may accept these commercials, depending on the format, time of day, and target demo. The PD, working in conjunction with sales, writes policy on how many commercials a sponsor can buy in one hour, in one daypart, and in one day. PDs, working with sales, establish the rules for product separation, meaning how close together two commercials in the same product category (such as restaurants) can be placed. These are written policies that account executives are required to follow when selling commercial time or inventory.

PDs from time to time butt heads with the sales department. If the radio station is doing particularly well and is sold out in several dayparts, the sales department may be tempted to ask for, and in some cases demand, more inventory each hour. The sales group may claim they have clients waiting to buy the time, and they likely do. Usually, these battles are fought in the general manager's office between the sales manager and the program director, with the general manager being responsible for making the final decision. Typically, when a station is sold out in a daypart, with clients waiting to get on the air, the laws of supply and demand dictate that the price of a commercial goes up rather than making more inventory available and risking listener tune out.

Comparison Shopping a Format

Radio stations that stream their on-air signal on the web make a PD's job of comparison-shopping music formats much easier. PDs can check out websites and listen to similar formats in other markets around the country for ideas. This is not necessarily a good thing, however. Taken to the extreme, a PD risks losing the originality and unique personality of the radio station if he or she tries to copycat a format from a radio station in another market. PDs who survey several radio stations playing the same format generally find that the format has a different sound from market to market. This is because each PD has, in theory, fine-tuned the format to fit his or her market.

Coordinating All of the Program Elements

Overall, the PD uses all of the elements discussed so far to assemble a living, breathing radio station. Just like a person, the radio station develops a personality and a distinctive style in everything it does that listeners identify with and relate to. This is called positioning the radio station so that listeners identify your radio station as *the* home, or *the* source, for their lifestyle and their kind of music.

The PD serves as the coach of a talented team of on-air professionals brought together to create the product that the sales department markets and sells to advertisers. Within the radio station's structure, the PD supervises the on-air personalities and the news, sports, and production departments. Typically, he or she works with the on-air talent and support staff in the station to make sure each person is focused and on goal.

PDs not only keep track of everything occurring in their radio stations, but it is also their job to find out and keep track of everything the competition is doing. PDs supervise and oversee competitive intelligence gathering. It is quite common for various staff members of a radio station to be assigned a competitor to track and monitor.

Format Delivery

After resolving all the issues described in this chapter, the program director must resolve one final issue with regard to the radio station's format presentation. His or her decision is usually made based on the projected profitability of the station format. The question is: Should the presentation of the format be live, live-assist, automated, or syndicated?

Live

A live format is based on air personalities in the studio working their shows, following the format, taking phone calls from listeners, posting to social media, and working creatively with newspeople, traffic reporters, and the weather service. There might even be a program producer doing a lot of these tasks for the talent during the show. There is a lot of studio activity and interaction between the various people appearing on the show (see figure 14.9).

The advantage of a live format is that it allows talent to stretch their legs and be creative. Tremendous creativity occurs as the different people who make up a

Figure 14.9 A live format is based on air personalities in the studio working their show, following the format, taking phone calls from listeners, and working creatively with newspeople, traffic reporters, and the weather person. Often one person serves as a producer to keep the show moving and handle technical details.

show play off one another and entertain the audience, adding a great deal of spontaneity to the show.

The downside of a live show is that all of these people cost money. Also, if the personalities do exceptionally well in the ratings, some other station may try to steal them with even greater monetary compensation. For all these reasons, live radio is more expensive to produce and thus requires a lot more station revenue to support.

Live-Assist

A live-assist format is based on a live air personality in the studio, with computers and outsourced talent doing much of the additional work. Preselected music is on one computer, and commercials and imaging are on another; in some systems, everything is loaded into one server (see figure 14.10). The music and commercials have been loaded into the computers based on music logs generated by the music director and program logs created by the traffic director. Basically, all the announcer is doing is hitting a space bar on a computer keyboard to start each music segment or commercial stop-set. News may come from a local TV news anchor over a digital audio link to the radio station. The air personality and the TV

Figure 14.10 Even small-market radio stations have adopted live-assist formats. Although dated, thanks to the two computers in this control room the announcer has more time to be creative. The announcer does not have to worry about a lot of minor details during his show because many of the tasks are automated.

anchor banter back and forth just before or after the news and, as far as the listener is concerned, the two are sitting in the same studio. Weather may come from a service like Accu-Weather, and traffic reports usually are outsourced as well.

An advantage of the live-assist format is greatly reduced talent costs. The station is only dealing with one or maybe two people for the morning show. A live-assist show gives the PD considerable control over every element of the station's sound, since the music and program elements are all programmed into a computer. Because many of the tasks are automated, the on-air talent has more time to be creative. Thanks to computers, announcer errors due to "brain fade" are reduced, and the format can be executed very efficiently.

There are not a lot of downside issues with live-assist. In fact, a lot of live shows use elements of live-assist. If there is a downside to live-assist, it is that creativity may suffer if the talent see themselves as just button pushers. But then again, that is where the role of the PD as a talent coach comes in.

Automated (Voice Tracking)

An automated format involves live announcers prerecording their stop-sets into an automation system that has been loaded with the program and music logs; this is referred to as voice tracking a show, which was described in chapter 9. In smaller markets, the on-air staff is often made up of former announcers who have gone on to become salespeople or to some other position in the station.

The station retains all of the control over its music, play clocks, and format, just like a regular station. All of the music and program elements are stored in the system. Included in the format are stop-sets in which the announcer can talk, do weather and news, and announce the music. Each announcer stop-set is assigned a file on the hard drive for the announcer to record into, just like everything else on the program log that has been loaded into the computer (see figure 14.11).

The PD loads the music logs into the system, and the traffic director loads the program logs into the system. The automation system merges the two together to produce a complete day, ready to air unattended. The on-air personality receives a printout of the log prior to the show to see where the stop-sets are and what he or she needs to do or say during breaks. Using the production room, the personality records the stop-sets into the assigned computer files. It takes about 30 minutes for a personality to record and load a 4-hour show into the computer.

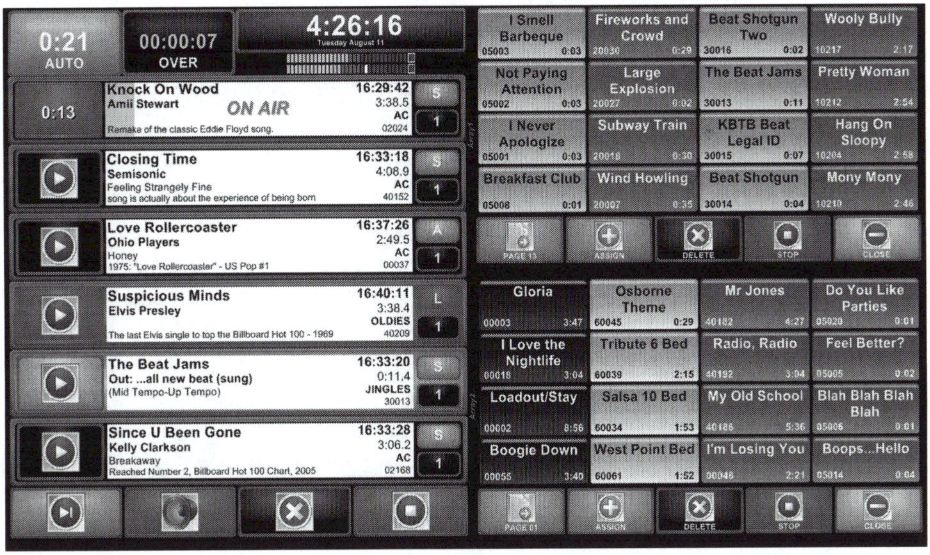

Figure 14.11 Computer automation and live-assist systems put radio station resources at the announcer's fingertips. On the left side of the screen are six playback decks into which the music, commercials, etc. are automatically loaded from the program log. Just above the playback decks are the station clock and audio meters. On the right side of the screen are 32 fast-fire buttons to play back selected program elements instantly. Announcers can configure the fast-fire buttons to suit their particular needs or show format. (Courtesy of ENCO Systems.)

This allows someone, such as a salesperson, to come into the station, record a show, and then be out on the street all day selling time for the radio station. It allows PDs to do their shows and then have the rest of the day free to work on station projects, rather than being tied to a control room for 4 hours every day. The morning show is likely to be voice tracked during the afternoon of the previous day. The midday and afternoon-drive shows are often recorded in the morning, and the evening and overnight shows are tracked before everyone goes home at 5:00 PM.

Taken to a higher level, automation is being used on a regional and national basis. There is no reason the air personality has to be physically in your radio station to record his or her stop-sets. A station equipped with a WAN, or satellite downlink, can use talent located anywhere on the planet. Likewise, with music-scheduling software, the program director does not have to be located in your station either.

The nation's largest chain of radio stations, iHeartMedia, uses this level of voice tracking and automation to bring major-market talent to smaller markets served by the company. The small-market radio station gets a major-market talent without all of the costs, such as a salary or benefits, normally associated with hiring a full-time person.

A local station staff member prepares information for the talent doing the voice tracking, such as pronunciation guides, local-event schedules, landmarks, scores from local sports teams, and the names of high schools and universities in the area. In some cases, the radio station sets up a website with all of this information so that the voice-tracking talent can access it any time he or she is ready to cut their tracks.

Because of the incredibly large talent pool such national media companies have to draw from, a local station's voice tracking for the morning show may come from Chicago, middays from Orlando, and the afternoon drive from Los Angeles. In fact, in addition to a live-air shift in the major market where he or she is based, voice-tracking talent can do as many as three or four voice-tracked shows a day for stations in the smaller markets. Despite what you might think, with a good production person looking after the system it is virtually impossible for local residents to tell where the talent is located, since the show has been localized.

Advantages of voice tracking include reduced costs, reduced costs, and reduced costs. Automation greatly reduces talent and operating costs, with most members of the station's staff pulling double duty. PDs have much greater control over the format, and it is very difficult for the announcer to make a mistake on the air, since a computer is running the show.

Thanks to the efficiency and savings of automation, a smaller-market station ends up with a top-notch program director and a better-sounding product than the local live show it might otherwise have produced. Automation frees up station and staff resources while retaining the local sound. Because computers are operating the radio station, it can run unattended, meaning that from 5:00 PM until 8:00 AM the station operates with the lights out and everyone home snug in bed. The same thing applies to weekends. If there is a problem, the computer calls or pages someone. Actually, the computer has a list of several people to call, in descending order of importance, and continues calling until someone answers and gives it a PIN number telling the computer to stop calling. Done well, it is very effective and very profitable, and local listeners will never have any idea where the talent is coming from.

A downside of automation is that it involves computers and attention to detail. Computers can fail; lightning and power surges can kill computers. However, when computers are properly protected and maintained, they operate for long periods (thousands of hours) of time without calling in sick or needing a day off to go see the dentist. Many stations operate what are called mirror computers. The mirror computer mimics everything the on-air computer does and is on standby, ready to go on air should the main computer fail.

Syndicated Radio

A syndicated show or programming service is similar to automation. Using such a service, an entire radio station can be located in a closet, and a small one at that. The syndicated format provider delivers major market-sounding talent and music. Syndicators spend a great deal of money on research and programming their format. The station receives the program service via satellite. The format has detailed play clocks, just like a regular radio station, with fixed-length local breaks each hour for insertion of commercial stop-sets, news, weather, traffic, or whatever the station wants to program in these local stop-sets (see figure 14.12 on the following page).

The satellite link is tied to a local automation system in which all of the local program elements are stored. The station traffic director loads the station's daily program log into the computer. A local production person is in charge of making sure the elements of the format, such as commercials, news, and weathercasts, are loaded into the computer and are ready for playback. The format provider uses digital commands to tell the local computer when to take a commercial stop-set or other local break.

What can really make a system like this come to life is how well the local station handles station identity. Each announcer from the syndicated program service provides a number of imaging liners that incorporate the local station's call letters and location or local station slogan. Each announcer's localized imaging liners are stored in separate files on the station computer. Just like the digital commands from the format provider that trigger the local stop-sets, digital commands tell the local computer which announcer is on the air and what image liners to have ready to play when the syndicated announcer sends a command to play one. Suddenly, the national announcer is a local announcer because his or her imaging, using your call letters, locations, and so forth, is played at appropriate times during the show. National announcers can even introduce the local news and weather. By keeping these elements up-to-date, your announcers can talk about local events and things going on in the community. These imaging liners are freshened by the national announcer as needed.

Advantages of the syndicated program format are even greater reduced costs, reduced costs, and reduced costs. Because the syndicated format is thoroughly researched and uses major-market talent, it allows a station to have a solid, well-constructed format that has strong listener appeal. If the local radio station does a good job of localizing the syndicated format by keeping imaging liners up-to-date and adding news, weather, sports, and so on, it is very difficult for a listener to tell what is going on. The major advantage of a syndicated format is a major-market sound at a very low cost to the radio station.

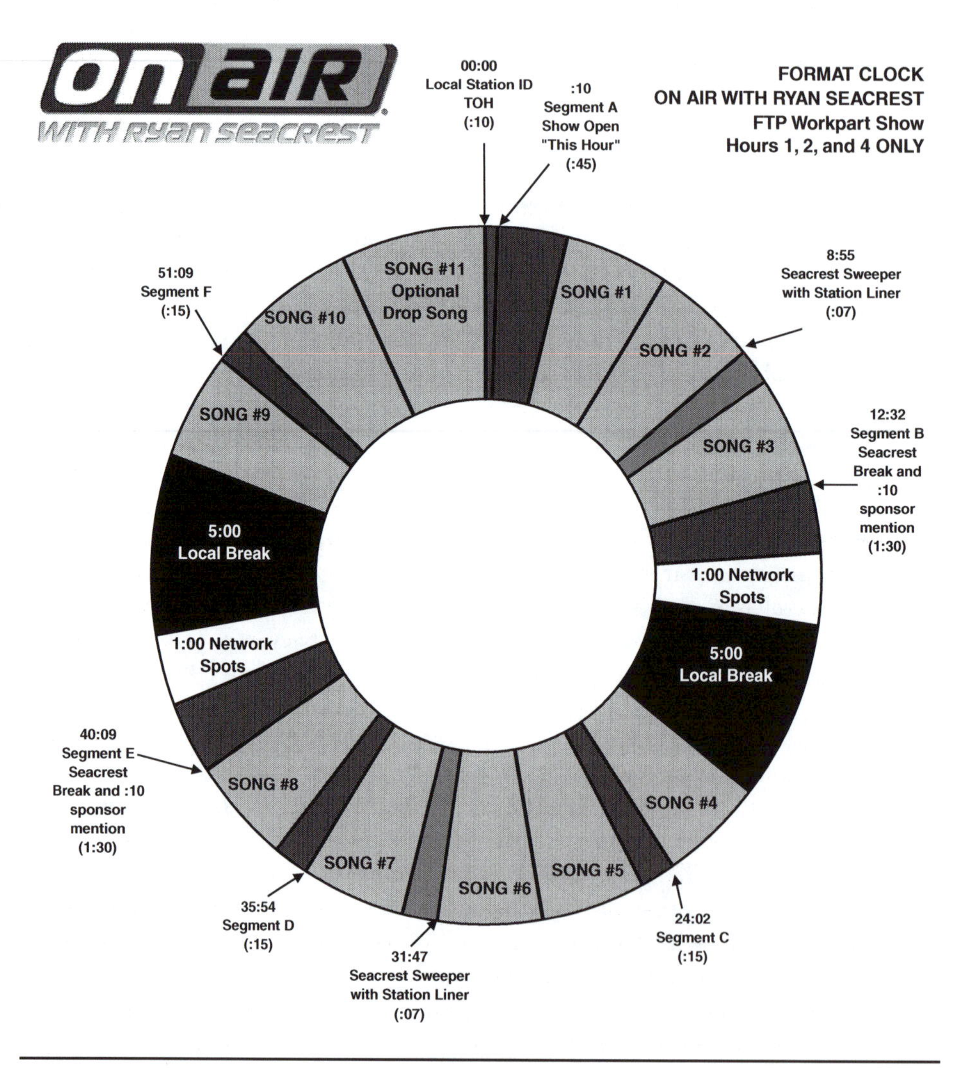

Figure 14.12 Syndicated formats have detailed play clocks, just like a regular radio station, with fixed-length local breaks each hour for insertion of commercial stop-sets, news, weather, traffic, or whatever the station wants to program in these local stop-sets. By creatively using the stop-sets, a station can truly sound local.

Although appearing similar to automation, the syndicated format is less flexible and requires more advance planning and attention to detail on the part of the local station. Another disadvantage is that the station has no control over the music or the format elements; those are determined by the national syndicator.

The station depends on the syndicator for the station's personality. The station plans everything in advance so that the syndicated talent can cut localized materials and promotional announcements. Basically, such a station has to work "inside the box." However, people who are skilled at working with a syndicated format can stretch the limits of the box with creative solutions to the restrictions they have to work with.

With syndication, as with automation, all of the observations made earlier regarding the reliability and the failures involved in using computers still apply.

The Format-Delivery Decision

The PD, in consultation with the general manager and the general sales manager, makes the decision on which method to choose to execute the format. As previously stated, this is a business decision based on the costs associated with the format and the projected return on investment that is required to support the station.

In some cases, a station may choose more than one of the various styles of presentation. For example, in a large market, a station might choose a live personality in the critical morning-drive rating period, with a producer, host, sidekick, and news and sports person. Midday, the station may shift gears and switch to live-assist with a live air personality producing his or her own show. The afternoon drive, another critical rating daypart, is likely to be a live personality show. As the station transitions into the evening hours, when ratings typically drop due to competition from television, the station might switch back to a live-assist personality. For overnights, from midnight to 5:00 or 6:00 AM (typically the smallest audience of the day), the station may pick up a syndicated show so that the station operates unattended, thanks to the computers.

However, even in a large market, a station may decide to go with one of the nationally syndicated morning shows. Don't think for one moment that a show has to be local and live to do well in the ratings. Such shows have proven many program directors wrong in the live versus syndicated programming war. Typically, a lot of news or sports-talk programming is syndicated, and stations do very well with it.

A smaller-market radio station might choose a live morning show and then automate the rest of the day. Another small-market alternative would be to use a live morning show and then switch over to a syndicated format for the rest of the day.

By availing themselves of the technology that is available, PDs can maximize their budget resources, putting live talent in the dayparts where it is needed the most to earn rating points, and at the same time, controlling costs and helping to generate the most revenue for the radio station.

The Production Person's Role in Programming

Although it may not be obvious, there is one common denominator critical to the success of each format-delivery method: the production person. Each format-delivery method requires a production person for it to be successful.

In a live-formatted station, the production person is focused on producing commercials, station promotions, and imaging for the station as a whole. The production person works closely with the PD so that the personality, style, and attitude expressed by these elements mesh with the format.

A live show likely has a producer who is responsible for creating many of the bits and show elements that make the show a success. Producers seek a production person's assistance on projects that may be beyond the producer's capabilities. A live-assist station likely asks more of production people, calling on them to produce special program elements for the live-assist personality, in addition to regular production duties. Although there are differences from the live format, they usually are not that great.

A local automated radio station uses the services of a production person slightly differently. Although he or she will still produce commercials, station promotions, and imaging, the production person's focus may shift to include responsibilities with regard to the automation system. The production person often is declared the "automation king or queen" and is responsible for making sure every element of the automation is loaded onto the hard drives properly. This includes making sure that logs are loaded properly, maintaining the system files, and being the gatekeeper for everything that gets loaded into or out of the system.

An automated station using regional- and national-level talent may require a production person to assume some of the duties of a program producer. Such duties might include supplying the distant talent with local information and timely details about the community to localize each show and verifying that the voice-tracking talent's stop-sets are in the system. In this situation, the production person's role becomes critical, since the producer is responsible for the presentation of the show on the local station in addition to his or her regular duties.

A syndicated format is almost entirely dependent on a producer to coordinate between the national format and the local elements necessary to make a complete radio station. In this instance, the production person takes on the role of running the radio station 24/7. Production people produce everything that is loaded into the system to interface with the satellite-delivered format, including news, weather, sports, and so forth. Very likely, the producer is also responsible for recording and producing all of the station imaging from the national announcers to localize the radio station. This requires advance planning on the production person's part to make sure that all of the elements are created and loaded into the computer.

As stated before, the method of format delivery is selected based on a business decision, as is the format selection. To say that one method of format delivery is not as good as another would be a serious error. Each method of delivery has a place and a purpose in the industry. As an example, there are many family-owned, small-market radio stations that would have gone dark years ago had it not been for nationally syndicated formats delivered via satellite.

Interestingly, it was the smaller markets that embraced the new automation technologies first, as a means of economic survival. By helping to reduce operating costs, these services have made it possible for small-market stations to operate on a profitable basis, producing good earnings for their owners. At the same time, the small-market operators broadcast a much better sounding product than if they were using all local talent.

There are those who feel that advances in technology are eliminating jobs from the radio industry. It is not so much a question of eliminating jobs as it is adapting job descriptions and creating new positions as the technology moves forward. Jobs that exist today may not exist in 3 or 4 years; however, jobs that do not exist today will likely be created thanks to the advances in technology. For example, who would have predicted that radio stations would have Internet and social media content managers? The end result of the advances in technology is a better, interactive, and quality radio product for the listener. What is ironic is that the intrinsic value of production people sometimes increases as the market size decreases, because their role in the final on-air product becomes so critical. Small market production people and their computers often are the radio station!

Suggested Activities

1. Go to www.westwoodone.com and look at the various 24-hour formats that are available from Westwood One. Do you recognize any that are on the air in your community?
2. Go to Nielsen Audio (http://www.nielsen.com/us/en/solutions/capabilities/audio.html) and look at all of the different kinds of listening data that are available. Examine the methods that Nielsen Audio employs to collect the audience-rating data used to formulate radio ratings.
3. Contact a local radio station and see if you can make an appointment with the program director or general manager to talk about their Nielsen Audio ratings and how the ratings affect the local radio market.

Websites for More Information

For more information about:

- *audience ratings*, visit Nielsen Audio at www.nielsen.com
- *music-scheduling systems*, visit RCS Sound Software at www.rcsworks.com
- *radio station automation systems*, visit Enco Systems, Inc. at www.enco.com
- *syndicated radio formats*, visit Westwood One at www.westwoodone.com

■ Pro Speak

You should be able to use these terms in your everyday conversations with other professionals.

focus groups—A qualitative market research tool in which a group of people with common demographics and similar attitudes are surveyed and questioned on a particular topic by a trained moderator.

psychographic—Determining how a person or a group of people think about a particular product or service; what their image is of your product and how they see themselves relating to your product.

target audience—A group of people that a radio station has selected to try to reach with the station's programming. Typically, this audience can be identified both quantitatively and qualitatively.

15

Getting Your First Job in Radio

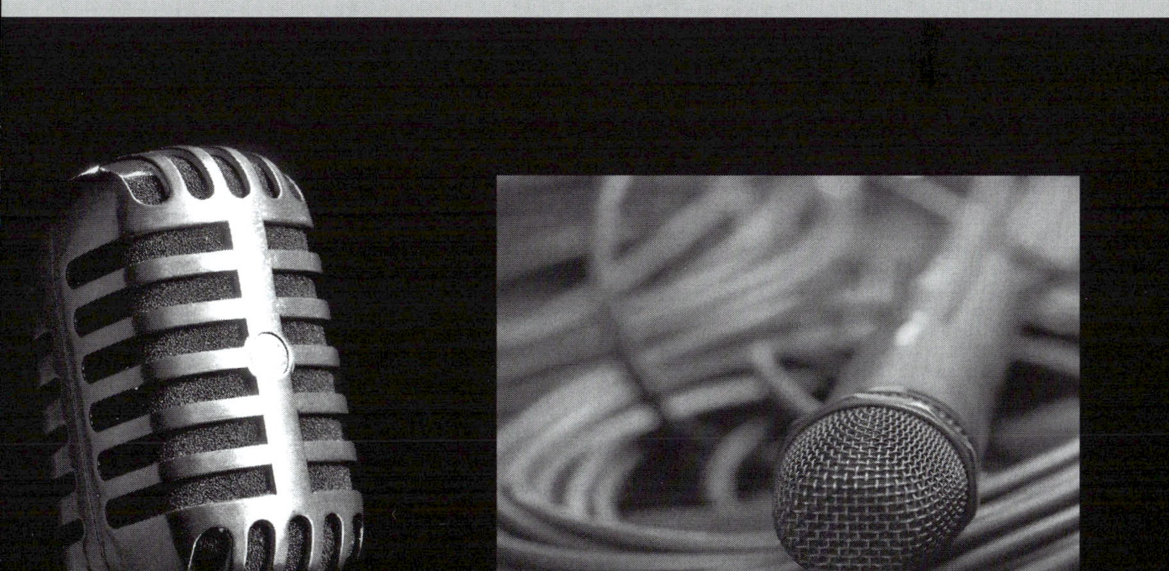

HIGHLIGHTS

Your parents have asked you a thousand times, maybe two thousand times: What's the job market like in radio? How much are you going to make when you graduate? Depending on whom you talk to, you may hear a story of doom and gloom for graduates entering any of the fields in media. You'll likely be told how convergence and consolidation killed the media job market. Yet the demand for creative, *qualified* broadcasters is as strong as ever. One program director recently told the author, "I am only interested in those who have a passion for radio. I want the best. I am not looking for someone who just wants to punch a time clock and put in their 8 hours." Employers no longer want or need those who are just mediocre.

My professional experience has included working as a program director, as major-market on-air talent, as a sales manager, and as a general manager. When I left the industry after more than 20 years to pursue teaching, I left a job as the affiliate relations director for one of the top three statewide radio networks in the country. My current experience working with stations reflects that. With regard to job openings, radio stations are always looking for aggressive, qualified, multitalented, creative individuals who seek to be the best in the market.

You will have to do some research to figure out what your starting salary may be. Salaries vary depending on the region of the country, the size of the market, and the size of the station. The best way to find out about salaries in your particular area is to talk to a program director or a human resources person at one of your local radio stations. He or she can give you some figures based on your experience level and the market you are interested in. Also, talk to your state broadcasters' association; these groups generally can give you a good idea of pay scales for the various areas of your state.

Getting your first job in radio is entirely up to you. Radio stations don't show up on campus and beg people to fill positions or offer lucrative starting salaries because there is a talent shortage. There is a great line from the movie *Barefoot in the Park*, "There are watchers in this world and there are doers. And the watchers sit around watching the doers do." You have to be a doer because no one is going to hand you a job. You are going to have to walk up and take it.

It might be tempting to bypass this chapter thinking your university's career services office can take care of everything for you. That might not be such a good idea. Although that office is prepared to help most students land a first job, it is not typically prepared for talent-based industries like radio, television, and acting. The purpose of this chapter is to give you some *advice* on how to land your first job in radio. It is advice because, as you are going to learn, there are no hard-and-fast rules when you are dealing with creative people. The information in this chapter is based on the professional experience of many in the broadcasting industry.

Internship: Get a Jump on the Competition

If you wait until the semester before you graduate to begin your job search, you're late. About a year late, to be exact. Early in your junior year in college, at the latest, you should begin taking the first steps necessary to land a full-time position in radio.

The Planning Process

A job search starts with the planning of an internship at a radio station or station group. Although some universities require an internship in the media to get a degree in communication, many do not. However, you need an internship, whether it is required or not. If you have to, set one up as a special topic in an independent study class with a professor (see figure 15.1). And to clear the one question on the top of many potential interns' minds: most internship opportunities are unpaid.

Start planning your internship at the beginning of your junior year so that you can schedule the internship, ideally, in the summer between your junior and senior years. Summer is a good time because, depending on where your internship is located, you can possibly lower your expenses by living with family, friends, or relatives. In some places, however, depending on the market and the hiring trends in your particular area, you may find that your second-to-last semester or last semester is a better time than the summer, because in many markets, stations offer positions to outstanding interns at the end of the internship. This is something to discuss with your professors. Regardless of the timing of your internship, plan for it well in advance.

Depending on the number of hours you devote to the internship each week, you might be able to hold a paid, part-time job as well. You should plan your internship to give you a total of at least 160 hours of professional workplace experience.

Figure 15.1 An internship can be a valuable experience and provide a student the opportunity to explore a career. A radio station's staff are glad to see someone express an interest in their field and are willing to help you learn the industry when you ask for their help. The student sitting at the console is getting hands-on experience under the guidance of a member from the morning show team. (Courtesy of 102 JAMZ, Greensboro, North Carolina.)

Selecting a Station

Selecting a good station or station group is important to a successful internship. First, the station should be a commercial station, unless you want to work in public radio. Second, the station should be in the largest market that is reachable from your home or the university so you don't have to spend a fortune on living expenses. The larger the market, the better your resume reads. And third, you should do some research and select the top-rated radio station in the market, even if it's not your favorite music format.

In addition to how it reads on your resume, why should you select the top-rated station in the largest market? Through your internship, you will have the opportunity to actively seek out someone in the industry to become your mentor, or learning partner. You're looking for someone to help guide you and to share his or her professional knowledge with you. Doesn't it make sense to work with the best and find out how the best do what it is they do? The professional you are seeking as a mentor or learning partner did not achieve his or her position by accident.

Another reason to select the best station in the market is so that you can network with top industry professionals. These experienced pros have contacts that could be valuable to you later during your job search. For example, if you have an excellent internship experience, your supervisor may be able to offer you a job when you graduate. If not, that person may be able to call friends at other stations and help you find another opening. That's the kind of result you want. Networking is discussed later in the chapter.

Asking for an Internship

No one is going to hand you an internship. You have to ask for one. Most stations, especially near universities, have someone who coordinates student internships. If you're planning a summer internship, make that first call to the station in early December, well before the holidays. Your first call is for surveillance purposes: your only task is to get the internship coordinator's name, title, and proper address.

Wait a day and then make the second call. You have a better chance of getting through to the internship coordinator because you are confidently asking for that person by name. The internship coordinator is likely to tell you that you are calling early for a summer internship and that you should call back in January or February. That's fine, because no matter how the call goes, it gives you the opportunity to demonstrate how professional you are by sending a follow-up letter. Remember: be aggressive, creative, and professional.

The follow-up letter or email should arrive within a day or two of your call. Thank the person for taking your call, briefly introduce yourself, and tell the person you *will* contact him or her again, as instructed (see figure 15.2). The letter must be professional in every respect, from the grammar and spelling to the 24-pound, 25 percent cotton-fiber paper and matching envelope. Under no circumstances can there be spelling or grammatical errors in that letter. Other than your phone call, the letter is the first tangible impression you are making on this person. How do you want to be remembered?

The letter should never be more than one page long. Under your signature and typed name, place your contact information, including your address, email address, and phone number. A word about email addresses is in order here. Some people have "unique" email addresses like bubba-redneck@quickmail.net. You need a professional-looking email address that includes your name, or a portion of your name. This makes it easier for the person to associate the email address with your name and to remember you.

With regard to the phone number you provide, make sure you have voice mail or an answering machine at that number and you have a professional-sounding outgoing message that includes your name. If a potential employer were to call you about an interview and heard a recorded message sung by you and your roommate about "drinking beer and having sex" (true story), chances are that he or she would not leave a message!

The envelope for your letter should have the same professional appearance as the letter. With the information you obtained earlier, address the letter to the specific person you talked to. It should be *hand addressed*, making sure your return address matches the one on the letter inside. Last, but not least, put a stamp on the letter. *Never* use a postage meter to send a thank-you letter. It makes it look impersonal and like all the other junk mail the person receives. Hand-addressed letters with stamps draw people's attention.

After the letter arrives, demonstrate a little aggressiveness and make your follow-up call to the station three to five days earlier than you said you would. Continue to maintain contact with this person every three to four weeks, with a phone call and a follow-up letter each time.

At this point, you already have a much better chance of landing the internship you want, thanks to the professional nature with which you have approached the internship coordinator and your persistence in following up. Your actions demonstrate that the internship is important to you and that you value the opportunity he or she may extend to you. Ask and you get; don't and you won't.

Jeff Davis
Operations Manager
WRPC
26 Radio Lane
Charlotte, NC 28723

Dear Mr. Davis,

Thank you for taking my call today and explaining WRPC's summer internship program. I was particularly impressed with the training program you offer interns over the summer. I have no problem with your 15-hour per week minimum, as I must complete 160 hours at a radio station to receive course credit. I will be going online in the next few days and completing the WRPC internship application and forwarding it to you next week.

This spring I will have completed both my radio production and radio programming classes as you suggested. In addition to Adobe Audition CC, I am becoming skilled in Pro Tools as well. I would really like to spend some of my internship time in production as well as in the promotion department.

Again, thank you for taking time with me and I will plan on calling you the second week in March to set up an internship interview appointment.

Best Regards,

Fern Lulham

Fern Lulham

12407 East 46th Street
Charlotte, NC 28723
828-555-1212 Home
877-555-1212 Cell
flulham1@peoplelink.net

Figure 15.2 Sample follow-up letter.

How to Gain More from Your Internship

An internship is an opportunity to get professional experience and can be extremely valuable. It is an opportunity to explore your desired career in a professional work environment. How much you gain, though, is up to you. Be up front with the internship coordinator about what you want to learn and experience while doing your internship.

Your approach to a successful internship must be a professional one. For several weeks, you are going to get the opportunity to work in a radio station. There are a number of things you can do to make your experience much more rewarding.

Before arriving for your first day of work, make a phone call to your supervisor. Ask that person when he or she would like you to report and also ask what the standard business attire, or dress code, is for the area you are working in. Follow the supervisor's lead. Regardless of how the other employees look, avoid excessive jewelry and forget the tongue piercing, extreme hairstyle, or heavy-duty perfume or cologne.

You are entering a professional work environment. Time is money, time is money, time is money! Set your watch and adjust your lifestyle to the station's master clock. *Never* be one second late for anything. Vince Lombardi is regarded as one of the greatest football coaches of all time. The Super Bowl trophy is the Vince Lombardi trophy. He became legendary for Lombardi time, demanding that players be 15 minutes early for everything. Fifteen minutes early was on time, and on time was late. Late is a four-letter word.

Make it your goal to remember the name of everyone you meet and, likewise, make sure everyone knows your name. Your internship is an excellent time to network with people and to pair up with someone similar to you to become your mentor. The guidance of a mentor can pay big dividends later when you are seeking your first position.

Take advantage of every opportunity offered to you. Radio is a 24/7 business, and many opportunities may come outside the regular 8:00 to 5:00 business day. If someone extends an invitation to you to participate in a station event or promotion after office hours or on a weekend, take it (see figure 15.3). Make yourself available to do anything that is asked of you. No matter how simple the task may seem, do it better than anyone has ever done it before. Don't be afraid to try new things.

Above all else, ask lots of questions. That's why you are there—to learn. There is no such thing as a dumb question. You'll find most industry professionals are as interested in you as you are in learning how to do things in a professional manner. The most reliable way to get an answer to any question you have is to ask the person for his or her opinion. For example, "What do you think is the best way to . . .?"

When asked for an opinion on the best way to do something, most pros are more than happy to share their knowledge. When you are unafraid to ask and answer questions and when you establish good communications with the industry professionals you're working with, unexpected things can happen. For example, during a morning sales meeting, an intern at a major-market radio station asked a question about a promotion that was in the planning stages. The sales manager turned the question around and asked her how *she* would handle the promotion if it were her project. She answered with an idea she had picked up in a sales class. It was a fresh approach to the promotion that the sales manager had not heard of. Imagine her shock as the sales manager adopted her approach to the promotion on the spot! By the way, that question landed her a major-market sales position straight out of college. The key to her success was that she was alert, observed what was going on, and when *asked* for her opinion was ready with an intelligent answer.

Another student who did an internship with a top-rated, large-market morning team really wanted to be a morning-show producer and worked very hard at the craft for several weeks alongside the show's producer. With about a week left in his internship, he asked if there was any way the producer would consider allowing him to produce just one show. He was stunned when the show's producer stood up and said, "Take it, it's yours" and left the studio. The producer reasoned that if the

Figure 15.3 As an intern at a radio station, take advantage of every opportunity offered to you. Radio is a 24/7 business, and many opportunities may come outside the regular 8:00 to 5:00 business day. If someone extends an invitation to you to participate in a station event or promotion after office hours or on a weekend, take it. Here an intern is picking up tips on the use of wireless remote transmission gear.

student had enough confidence in the skills he had learned to ask the question, the producer would give him a chance. As a result, for the last week of his internship the student produced the show under the guidance of the regular producer. Thanks to a good, radio-oriented resume and an outstanding internship experience, the student started work at that station two days before he graduated.

While doing your internship, keep detailed notes on any work you do or projects that you contribute to. Ask your supervisor for permission to make copies, as some materials may be copyrighted or it may be a violation of station policy to copy and remove such materials from the station. These materials are valuable for your audition to demonstrate professional experience. For radio, air checks and any production you worked on are great. Even a sample of your voice-tracking abilities can be a big help. After clarifying it is not for air, ask if you can voice track a show and have the program director critique your work.

Although you will probably be required to work a minimum number of hours to satisfy the requirements of an internship anyway, this author strongly suggests that you become as involved in the station as your supervisor allows. Network with everyone you meet and do your assigned job better than anyone who has ever done it.

Networking. At least 80 percent of the job opportunities in radio are discovered through networking. Less than 10 percent of radio jobs ever show up in a job search engine. Networking is all about making contacts. As you meet people, begin to build your network of contacts, communicating with them on a regular basis and adding more people to your network over time. Be selective about whom you add to your network. You should be able to get along well with each person and, quite frankly, each one should like you. During your internship would be a good time to establish a LinkedIn page to keep track of your "network."

If you do not want to use social media to build your network, organize your network by keeping a business card from each person you meet or by adding him or her to your smart phone, because a good network has too many names and numbers for you to keep in your head. Include the person's name, addresses, phone numbers, email address, and any bit of information you think is important about this person. You can also trade contacts and information with others in your network. A quick note about keeping everything in a smart phone: make sure you have the backup service offered by your cell phone provider. Things happen to phones and you may want to keep your contacts for years. If you truly desire to be a pro at networking, send each new person you meet a short handwritten thank-you note or letter. The fact that you took the time to send a note puts you one step above everyone else professionally (see figure 15.4).

Networking is centered on the exchange of information and advice. When the time comes to find employment, *do not* pick up the phone and call everyone in your network asking for a job. The best way to approach someone in your network when you are looking for a position is to ask for some type of help, or advice. You can ask someone in your network for job-searching advice, for the best places to find information and position listings, career-planning advice, state-of-the-indus-

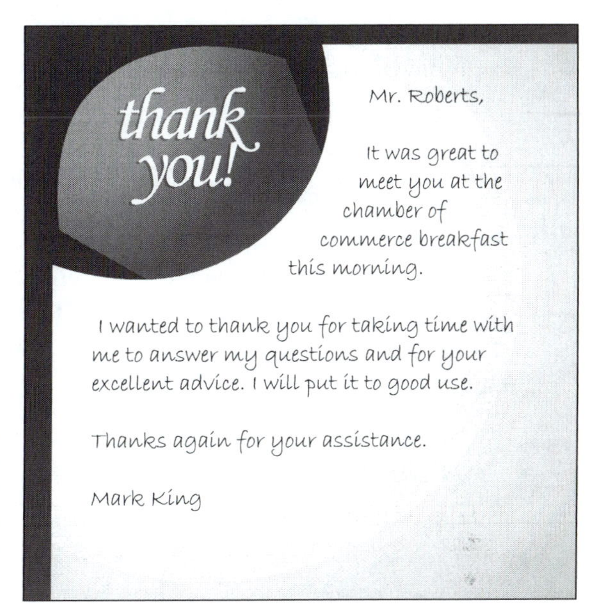

Figure 15.4 These days most of us (especially men) tend to neglect the art of handwriting a note, but a handwritten thank-you note always makes a good impression.

try advice, or advice about a particular company. You need to be seen as seeking guidance from the people in your network, not asking them for a job.

By using this approach, you are positioning yourself as someone who is seeking to better herself or himself through knowledge. When you ask for someone's advice, you are telling that individual that you respect his or her opinions and thoughts. People are much more willing to help someone asking for advice; it makes them feel like their opinion is valued and respected. Which of the following approaches would work better for you?

> Aaron, I met you at a station promotion and I am about to graduate from college. I've waited a little late to start my job search and I am desperately looking for a position so that I won't have to move back home with my parents. Aaron, you got any jobs open right now?

Or . . .

> Aaron, Jeff Hall at iHeartMedia in Asheville suggested I contact you. Jeff was my internship supervisor and he mentioned that you work in radio production. I am just starting a career in production and I am looking for someone who can give me some good advice about the best way to get started. Could we get together sometime next week? I would sincerely appreciate any time you could give me.

When you meet, ask for advice on seeking that first position and how to get a career off to a good start. Often, people are happy to share contacts and leads with someone asking for help. If they don't know of anything, they may refer you to someone in their network of people who might have something. Just as before, send a handwritten thank-you note or letter after your meeting. You will get a call the next time they hear of a job opening.

The best example of successful networking is the following unsolicited email this author received from a student.

> Ultimately, it was my internship with XXXX that sealed the deal for my new job. This is no joke; it took about a month to get hired. So pass along to graduating students that they will have to be patient and aggressive at the same time or they'll get discouraged very fast. I can't stress the importance of the internship. People always talk about that first position as "getting your foot in the door," well, the internship is ultimately getting your foot in the door. You just can't beat the importance of all those contacts you meet with during your internship.

The student landed a job in a top 50 market. Networking is a powerful means of finding a job.

Social Media

Social media outlets are a force to be reckoned with in seeking and keeping a job. Three of the most popular sites are Facebook, LinkedIn, and Twitter. Traditionally, Facebook is a personal social media site and LinkedIn is a site aimed at professionals. Twitter can be an excellent relationship builder and often allows individuals to have direct contact with major players within a corporation. Such sites are moving targets and change constantly as they evolve.

Employers use these sites as powerful tools to reach candidates seeking jobs. Likewise, employers use social media sites to check out potential employees and to keep tabs on current employees. There are many horror stories about careless postings by potential and current employees. At the same time, social media can be a valuable resource for students seeking employment and desiring to build their professional contacts. Social media turns you into a publisher. What are you going to publish about you? How will you build your image or "personal brand" online?

As creative as social media are, there are some common pitfalls to avoid. If you are a Facebook user follow a simple rule. Do not put anything on your Facebook page you don't want an employer to see or read. If you cannot publish what you're putting on Facebook on the front page of your local newspaper, don't!

Many career advisers suggest having two Facebook pages, one for personal use on which the privacy settings are "friends only" and another for professional use. Facebook also allows you to approve any posts where someone else has tagged you. This gives you the opportunity to review the post. If you do have a personal and a professional site it is up to you to make sure the two do not intermingle.

There are applications such as BranchOut or BeKnown that plug into your Facebook profile and allow you to have both a public-professional profile and a completely separate professional network.

Another way to use a professional-oriented Facebook page is to post a link to your LinkedIn page where all of your professional information is located. Use one site to drive viewers to the other. LinkedIn is an excellent place for students to post a resume as well as educational and professional background materials. It is also a good place to post links to writing, audio, and video projects. LinkedIn is a place where you can post an electronic portfolio of your work.

So how do you avoid some of the pitfalls and horror stories you have heard about social media? Here are some tips.

- Posts are not private. If you have any kind of social media site assume your potential or current employer will see what you have posted.
- Raise your privacy settings. Privacy settings only prevent the *average* user from getting at your information. Assume someone will come looking to see what you have protected.
- Do not disclose personal information you are not comfortable with everyone knowing. This includes information you would not normally reveal to an employer in an application or job interview. Never post your birth date. With your name and birth date you would be shocked at the information that can be revealed about you.
- Do not reveal confidential information, be it personal or business, you have been entrusted with. Always assume someone is monitoring your site.
- Monitor yourself. Google your name often and see what comes up. Over 80 percent of employers Google prospective employees before a background check. The first page of listings with your name that comes up is known as your Google resume; make sure all the listings are good stuff.
- Set up alerts. Tell your friends what your expectations are with regard to personal information and privacy. Best to have a clear understanding than an oops! One of the biggest red flags to draw attention are photos that can place you in a very bad light. Likewise, be considerate and respect others when posting. Don't be afraid to edit or ask others to remove posts about you.
- Don't mix personal and business contacts. Maintain a separation between the two and be more conservative with your business contacts.
- Do not post anything that can even remotely be construed as a crime or just inappropriate. What you may think is a funny joke may be seen as demeaning, criminal, or unethical. The police or your employer could use this information in ways you have never imagined.
- Be careful what you post about business events. Everyone has heard of the office party photo that caused someone's termination. Be warned; photos of you taken in a public place are fair game.
- Don't engage in social media at work unless it is your job and it can earn you a raise. Many companies (about 45 percent) use software that records every keystroke on your computer. In surveys over 50 percent of employers admit to terminating an employee for unauthorized computer use in the workplace.

Some students feel that once their resume or information is posted online that their job is to sit back and wait. Nothing could be farther from the truth. You cannot hide behind the screen. You have to network and talk to people. When you meet someone online, follow up with some other form of communication, such as a phone call or email.

There is no question that companies are relying on the social media for job searches and for checking on candidates. Your job is going to be to adapt the current social media trends and job search practices in order to refine and develop your "personal brand" or image.

Resume

Many students assume that resumes are generic, one-size-fits-all things that the counselors in your college career services office will do for you. But stop and consider this: it is estimated that 98 percent of the resumes submitted to potential employers do not even begin to describe adequately or do justice to the candidate whose name and personal information appears on it. Writing a great radio resume is just like writing a great radio commercial. Focus your resume on showing how you can solve a particular employer's problems and needs, and you've likely got an interview. Forget the advertiser (you) and focus on solving the customer's (employer) problems and needs, and the product (you) will sell.

With regard to your resume, there is only one rule set in concrete, steel reinforced, and armor plated. Under no circumstances can it include a single spelling or typographical error. More employers than you can imagine toss resumes at the first grammar or spelling error. The best resumes are well organized and often professionally prepared. After you have met the requirements of rule number 1, then you can get creative and use some color, graphics, and creative packaging. The industry buzzword for a professionally designed and prepared resume is career design. Career designers are people who specialize in preparing resumes and graphically marketing people.

Resume Style

When it comes time to write your resume, you need to select a resume style. Two of the most popular resume styles don't work particularly well for starting out in radio. The first kind is a chronological resume. As the name implies, this is a chronological listing of the jobs you have held, working backward from the present. Under each job you list your skills and achievements. Although very informative, a chronological-style resume is not very creative. The focus of the format is on personal growth and job-to-job continuity. Employers tend to look at these historically, to see whether you have had steady upward growth in your career development, but this is pretty hard to do if you are a student just starting out and don't have years of experience to list on your resume.

The second most popular resume style that won't work so well for your purposes is a functional resume. This is one that begins with a couple of paragraphs centered on a broad description of your qualifications, abilities, certifications, skills, and how great you are. Your employment history is secondary in importance to the opening and provides few details about the positions you have held or what you may have accomplished at each job. The focus of a functional format is on what you have done, but not on where or how you did it. Although functional resumes appear very creative, employers tend to be cautious when someone presents this type of resume, particularly someone fresh out of college. The general thinking is that if you can't list your jobs in chronological order and what you accomplished at each one, you might be trying to hide something, like gaps in employment or frequent job changes.

An ideal resume for someone starting in radio combines these two styles. The resume opens with a section detailing your market value and why you are worth considering. This is the "Hey, look at me" section of your resume. You present your skills, qualifications, and accomplishments just as you would in the functional resume. The creative opening is then followed with a typical chronological listing of the positions you have held, working backward from the present. Under each job, list your skills and achievements, just as you would in a chronological resume (see figure 15.5 on the next page).

The benefit of the combination-style resume is that it allows you to sell your assets as a potential employee right up front, to get the employer's attention. You start by showing how you can solve the employer's problems and needs. The detailed chronological section supports the opening. Likewise, the opening adds a creative touch to the rather dull chronological listing. It allows you to add some bulk to your resume that's not fluff or nonsense. The combination style appeals to employers because it demonstrates creativity that is backed up with a job history the employer can track. In essence, it is the best of both worlds.

Resume Layout

How your resume appears is important. A typical human resources person takes about 20 to 30 seconds to look at the first page of a resume. If he or she does not see anything that interests them, the resume is tossed into the "no thanks" pile. The best resumes are laid out matching the normal viewing pattern of someone's eyes when reading. Typically, people read a printed page from left to right and top to bottom.

Take a piece of paper. Vertically fold the paper in half and crease the fold. Now fold the paper in half horizontally and crease the fold. When you open the paper you will see that the page has been divided into four quadrants. Number the upper left section number 1; the upper right section number 2; the lower left section number 3; and the lower right section number 4. As you write and layout your resume each section should have an equal amount of copy and white space. This will give your resume a balanced layout. Logically, the most important information goes in sections 1 and 2.

You can balance the layout by putting the most important information on the left side of the page, such as your education information, employment history, etc. Place the dates and locations on the right side of the page (see figure 15.5). Essentially, you are creating two columns on the page and ensuring a balanced layout.

Mark R. King
Cell: 704.223.1234
mrking1950@supersmail.com
4188 Utopia Drive • Charlotte, NC 28226 • 704.828.4321

Objective

To excel in broadcast sales using my team-building skills and exemplary verbal, adaptive, listening, and organizational skills.

Summary of Qualifications

- Radio Advertising Bureau, Radio Marketing Professional, certification #R-618923
- Interpersonal skills that build relationships based on trust
- Underwriting sales coordinator for WWCU FM at Western Carolina University
- Skilled with industry software such as ACT!, Pro Tools, and Adobe Audition CC

Career Profile

Underwriting Sales Coordinator, WWCU-FM—Cullowhee, NC **2016–2017**

- Developed sales projections, conducted meetings, established goals and budgets
- Supervised two underwriting salespeople
- Consistently exceeded sales projections and met budgets

Intern, Welk Music Group: Vanguard and Sugar Hill Records Corporate Offices—Los Angeles, CA **Summer 2016**

- Assisted promotion, sales, marketing, new media, and publicity departments
- Worked for two labels and approximately 15 releases
- Promoted artists in specific markets by providing retailers with materials

Education

Western Carolina University, Cullowee, NC **May 2017**
Bachelor of Science Degree in Communications **GPA 3.87**
Major: Electronic Media • Minor: Marketing

Community Service and Volunteer Work

Fundraising events for various community organizations

- A.W.A.K.E. Foundation for battered women and children **2015–2017**
- Ronald McDonald House for Children's Charities **2015–2017**

Sigma Phi Epsilon ΣΦΕ

- Fundraising Chair **2015–2017**
- Vice President of Member Development **2016–2017**

Software Microsoft Windows & Office • ACT! • Adobe Audition CC • Pro Tools

Letters of reference provided upon request

Figure 15.5 Sample resume.

Targeting Your Resume

Unlike a generic resume intended for general applications, a radio resume needs to be focused and targeted at the specific station and market you are applying to. Instead of a one-style-fits-all resume, you need one that presents your talents, skills, abilities, and experience in a manner that matches the needs of a specific station. Targeting a station involves carefully looking at the requirements for the position. The requirements are, in essence, the program director's statement of the station's problems and needs. After you analyze the problems and needs, adjust the copy on your resume item by item to answer as many of the stated problems and needs as you can with your skills and abilities. Look at the example resume in figure 15.5. Can you tell what kind of position this student is seeking? Where is the most important information?

The other aspect of targeting a radio station is how you go about presenting yourself and your accomplishments. Most people tend to just list their former positions and responsibilities. Big whoop—most employers don't care about your prior responsibilities. Employers want to know what you accomplished, what you achieved for your previous employers. Instead of listing the responsibilities of each position you have held, list your accomplishments. For example, look at the two following statements. Which statement looks, reads, and sounds better? Which candidate would interest you?

1. "I was responsible for the production department at WWCU at Western Carolina University and managed a staff of five students."
2. "As production director at WWCU at Western Carolina University, I streamlined production workflow and increased efficiency, resulting in a 20 percent reduction in the time needed to get public service and underwriting announcements on the air. I was honored three times with local ADDY Awards for public service campaigns. As a result of the new procedures, outstanding quality, and fast turnaround time, my department operated more efficiently, resulting in a significant cost savings for our university station."

Analyze your previous jobs by what you have accomplished, and translate those achievements into positive and descriptive language. Learn some of the broadcast industry buzzwords and phrases, such as:

- inventory control
- improved profitability
- turnaround
- implemented new procedures
- identified and resolved problems
- increased return on investment
- introduced new technologies
- improved productivity

Keep in mind that responsibilities have never solved a problem or met an employer's needs—accomplishments have. Make sure everything on your resume accurately represents your accomplishments and answers the specific job qualities or skills requested in the ad. Look again at the student resume in figure 15.5 and pick out the achievements and buzzwords.

Resume Contents

If you have not figured it out by now, resume writing is critical to your successful employment. There are a number of good books and publications on resume writing; use them for ideas. You should also take a look at the materials in your university career placement office for ideas. Keep in mind, though, that you are applying to a creativity-based business. What follows is a very basic listing of what should and should not appear in your resume for a radio station, or any other business, for that matter.

Information That Does Not Go on Your Resume. There are a number of personal information items that you do not put on your resume. Although your name may give the employer a clue as to your gender, do not list yourself as male, female, or the personal gender pronoun you use to identify yourself. Never list your date of birth, social security number, age, height, weight, health, race, citizenship, ethnicity, religion, sexual preference, or whether you are single, with a domestic partner, married, divorced, have children or dependents other than children, or are a single parent. You should not state or imply any family plans you may or may not have regarding marriage, domestic partners, or starting a family. Statements of religious beliefs are also not appropriate on a professional resume.

Many employers will not review a resume if any of the above information is included as the employer does not want to be seen as giving preferential treatment to anyone. If you have a question about what information an employer can require you to divulge, visit the website of the US Equal Employment Opportunity Commission (www.eeoc.gov).

The Header. At the top of your resume, a header contains all of your personal contact information. The header should have your name, address (both university and permanent, if you are still in school), and appropriate phone numbers, including your cell phone, and email address. A word of caution: *do not* use a current work phone number, fax number, or email address. This is a red flag to a potential employer that you abuse company resources for personal gain.

There are two places to put the header at the top of the resume: in the center, or somewhere else. A centered header is considered the traditional style, and any other position is considered a creative style. Using publications and online resources, examine a number of creative resume samples and see if there is a style that looks professional and fits your personality. Radio people tend to be creative and like to see creative approaches to projects. Remember: keep it professional and creative with neat, clean lines, not avant-garde.

Goals and Objectives. Below the header is the first opportunity to sell yourself, your first opportunity to land an interview. Imagine yourself as an employer for just a moment. As you begin to read a resume, what do you want to see? If you are like most employers, you are asking a few questions: Why am I reading this? What does this person have that I am interested in? What can this person do to help solve my problems and make our company more profitable?

You need a specific goal, or objective, that clearly demonstrates to a station how you can solve its problems and needs. Prove to the reader that you are responding directly to them. Most resumes have incredibly vague and meaningless career goals or objective statements. For example, one that sounds like it came out of a can, is overused, and means nothing is, "I am seeking a position with a market-leading radio station." Many employers will simply toss resumes with those words into the no thanks pile. Resumes that are even more likely to hit that pile contain the words, "I am open to any position in your station you might wish to consider me for." If you don't know what you need, where you are going, and how you are going to get there, then no one is going to be interested in you, period.

As someone entering the job market for the first time, your goals and objectives are going to be different from those of an experienced pro. You need to target personal attributes and qualifications that meet the stated needs of the prospective employer, and then back them up with any related experience you might have. You need to let the reader see that you have potential.

Start by making a list of your skills, talents, and outstanding achievements in college. Analyze your internship for things you accomplished or excelled at. Select items from your list that address the stated qualifications of the position you are seeking. A good way to start off your resume is with a quote from your internship supervisor, or other industry professional, followed by your goals and objectives. Here is an example from a recently graduated student:

> "It is hard to believe that Lashea is a college student. Her creativity, writing skills, and mastery of digital audio is outstanding and helped her fit right in at WXXX. In fact, the quality of her work was so good several in our sales department thought she was one of our full-time production people."
>
> Charlie Herndon, Program Director
> WXXX, Knoxville, Tennessee
>
> **Goals and Objectives**
> I am seeking a full-time production position with a demanding, creative, radio production department in the Pensacola, Florida, area. I want to be involved with a station that needs a team player with an excellent track record of productive work relationships and on-time production delivery.

Notice that rather than leaving the goals and objectives as a single sentence, this student took the opportunity to address a specific qualification or requirement that had been listed for the position she was seeking. Just as a commercial has to attract the listener's attention in its first few seconds, the person who reads your resume needs to see something in the first 10 to 15 seconds to cause him or her to read further. A resume is a marketing tool for you, and your choice of words and the messages you deliver are critical. The reader needs to see something that lets him or her know that you have potential and deserve some attention. A resume has two basic purposes. The first is to get the employer's attention and an interview. The second is to build your confidence and pump you up in the eyes of the employer through your own personal inventory of skills and the realization of what you really do have to offer an employer.

Sometimes, in doing a personal inventory, a student realizes he or she does not have a lot of professional experience to put on a resume. If you think this is a big negative and will not get you a job, you are mistaken. For example, a student was interviewing in a top 30 market for a sales position. He was worried he would be asked about his lack of professional experience. He received the following good advice as to how to turn an objection into a benefit. If you are told that you have a great resume but that you lack professional experience, tell the interviewer, "You are correct and this is to WXXX's benefit because I don't bring years of bad work habits with me. You can help me develop my professional skills and work ethic in *exactly* the manner your company desires." In this example, the interviewer agreed, and the student was hired into a top 30 market. If someone asks you what you have to offer his or her station, be prepared to tell them.

Career Profile. Employment and employment history are very dull words. Your career profile sounds a little more interesting, and works well as a section following the introduction, your goals and objectives, and a summary of your outstanding qualifications. Although you design your goals and objectives to catch an employer's eye, your career profile is typically what draws the most attention from a prospective employer. This section of your resume begins with your current job and works backward, chronologically. Include your internship (clearly listed as such) under your career profile. Remember that any gaps in the timeline of your career profile, without an adequate explanation, are an automatic red flag to an employer.

If you have gaps in your career timeline, explain them in a simple, straightforward manner. For example, attending college is a perfectly acceptable explanation, as is a medical leave of absence or a personal leave of absence. Although employers can ask you about things like your college experience, they may not legally ask you about medical or personal issues. Because of the variety of situations, how to handle a specific gap on your resume is something best discussed with a career placement counselor at your university.

For each employer you should list:

1. The name of the corporation, the station call letters (if applicable), and location (city, state).
2. Dates of employment (rounded off to the nearest month).
3. Title or position.
4. Specific things you accomplished for the employer.
5. The experiences and skill sets required to carry out the duties of the position.

You should *not* list your supervisors' names, company phone numbers, or reasons for leaving.

Be sure to put most of the emphasis on your skills and how you used them to accomplish things or on your achievements while holding the position. One of the jobs some people shy away from listing on their resume is waitress, waiter, or server. Waiting tables, however, is actually a good example of how you can find applicable skills and abilities in just about every field. A server has to deal with a number of different restaurant staff members. That is called "excellent employee relations" or "the ability to work with people from a diverse number of backgrounds." And also: you did not wait on tables, you waited on restaurant customers. Make one serious mistake, and customers never return. That is called "great customer relations skills" or "the ability to work with a variety of customers simultaneously." Employers know that being a server in a restaurant for any length of time demonstrates excellent people skills.

Your present employer receives the most attention and space in your resume; give progressively less space to each employer as you work backward. Normally, you need to go back only 10 years and, in the case of college students, not even that far, since your employment history is probably not that lengthy.

As you write each employment description, use positive power words or buzzwords, as they convey your professionalism and knowledge. However, do not hide behind the buzzwords. For example, everyone is "creative." Tell your potential employer why or how you are creative. Everyone has "organizational skills." Tell your potential employer how you organized a project for a large client or a remote for a popular local business. Keep the focus on your accomplishments and how they contributed to the profitability of the business. For example, look at the following statements and decide which one you think would sound better to a potential employer.

1. "I helped install the new production computer software."
2. "I worked on the installation of new and enhanced production computer software, which resulted in increased productivity and improved workflow in the production department."

Notice how the second statement is a positive statement that includes accomplishments and benefits.

Take a look through broadcast trade publications, websites, and the professional want ad section of a major Sunday newspaper. Look at the buzzwords employers use to describe the positions they are hiring for. Learn to speak their language!

Education. The education section of your resume can go either toward the end of the resume, if you have a significant work history, or near the beginning of the resume, if you are a recent graduate with a more limited work history. Where you place the education section is your choice.

The education section begins with your most recent degree, the year you earned it, and the institution where you earned the degree. If you graduated magna or summa cum laude, this should also be listed with your degree. If you had a GPA of 3.0 or above, list it along with your degree. Do not list your high school diploma.

In addition to your degree, you may also choose to list significant honors, awards, or scholarships you have received, along with any special licenses or certifications you have earned. In particular, list any college classes that have direct applicability to the job you are applying for. If you did a major research project that directly applies to the field or position you are applying for, you should list it as well. List your membership in any honorary, professional, fraternal, and social organizations that relate to the position you are applying for.

Miscellaneous Sections. There are many things that distinguish you as an individual that you may wish to include toward the end of your resume under various headings. If you served in the military and were honorably discharged, briefly mention this under a heading of Military Service. However, keep the listing simple, using only terms a civilian can understand. If your military service was directly related to the position you are applying for, it is appropriate for you to go into detail about your military career. For example, if you are applying for a position at a radio or television station and you served with Armed Forces Radio and Television, the details of your military service could be beneficial. Again, any details about your military career need to be written in plain English so that a civilian can easily understand what you did and where you did it; stay away from military terms and abbreviations.

Do you speak a second language fluently? You might not think this is important, but in a global economy it is an increasingly valuable job skill. Not every radio station client speaks English well. In some markets, this skill is critical enough that it should be listed at the very beginning of your resume. For example, you are a valuable employee if you speak Spanish in any major metropolitan area, particularly in cities like Miami, Atlanta, or New York.

Certifications and membership in student and professional organizations should be listed on your resume if they relate to the position you are seeking. For example, a salesperson who has earned the Certified Digital Marketing Consultant certification from the Radio Advertising Bureau should list this information on his or her resume. Engineering students should list any Society of Broadcast Engineers certifications held. Certifications and memberships are listed under a heading such as Professional Memberships or Professional Certifications.

Community service and volunteer work is an excellent indicator of character. Many employers like to see a history of community involvement. In particular, radio stations require that on-air staff become involved in community projects and maintain a high profile in the community. If you serve on any community boards or service group boards, this information should be included on your resume (see figure 15.5). Community involvement is listed under a Community Service heading. A word of caution: if your community service is related to your religious beliefs, you might want to carefully write it to emphasize what you did rather than your particular religious beliefs.

Most radio production people are capable of operating a number of different digital audio software programs and various pieces of equipment. Tell your potential employer about the abilities you have in a section of your resume entitled Software or Technological Skills. List the hardware and software you are proficient with. List whether you are Mac, PC, or dual-platform proficient, and what major computer operating systems you are familiar with (see figure 15.5). Employers want to see that your computer and software skills are current.

And finally, there are your personal interests, which can spark conversation and show a potential employer you have a life outside work. These are particularly good to list if they relate to your work. As an example, consider the radio person who collects vintage microphones and radios. The radio person is also an advanced, open-water scuba diver and enjoys sports car racing. Collecting microphones and radios shows a historical interest in radio work. The scuba diving reflects taking calculated risks on a regular basis. The sports car racing reflects that the person is competitive—a good quality to have in one of the most competitive industries in the world. You should devote only a simple, brief listing for each item to give your potential employer an idea of who you are outside work.

References. Normally, you state "References furnished upon request" at the end of your resume. References are not usually listed on the resume, but they certainly deserve some attention as they are a critical part of the hiring process for most employers.

The reason you do not put the names of references and their phone numbers on your resume is that most former employers will only confirm dates of employment during a telephone reference request. Yet your potential future employer will want more information than that, and this is when letters of reference are a good idea.

Most employers want to see and verify references. As a student starting out, you can enhance your professional status with a prospective employer by having five or six current reference letters ready for examination when requested. Rule number 1 with respect to references is to use only professional references, not family members or friends. References from people you worked with during your internship are very appropriate. As a recent grad, academic references from professors whose classes have direct applicability to the position you are applying for are also appropriate. Rule number 2 is that reference letters *must* be on company letterhead. There are no exceptions. Rule number 3 is that reference letters should not be more than one page long.

Prepare to secure your reference letters as you are putting together your resume. Make a list of five or six individuals who will provide you a favorable reference, and contact them. Tell them what you are doing and ask if they would help you in your job search by providing a favorable reference letter. When asked, most people are pleased to help. If your resume is ready, send each individual a copy so he or she will have a good picture of what you are seeking to accomplish.

Typically, one of two things will happen when a person agrees to give you a reference. The person either asks you to suggest specific things to be covered in the body of the letter, or the person asks you what you want the letter to say in general. In either case, you need a letter that supports the critical elements of your resume and highlights your specific accomplishments. It is appropriate to tell the person what the letter needs to say, or to coach the person on how you want the letter to be written.

When you receive a reference letter that has been written for you, promptly send a thank-you note to the person expressing your gratitude for his or her support of your job search. Along with your thank-you note, you should send the person a copy of your resume if you haven't already done so. Explain that it is for reference purposes should a prospective employer call him or her for more information. Keep in touch with your references and contact them after each interview so that you can keep them up-to-date on your job search. Let them share in some of your excitement of finding a new job.

Once you have your reference letters in hand, take them to a professional copier service and have color copies made. Create a title sheet with the header Professional Reference Letters. On the title sheet, list each reference by name, followed by the company he or she represents, their address, and phone number. Place the copies of the reference letters behind the title sheet and have them bound at a professional copier service or office supply store. You should also have your reference letter package scanned and converted to a PDF file that can be delivered electronically. When submitted with your resume, such a presentation gives you a great deal of credibility. In effect, you and five business leaders are applying for the position.

Keep three copies of your resume and references with you at all times when interviewing or networking. You need three so that you have two backups in case anyone else asks for a copy during an interview or while you are talking to someone. Remember, broadcasters typically need someone yesterday. They work in minutes and seconds and are accustomed to things happening fast. If someone you meet asks for a resume and references—hand it to them on the spot! If you are attending a job fair, carry a dozen resume/reference packages with you.

Audition

Now you have an application letter, a resume, your references, and you are ready to apply for your first job at a radio station. Right? Not quite. Remember, this is a talent-based industry, and potential employers want you to demonstrate your skills for them in the form of an audition. Almost all radio positions require an audition that showcases your best work.

Your audition should not be more than 4 (preferably 2 to 3) minutes of your very best work. If you are applying for a morning-drive position, you can extend the time frame up to 10 minutes, if you feel it is necessary. Something to keep in mind is that the person who listens to your audition has likely never met you. Your audition is both your introduction and who you are, to some degree. A prospective employer interviews the best-sounding audition.

Creativity is the key when it comes to the audition. Your audition needs to stand out from the pack and deliver the message that you have all the necessary skills to create great radio. At the very most, you are going to have 15 to 20 seconds to catch the ear of the person reviewing the audition. The first 15 seconds need to be bright and expressive and need to reach out and grab the listener by the throat and scream "listen to me!" If it does not, your audition will get deleted or tossed into the round file.

Start the audition with your shortest, *best* work, and follow it with a series of fast-paced segments and cuts that move from one aspect to another of your production skills. The segments should not be longer than 15 to 20 seconds long. Think of your audition as telling a story with an incredible beginning, a good middle, and a great ending. Save your *second-best* work for the end of the audition because you need to leave the person impressed. It is important to have a high wow factor at the end of your audition because this is what motivates the program director to pick up the phone and call you.

For a production person, the content of your audition needs to include at least five commercial segments or complete commercials. You need to include as much demo material as possible. Show the extremes to which you can go: from a soft, caring, funeral home commercial to a rock concert commercial. Include a couple of station promos, sweepers, and liners. Once your audition starts, there should be no spaces or gaps. It's a steady barrage of great production work with a super ending. Basically, your audition is a production about your production.

Listen to audio demo, cut 15-1. The audition you will hear was a class assignment and was produced entirely in a studio by a college junior who had never held a position in radio. His target: an edgy Alternative station. Although it may not be the style of music you prefer, listen carefully to the production. Creatively mix your segments so that the listener is teased and wants to hear more, rather than just lining up one commercial after another. Don't be afraid to develop a short imaging sound effect, such as the record noise and static at the beginning of the demo or the sound of a radio being tuned, to go between segments. This same create-it-in-a-studio style can be used for an announcer looking for his or her first position.

For an experienced announcer seeking an air shift, the content of the audition needs to be focused on your on-air personality (see audio demo, cut 15-2). The student on this demo had extensive on-air experience before reentering college. If you are working, even on a part-time basis, you should be air-checking yourself every day and saving the audio files on a flash drive. From these air checks, edit your best work together, keeping in mind that the focus is on you. Although you may talk over music or talk about upcoming artists and station events, edit out as much of the music as possible. For example, as soon as you finish talking over the intro to a song, cut to the next segment. The program director does not need to hear any music.

Audition Submission

Radio is a got-to-have-it-yesterday business. Most auditions are delivered electronically (MP3), along with a PDF of your resume via email. Although a word processor file can also be attached to an email message, it is not protected from alteration or accidental modification through error as the PDF is. One email can take care of it all. Although a little unusual, you may find a PD that still prefers a hard copy of the audition and resume. A couple of options are a CD or an inexpensive flash drive. The physical presentation of your CD is just as important as what is on the CD. Use a quality CD and, after you have burned it, listen to it in a CD player other than the one on your computer. The presentation of your flash drive or CD needs to demonstrate to your prospective employer that you are a pro. The CD must be clearly labeled. There are a number of software/label products that can create a professional label on your computer. The label should appear as neat, clean, and professional as possible.

Be sure your CD label has your name, address, phone numbers, and email address clearly displayed (see figure 15.6). If you are sending a flash drive you can attach a business card sized tag to the flash drive with the same information. Believe it or not, auditions sometimes end up in some pretty strange places in a radio station, and if they are not labeled, they are history. Often, they get passed around among the individuals making the hiring decision.

Figure 15.6 Sometimes simple is best. This CD label clearly displays the applicant's name, address, two phone numbers, and email address. Keep in mind when you label your CD that it may become separated from your resume, and you never know where it will end up. It is important to have your complete contact information on the disc.

Remember, an audition is the first impression you make on a prospective employer. If your materials look sloppy, you are perceived as sloppy. If your application email or letter looks poorly done, you are perceived as someone who cannot handle details and lacks professional standards. First impressions are lasting impressions.

Packaging Your Application

If the employer wants you to send your audition and resume electronically you still have some work to do. Your email cannot look like a text message. It needs the same attention to detail that you would give to any letter of application. No spelling or grammar errors, a clear email address that includes your name in some form, and don't forget to include your contact information. One area that needs attention is the subject line in the email. Believe it or not there are people who open with the word "Hey" in the subject line. Do you know how many junk emails do this to make it look like they know the recipient? In the subject line put the reason you are writing: "WXXX Production Person Position." Which of these two subject lines would tempt you to open an email?

On the other hand, if the station wants a hard copy of everything you need to prepare a "package" to sell yourself. With your resume, professional references, and audition you need to write a cover letter and ship your package to a prospective employer. The cover letter and the package should be addressed to the individual doing the hiring. Never send a package to the attention of the program director or station manager. If you do not know the name of the program director, check the station's website or call the station to get the person's exact name, title, and mailing address. If you are given a post office box number, ask for the station's physical mailing address. Even the label on your envelope says something about you. Carefully handwrite the address and return address.

Rule number 1 of sending materials: send them flat, never folded. Rule number 2: *never* send your materials in a brown or manila envelope. Junk mail comes in brown or manila envelopes. The large padded envelope you select to send your materials in is critical. Distinguish your package so that it stands out from all the rest. At the very least, purchase some large, bright white envelopes to send your materials in. Sending your materials via priority mail from your local post office is an even better way to distinguish your package from the rest. Use the bright, colorful, free envelope the post office provides for this service (see figure 15.7). Even if you are mailing your package to a station in the town where you live, send it via priority mail. Priority mail attracts attention and gets opened. It also subtly says to the employer that you value the application and the opportunity.

Figure 15.7 How you package your resume and materials is important. If you were a program director, which of these envelopes would you open first?

If the potential employer is located out of town and you are really interested in the position, send your materials via a shipping carrier and require a signature. This lets the employer know you are a professional and you are serious.

Tracking an Application's Progress

Once your carefully prepared materials have been delivered to the appropriate person, what do you do? Wait. Then wait some more. Remember: radio is different when it comes to getting a job. If you answered an ad in a broadcast trade publication, very likely the ad said "*No* calls." There is a reason for this. It is not unusual for a trade publication ad to generate more than 100 responses for a program director to consider. Some of the applicants are qualified; the vast majority, however, are not.

Program directors are human beings. In today's marketplace, program directors often handle the programming for more than one station. Their time is very valuable. Do you really think they have the time to take over 100 phone calls from potential applicants, many of whom are not qualified to begin with? Resist the urge to call if the ad says "*No* calls." You'll only be annoying the person you want to impress, and besides, the chances are very high that you won't get through anyway because the program director has told the receptionist (gatekeeper) to block all calls regarding the position.

If, on the other hand, the ad did not say "*No* calls," then go for it. It may not be easy to get through to a busy person. No matter how frustrated you are that the person has not returned a call or you have not gotten to talk to that person, *never* challenge the receptionist/gatekeeper. In fact, make some small talk and tell the person you understand he or she has one of the toughest jobs in the station. Thank the person for any help he or she can give you. If the person takes a liking to you, you may get through. However, if the person feels you do not appreciate his or her position, your calls will never get through. Treat people the way you would like to be treated.

Is there any other way to get to the program director? Try a simple email. And if or when you do get through to the program director with a phone call or an email, keep it short and simple. Tell the program director you know he or she is swamped but that you just wanted to confirm that the package you sent had arrived and thank him or her for consideration. *Do not* try to advance the call or email past this point. All you want to do is confirm that your package arrived and say thank you. If *the program director* wants to advance the conversation, answer any questions politely, but don't get pushy.

What you have to understand about radio is that program directors *do* call the people they are interested in, without hesitation. They are constantly searching for the best talent to fill the positions they have. During the career of this author, I once delivered a resume and audition to a major-market news/talk station on Wednesday morning, was called Wednesday afternoon, interviewed Thursday morning, and reported for work on Friday. The process does not always happen that fast, but radio stations *do* move when it's time to move.

Suggested Activities

1. Go online and visit the website for three major radio companies. Look at the ads and the positions listed, paying particular attention to the language used in the ads. Look for words and phrases you can incorporate into your cover letter and resume.
2. Visit your state broadcasters' association website. Many have employment clearing houses or job banks that list available positions around the state. Search for something that appeals to you. Review your resume to see what changes you would need to make to fine-tune it to apply for a particular job.
3. Go to the library and see if your university or school has online access to the Standard Rate and Data Service (SRDS). This service provides detailed current listings for all of the radio and television stations in the United States. Learn how to read the listings so that you can find out valuable information about a station, including names of personnel, phone numbers, and addresses.
4. Visit the National Association of Broadcasters Education Foundation Broadcast Career Link website (www.broadcastcareerlink.com). Broadcast Career Link has job listings for radio and television stations all over the country. This is an incredible resource. Explore the site and become familiar with all of the services offered for the job seeker.
5. Do a Google search for "career designers" and explore the many different aspects of this interesting field.

Websites for More Information

For more information about:

- *broadcasting job banks and career development,* visit the Broadcast Career Link at www.broadcastcareerlink.com
- *employment opportunities with each of the major station chains,* visit each of these corporate websites:

  iHeartRadio and Entertainment at www.iheartmediacareers.com;

  Cumulus Media at www.cumulus.com; and

  CBS Radio at www.cbsradio.com
- *personal information that is required and is not required on a job application or resume,* visit the US Equal Employment Opportunity Commission at www.eeoc.gov
- *using social media in your job search,* visit Joshua Waldman's Career Enlightenment website at www.careerenlightenment.com

Appendix
Digital Radio Production
Audio and Music Demo Cuts

Audio Demo Cuts

Each audio demo cut has a narrative explaining what you will hear and what key element you should listen for.

Chapter 1

Cut 1-1 A sample commercial with the music and copy mismatched.

Cut 1-2 A sample commercial with the music properly matched to the copy.

Chapter 2

Cut 2-1 Audio frequency sweep from 100 Hz through 20 kHz, with the major frequencies announced.

Cut 2-2 Audio frequency sweep from 50 Hz to 5 kHz at a fixed sound level to demonstrate perceived sound-level changes.

Cut 2-3 A fixed frequency audio signal with the sound level varied to demonstrate a perceived frequency change.

Cut 2-4 Shuttle launch recorded at the Kennedy Space Center press site to demonstrate how it takes nearly 15 seconds for the sound to arrive there from the launch pad, which is 3.1 miles away.

Cut 2-5 Demonstration of music recorded 180 degrees out of phase.

Chapter 3

Cut 3-1 Vocal demonstration of the proximity effect using an AKG 414 microphone.

Cut 3-2 Demonstration of a ribbon microphone's open, natural sound.

Cut 3-3 Demonstration of the proximity effect with a ribbon microphone at various distances up to 3 feet.

Cut 3-4 Vocal recording with an AKG 414 at 3 feet, 2 feet, 1 foot, and 6 inches from the face of the microphone.

Cut 3-5 Omnidirectional microphone pattern demonstration with an AKG 414 microphone.

Cut 3-6 Bidirectional microphone pattern demonstration with an AKG 414 microphone.

Cut 3-7 Cardioid microphone pattern demonstration with an AKG 414 demonstrating 0-degree versus 180-degree sensitivity.

Cut 3-8	Hyper-cardioid microphone pattern demonstration.
Cut 3-9	A demonstration of polar frequency response by rotating an AKG 414 while in the super-cardioid pattern and speaking into it.
Cut 3-10	Demonstration of an AKG 414 microphone, without a shock mount, and the building noise and rumble it picks up.
Cut 3-11	Recording of birds and mountain stream in the Joyce Kilmer Forest.
Cut 3-12	Recording of white-water rafters on the Nantahala River.
Cut 3-13	A recording of a direct and an ambient microphone demonstrating the various mixes that can be created from the two tracks.
Cut 3-14	***P*** popping demonstration.
Cut 3-15	Demonstration of a Shure SM 81 with and without a pop filter.
Cut 3-16	(A) Demonstration of a Sennheiser 421 II with a person speaking directly into the front of the mic, popping their ***p***'s. (B) The same person speaking into the same microphone, turned 90 degrees to the side.
Cut 3-17	(A) Demonstration of a Sennheiser 421 II outside, without a windscreen, on a windy day. (B) The same microphone with a windscreen, under the same conditions.
Cut 3-18	Demonstration of a person speaking directly into a lavalier microphone placed at the lips.
Cut 3-19	Demonstration of a shotgun microphone's audio pickup compared to an omnidirectional microphone at the same distance from the sound source.

Chapter 4

Cut 4-1	A demonstration of equalization by placing one sound above or below another carving an equalization hole (i.e., to demonstrate height in a mix).
Cut 4-2	A demonstration of panning to help provide width in a mix.
Cut 4-3	A demonstration of reverb to help provide depth in a mix.

Chapter 6

Cut 6-1	An example of an announcer's voice with and without a voice processor.
Cut 6-2	An example of an announcer's voice with and without compression over music to demonstrate how compression raises the voice above the music.
Cut 6-3	A whisper that is compressed and becomes a loud whisper.
Cut 6-4	An example of an announcer's voice that compares a 4:1 compression ratio to a 15:1 compression ratio.
Cut 6-5	An example of an announcer's voice demonstrating a compressor pumping.
Cut 6-6	An example of an announcer's voice with limiting applied.
Cut 6-7	An example of an announcer's voice with expansion applied.
Cut 6-8	An example of an announcer's voice using a noise gate.
Cut 6-9	A demonstration of a high-pass filter cutting or removing building rumble.
Cut 6-10	A demonstration of a low-pass filter cutting or removing the high end from a vocal.
Cut 6-11	A demonstration of a notch filter removing pops and clicks.
Cut 6-12	A demonstration of a de-esser on an announcer's voice.
Cut 6-13	An equalization demonstration to show the effects of boosting and cutting frequencies.
Cut 6-14	An example of an announcer's voice with 40 milliseconds of slap delay.
Cut 6-15	An example of an announcer's voice with 40 milliseconds of slap delay and feedback.
Cut 6-16	An example of an announcer's voice with less than 35 milliseconds of slap delay.

Cut 6-17 An example of an announcer's voice with reverb, demonstrating three different-sized rooms or spaces.
Cut 6-18 A recording of a piano without reverb, compared to a piano with small reverb and short delay to make it sound bigger than life.
Cut 6-19 A recording of a music bed and how to move it behind or in front of a vocal using reverb.
Cut 6-20 A demonstration of a 20- to 35-millisecond chorus effect on a piano recording.
Cut 6-21 A demonstration of the phase-shift effect.
Cut 6-22 A demonstration of a 10- to 20-millisecond flanging effect.
Cut 6-23 A demonstration of the re-amping effect.
Cut 6-24 A demonstration of normalizing a vocal.
Cut 6-25 A demonstration of adjusting the input on a PreSonus voice processor.
Cut 6-26 A demonstration of adjusting the compressor controls on a PreSonus voice processor.
Cut 6-27 A demonstration of adjusting the expander on a PreSonus voice processor.
Cut 6-28 A demonstration of adjusting the de-esser on a PreSonus voice processor.
Cut 6-29 A demonstration of adjusting the shelving filters on a PreSonus voice processor.
Cut 6-30 A demonstration of adjusting the peak limiter on a PreSonus voice processor.
Cut 6-31 A demonstration of the Eventide Harmonizer and its effect on voice.

Chapter 7

Cut 7-1 An example of a commercial in which the advertiser talks only about his business and uses industry jargon (Amy's Cookie Factory).
Cut 7-2 An example of a commercial in which the advertiser stresses customer benefits (Amy's Cookie Factory).
Cut 7-3 An example of a commercial that uses "we" and "us" versus one that uses the business's name instead (Amy's Cookie Factory).
Cut 7-4 An example of a good wrap at the end of a commercial.
Cut 7-5 An example of how confusing telephone numbers can be in a commercial and a better way to present them.

Chapter 8

Cut 8-1 A demonstration showing how too many sound effects ruin a commercial.
Cut 8-2 Two different reads of a commercial to demonstrate the differences in presentation of the same material.
Cut 8-3 An original, unedited, location recording of the historic Baldwin steam engine number 722 owned by the Great Smoky Mountains Railroad, Dillsboro, North Carolina, for sound effects editing purposes.
Cut 8-4 An "OK" read of a commercial with the announcer doing several pickups so that the production person can edit to the read he or she likes.
Cut 8-5 A 60-second music cut created with music loops.
Cut 8-6 A comparison of an unprocessed voice track and the same voice track with mild compression and reverb, demonstrating how the processed voice blends with the commercial.
Cut 8-7 A 60-second agency commercial with space for a 10-second tag at the end for local copy.
Cut 8-8 A 60-second agency commercial with a donut in the middle for local copy.
Cut 8-9 A news story that has been unethically enhanced using sound effects and audio processing.

Cut 8-10 A person saying the numbers 1 through 10 in random order, with breath sounds between each; this is an editing exercise.

Chapter 10

Cut 10-1 Radio station sweeper demonstration.
Cut 10-2 A national, award-winning promo.
Cut 10-3 Morning show promo.
Cut 10-4 Promotional voice demonstrations.
Cut 10-5 Promotional voice doing sample liners.
Cut 10-6 Promotional voice doing sample sweepers.
Cut 10-7 Highly produced sweeper sample.
Cut 10-8 Transition sweeper from music to news.
Cut 10-9 Sample sweeper with station ID buried inside.
Cut 10-10 Demo of jingles from TM Studios.

Chapter 11

Cut 11-1 Thirty seconds of a remote broadcast on a POTS line.
Cut 11-2 Thirty seconds of a remote broadcast on a POTS codec.
Cut 11-3 Thirty seconds of a remote broadcast on an ISDN codec.
Cut 11-4 Thirty seconds of a remote broadcast on a Marti VHF remote pickup transmitter.
Cut 11-5 Thirty seconds of a remote broadcast on a satellite phone.

Chapter 14

Cut 14-1 An audition of imaging, including liners and sweepers from Jeff Laurence.

Chapter 15

Cut 15-1 Actual student production audition CD.
Cut 15-2 Actual air talent's audition CD.

Music Demo Cuts

Welcome to the real world of radio production! The music demo cuts are unique in that they have been written and produced for radio production students. All the music cuts have been composed and produced by Dr. Bruce H. Frazier, the Carol Grotnes Belk Distinguished Professor in Commercial and Electronic Music at Western Carolina University. His many associates in the television and motion picture industry know him as the owner of Sallycrab Music, a Broadcast Music, Inc. publishing firm.

In this appendix you will discover an outstanding selection of 40 production music cuts and suggested activities, including a custom studio-tracking session written specially for this text. You are free to record, rerecord, edit, and create new music editing these cuts as a part of your production class. However, under no circumstances are you permitted to use the music for commercial purposes of any kind.

As you explore the music, you will see that some of the cuts are odd lengths. This is because they are filled with multiple edit points so that you can experiment with editing and creating new music. For example, if a cut is 1:35 in length, chances are that you can easily get a 60-second and a 30-second cut from it. But it's also just as likely that in addition to those two cuts, you could also create another 60-second cut and still another 30-second cut from the same original piece of music!

What is really amazing about the produced cuts of music is the range you have to work with and what you can create from the cuts themselves. Below you will find a brief, subjective description of each cut and its suggested uses and applications. Listen carefully and I am sure you will think of many more creative uses.

To help get you started on the production process with the music cuts, refer to chapter 5 of your *Digital Radio Production* textbook.

Have fun!

Donald W. Connelly

Chapter 5

Cut 1 **Carolina Blues—full mixed version (1:35)**
Cut 2 **Carolina Blues—piano track (1:35)**
Cut 3 **Carolina Blues—bass track (1:35)**
Cut 4 **Carolina Blues—drum track (1:35)**
Cut 5 **Carolina Blues—rhythm guitar track (1:35)**
Cut 6 **Carolina Blues—lead guitar track (1:35)**

Description

These cuts have been composed and recorded as an editing exercise. Carolina Blues is a simple blues progression including a piano, bass, drums, and rhythm and lead guitars. The piece was written with a multitude of built-in edit points for you to experiment with. In addition to the full mix version, you get each of the separate tracks from the recording session. What is unique is that each of the tracks can stand alone behind a commercial and, when listened to individually, take on a whole new flavor.

Suggested Uses

1. Start by laying down the five individual tracks and doing your own full mix of the piece. Or, you can mix and match the tracks to form completely new music. The various tracks will need to be synchronized. Maybe you only want to use the piano and drum tracks, or the guitars and drums.
2. Because the 12-bar blues pattern is repeated four times, parts from different sections can be layered over each other. For example, try cutting the guitar solo over music to the first chorus.
3. Try using Carolina Blues for barbecue sauce, a riverboat cruise, a New Orleans restaurant, shrimp and crab dip, a Memphis nightclub, or anything that contains pork fat, or requires a napkin.
4. With these five tracks you can create over 100 new pieces of music. It's up to you.

Cut 7 **Rock 1 (1:30)**
Cut 8 **Rock 2 (1:00)**

Description

Rock 'n' roll! Rock 1 has an up-tempo aggressive sound, while Rock 2 is a more laid back traditional sound to it. Note that the cuts are filled with natural edit points, so feel free to experiment to create your own unique signature piece.

Suggested Uses

1. These cuts would fit well behind a car dealers ad, a club, or any business that wants to stand out with aggressive music.
2. Rock 1 and 2 also can work for sports themes.

Cut 9 Swing (1:00)

Description

Swing, the popularized version of jazz that came into vogue starting in the 1930s, is a holdback to an earlier time. While it might sound out of place to you, keep in mind there are a lot of businesses that want a light, easy-going sound behind their commercial.

Suggested Uses

1. How about creating something for a woman's dress shop, a restaurant, a gift store, etc.?

Cut 10 Funky (0:59)
Cut 11 Funky Too (1:00)

Description

Funky One and Too are just what their name implies—funky. Both cuts illustrate a unique, identifiable, style of guitar playing known to Funk. This is music you can have fun with and try out behind a variety of commercials as an alternative sound.

Suggested Uses

1. These might work behind restaurants, music stores, women's beauty and men's styling salons, or a sub sandwich shop.
2. Play around because there are many edit points in these songs.

Cut 12 Smooth (1:00)

Description

The name says it all, Smooth. Smooth features a standard rhythm section with additional musical colors from electronic keyboards All together, this produces a silky, light, soothing sound.

Suggested Uses

1. This versatile piece of music will fit behind so many types of businesses it is amazing.
2. This might work behind restaurants, clothing stores, florists, dentists, doctors, gift shops, book stores, beauty salons, and antique stores just to name a few. See what I mean?

Cut 13 Sweet (0:59)

Description

Very much like Smooth, Sweet goes behind a lot of business types. Think of Sweet as the best quality vanilla ice cream, it goes with everything!

Suggested Uses

1. Think veterinarians, hospital and health care ads, restaurants, floral shops, funeral homes, bridal shops, or any kind of ad that needs an "awww" at the end!

Cut 14 Pretty (waltz) (1:00)

Description

In terms of composition, Pretty has a soft building sound.

Suggested Uses

1. OK, this is the bridal shop winner!
2. Like Sweet and Smooth, Pretty goes well behind so many softer sell commercials.
3. Listen carefully and see if Pretty could be used to tell your public service groups story.

Cut 15 A Bit of Bach (2:00)

Description

Revel in Johann Sebastian Bach's Prelude in C, from the Well Tempered Clavier.

Suggested Uses

1. Every radio station has a few high-end clients that want to sound classy.
2. Use this for the jewelers, the Mercedes dealer, and the five-star restaurants in town.

Cut 16 Mozart (1:00)

Description

An excerpt from W. A. Mozart, Sonata in A, K331, 3rd movement.

Suggested Uses

1. Again, just like Bach, well-known classical music for some of your upscale or otherwise classy advertisers who want that exclusive sound.

Cut 17 Far East Loop (1:00)

Description

With just the first few notes this music identifies itself. Can you hear the koto-like sound and see the shimmering bells? (This piece of music was created with loops from the Adobe Audition 1.5 Loopology Sampler.)

Suggested Uses

1. The obvious advertiser is an Asian restaurant, but don't overlook the upscale health club/gym, health food store, massage therapist, or the annual jade sale at the jewelry store.

Cut 18 Reggae Loop (1:00)

Description

What can you say, it's reggae, but not over-the-top reggae. What makes this cut distinctive is the traditional back-beat rhythm from the guitar. It's a very flexible cut, with lots of neat edit points. (This piece of music was created with loops from the Adobe Audition 1.5 Loopology Sampler.)

Suggested Uses

1. Try this behind any commercial you want to liven up or give a completely different treatment to.
2. Try this behind an ad for an early morning breakfast or donut shop. This might also work behind a barbecue restaurant ad.

Cut 19 Garage Loop (1:00)

Description

This cut is like someone turned the Broadway play *Stomp* loose in your garage. Lots of metallic sounds combined with a seriously powerful bass line. (This piece of music was created with loops from the Adobe Audition 1.5 Loopology Sampler.)

Suggested Uses

1. My first inspiration was for a tattoo shop for this piece of music—it's that edgy.
2. Other candidates for this music are Alternative music and specialty stores appealing to a younger audience.

Cut 20 Dirty Loop (1:00)

Description

Dirty Loop is a mix of retro and new sounds starting off with what sounds like a Hammond electronic organ with a Leslie speaker. (Do a web search to find out what a Leslie speaker is!) (This piece of music was created with loops from the Adobe Audition 1.5 Loopology Sampler.)

Suggested Uses

1. Candidates for this music are bars, clubs, concert venues for local bands and Alternative music and specialty stores.
2. Watch out for the bass on this cut, it's clean, it's loud, and it's proud!

Cut 21 News and Public Affairs Bed (full version) (1:00)
Cut 22 News and Public Affairs Bed (short) (0:30)
Cut 23 News and Public Affairs Bed Bumper (10:15)
Cut 24 News and Public Affairs Bed Bumper (20:05)

Description

A news or public affairs bed is music that can go behind copy. It opens strong and then blends easily with copy. The News and Public Affairs Bed is a major-market sounder that stations typically pay several thousand dollars to have custom-created for their station. This bed was created especially for this text and you will not hear it anywhere else!

Suggested Uses

1. With multiple edit points, these cuts can be looped into a two-minute news bed or edited to fit a specific need.

Cut 25 Drum 1, Basic Pattern (0:10)
Cut 26 Drum 2, Pattern with Fill-A (0:14)
Cut 27 Drum 3, Pattern with Fill-B (0:10)
Cut 28 Drum 4, Pattern with Fill-C (0:10)
Cut 29 Drum 5, Fill Only-A (0:03)
Cut 30 Drum 6, Fill Only-B (0:02)
Cut 31 Drum 7, Ending Only (0:09)
Cut 32 Drum 8, Pattern with Crash (0:10)
Cut 33 Drum 9, Pattern with Ride (0:13)
Cut 34 Drum 10, Pattern with 8ths (0:10)
Cut 35 Drum 11, Pattern with 8ths + Open Hat (0:10)
Cut 36 Drum 12, Pattern with 8ths + Crash (0:10)
Cut 37 Drum 13, Pattern with Ride (0:13)
Cut 38 Drum 14, Pattern with 16ths (0:10)
Cut 39 Drum 15, Pattern with 16ths + Crash (0:10)
Cut 40 Drum 16, End Hit Only (0:05)

Description

There are so many things you could create with these drum patterns that it is mind-boggling. By the way, these are real drums played by a real drummer by the name of Moose. The drums include a seven-piece Premier XPT drum set, Sonor kick pedal, Sabian, Zildjian, and Paiste cymbals. This vintage drum set is made of maple, giving the drums a rich, full tone.

Suggested Uses

1. With these audio clips, you can design anything from beginnings to endings and patterns that are easily looped into longer segments.
2. Try importing these loops into Adobe Audition CC and experiment with combining the drum tracks with one of the other music demo cuts.
3. Edit your heart out to see what you can create.

Glossary

acoustical phase—A difference in time between two or more sine waves, measured in degrees. (chapter 2)

active crossover—A powered speaker crossover network with adjustable crossover frequencies. (chapter 4)

A/D—Analog-to-digital audio converter. (chapter 11)

AM—Amplitude modulation. The AM radio band extends from 535 kHz to 1,605 kHz. (chapter 13)

American Society of Composers, Authors, and Publishers (**ASCAP**)—A membership-based performing rights organization that protects the copyrights of members' works by licensing and distributing royalties for the performances of their copyrighted works. (chapter 1)

amplitude—Amplitude represents the strength, or loudness, of a sound or audio signal without regard for its frequency. The height of the sine wave can be measured either in decibels (for sound) or voltage (for audio signals). (chapter 2)

attack time—The fixed or adjustable length of time it takes for an audio processor to sense the presence of audio and react to the audio, measured in milliseconds. (chapter 6)

audio—Latin, "to hear"; sound that has been converted into electrical energy. (chapter 3)

balanced microphone cable—A microphone cable or line in which there is a positive wire, a negative wire, and a shield or earth ground surrounding the positive and negative wires. (chapter 3)

band-pass filter—An audio filter that has a high and low cutoff frequency, allowing only the frequencies in between to pass unaffected. (chapter 6)

bandwidth—With reference to a radio carrier wave, bandwidth is the space on the radio band occupied by the radio carrier wave frequency and the audio that has been superimposed upon it. The superimposed audio signal widens the radio carrier by adding sidebands just above and below the assigned carrier frequency. The carrier wave and the upper and lower sidebands of the signal compose the bandwidth of a signal. (chapter 13)

barker—A person who attempts to attract people to an event or place for entertainment purposes. He or she often talks about the uniqueness or novelty of the entertainment. A radio station can be a barker for its website. (chapter 10)

barometric air pressure—A measure of the weight of the atmosphere, or air, pressing down on the earth. It averages approximately 29.92 inches of mercury at mean sea level, which is the equivalent of 14.7 pounds per square inch. Barometric pressure varies with altitude, getting lighter as the altitude increases. (chapter 2)

bass boost—Another term for proximity effect. (chapter 3)

bidirectional—A microphone pattern that has maximum sensitivity at 0 and 180 degrees and maximum rejection at 90 and 270 degrees. (chapter 3)

BNC—An unbalanced coaxial cable connector with a bayonet-type locking system invented by Paul Neill and Carl Concelman, thus the name Bayonet Neill-Concelman. Typically found on video and digital audio cables. The center conductor of the cable is connected to a pin, and the shield is connected to a barrel surrounding the pin. A rotating ring outside the barrel locks the cable to any female connector. (chapter 5)

branding—A station's marketing plan that seeks to provide a clear differentiation from the other stations in the marketplace; based on the station's overall identity and attitude as defined by the listener. (chapter 10)

breakout box—A device designed to allow audio distribution at press conferences through multiple mic or line-level outputs. (chapter 11)

bright—A sound rich in high frequencies. (chapter 6)

broadband—An Internet connection capable of transmitting data at a speed over 200 Kbps. Typical broadband connections include cable TV modems, DSL lines, and wireless connections. (chapter 12)

Broadcast Music, Inc. (BMI)—A membership-based performing rights organization that protects the copyrights of members' works by licensing and distributing royalties for the performances of their copyrighted works. (chapter 1)

bus—A common point in an audio device into which several inputs feed and are joined together. (chapter 4)

butt-plug—A wireless transmitter designed for plugging into the base of a microphone. (chapter 3)

call-out research—Research conducted by a radio station or a research firm to ask listeners questions about how they perceive the station's music format; often used to keep a station's music format on track from week to week. (chapter 1)

cardioid—A heart-shaped microphone pattern with maximum sensitivity at 0 degrees and maximum rejection at 180 degrees. (chapter 3)

carrier wave—A high-frequency radio wave transmitted by a radio station that has been amplitude or frequency modulated. (chapter 13)

centerline—An imaginary line extending straight out from the face of a microphone, or on a 0-degree axis. (chapter 3)

codec—Coder/decoder; a digital audio device that codes analog audio into digital audio and decodes digital audio back into analog audio. (chapter 11)

coincident pair—Stereo microphone placement in which the two microphones are placed side by side facing the sound source. With the midpoint of the mic bodies serving as the rotation point, the heads of the two microphones are aimed inward nearly touching. One is aimed at the right side of the sound source, the other at the left side. (chapter 3)

combo mode—When the on-air radio talent serves as the show's engineer as well, running the control board and other elements of the show in addition to announcing. (chapter 4)

commercial inventory—The number of commercials that a radio station has to offer for sale. Commercial inventory can be broken down into weekly, daily, and hourly inventory. (chapter 1)

commercial inventory load—How many commercials are actually scheduled to run in a given time period, such as an hour, half hour, or quarter hour. (chapter 1)

compression—An audio processing technique in which the highest audio peaks are automatically turned down, reducing the dynamic range of the audio signal in a direct ratio to the audio input level; a form of automatic volume control. (chapter 5)

compression ratio—The fixed or adjustable ratio of the input to the output of an audio compressor. (chapter 6)

condenser microphone—A microphone that uses an electrostatic field to convert sound into audio and requires a polarizing voltage and a preamplifier. (chapter 3)

constructive interference—Occurs when two sine waves are just slightly out of phase with one another but are still together in their cycles of compression and rarefaction. Constructive interference tends to increase amplitude, reinforcing the sound waves. (chapter 2)

content delivery network (CDN)—A system of computer servers that distributes web content, audio, and video streams simultaneously to thousands of users regardless of their geographic location. (chapter 12)

core audience—The primary target demographic of a radio station. (chapter 10)

cost per click (CPC)—A method of paying for an ad on a website. The advertiser pays a fixed rate each time someone clicks on the ad and is directed to the advertiser's website. (chapter 12)

cost per thousand (CPM)—A universal media term for the cost to make 1,000 impressions. M is the Roman numeral for 1,000. (chapter 12)

courtesy lines—Communication connections that are available in the press area of a sports venue for visiting sportscasters to use without charge. Depending on the facility it could range from a POTS line, a broadband Internet connection for AoIP, to a WiFi connection. (chapter 11)

crispy—A sound with extended high-frequency response, like the sizzle of crisp bacon frying. (chapter 6)

critical distance—The distance from the sound source at which the ratio of direct-to-reverberant sound is equal, or 1; abbreviated d_c. (chapter 3)

D/A—Digital-to-analog audio converter. (chapter 11)

dark—A sound that is the direct opposite of bright; dull with a weak high-frequency response. (chapter 6)

daypart—A broadcast industry term that classifies a specific time period of radio listening. For example, 6:00 AM to 10:00 AM, Monday through Friday is generally accepted as the morning-drive daypart. (chapters 9 and 10)

dead studio—A studio or a location that has no natural reverberation or echo. (chapter 6)

dead-studio sound—A studio is said to be a dead studio when there is little or no natural reverberation—the less reverberation, the more dead the studio. (chapter 8)

decibels (dB)—A unit of measurement developed by Alexander Graham Bell to measure the loss of signal over one mile of telephone line. Because the bel is such a large unit, most measurements are made in tenths (deci) of a bel, thus the decibel. A decibel can be expressed as a volume of acoustic sound-pressure level or, for audio signals, as a voltage reading. (chapter 2)

de-esser—A specialty compressor that has a variable frequency range used to reduce sibilant "ess" sounds in speech between 6 kHz and 8 kHz. (chapter 6)

demographic—A market or segment of the population that is identified by demographics, which includes statistical data such as age, average income, marital status, and geographic area. (chapter 1)

destructive interference—Occurs when two sine waves are out of phase with one another and their cycles of compression and rarefaction occur at different times. Destructive interference tends to decrease amplitude and weaken the sound waves. (chapter 2)

digital audio workstation (**DAW**)—The term generally applied to a computer-based digital recording system that incorporates the ability to record, store, manipulate, transport, and deliver digital audio products. (chapter 5)

digital condenser microphone—A condenser microphone with a built-in preamplifier that includes an analog-to-digital converter and digital output only. (chapter 3)

Digital Millennium Copyright Act of 1998 (**DMCA**)—This act established that owners of a sound recording had the right to authorize digital public performances and to be paid for such performances. (chapter 12)

digital signal processor (**DSP**)—A device for the manipulation and modification of audio signals, such as reverb, delay, or dynamics processing. (chapter 5)

digital subscriber line (**DSL**)—An Internet connection made over a regular, twisted-pair telephone line capable of carrying data and a telephone conversation at the same time. Popular for home and small business use. (chapter 12)

diversity receiver—A wireless microphone receiver using two tuners and two antennas to eliminate noise and dropouts. (chapter 3)

domain—A unique name for a location on the web. Domains can be either generic (such as .com for commercial websites or .org for nonprofit organizations) or geographic (such as .uk for the United Kingdom or .fr for France). (chapter 12)

donut—A type of agency-supplied commercial that has a hole in the middle of the ad for the insertion of local copy. (chapter 8)

draw—Something to attract people to an event or a location; a gimmick or enticement. (chapter 11)

driver—An individual speaker within a speaker cabinet; each driver consists of a moving-coil transducer that converts analog electrical energy into acoustic energy. (chapter 4)

dry signal—An original, unprocessed audio signal sent from an audio console to an outboard audio processor, such as a reverb or delay unit. (chapter 4)

dubs—Slang for the word duplicate; a term used for a copy or copies of an original. (chapter 7)

duplex devices—Audio devices that allow bidirectional communications on a single line or radio-frequency transmission. (chapter 11)

dynamic microphone—A moving-coil microphone. (chapter 3)

dynamic range—The difference between the softest and loudest part of a sound, measured in decibels (dB). A dynamic range of 95 dB, for example, indicates that the loudest sound is 95 dB louder than the softest sound. (chapter 2)

effective radiated power (**ERP**)—When applied to an FM station, the term specifies in watts how much power the station can use to transmit its signal, as measured at the antenna. (chapter 13)

electret-condenser microphone—A condenser microphone that, although it does not require a polarizing voltage, does require a battery to run the preamplifier. (chapter 3)

electronic news gathering (**ENG**)—Location spot-news coverage. (chapter 11)

external marketing—Any medium or marketing tool outside the radio station (such as billboards, newspaper ads, and television commercials) used to attract new listeners to the radio station. (chapter 1)

fatter—A sound with a good, solid, low-frequency response with emphasis in the 100 to 300 Hz range. (chapter 6)

feedback—A condition that occurs when a microphone is left open near a monitor speaker, creating a loud howling or whistling sound. The microphone picks up the amplified sound from the speaker and sends it back through the system, regenerating the signal over and over. (chapter 4)

file transfer protocol (FTP)—A system for exchanging computer files with another computer using the Internet as opposed to the World Wide Web. (chapter 5)

FM—Frequency modulation. The FM radio band extends from 88 mHz to 108 mHz. (chapter 13)

focus groups—A qualitative market research tool in which a group of people with common demographics and similar attitudes are surveyed and questioned on a particular topic by a trained moderator. (chapter 14)

footprint—A satellite signal's geographic coverage area on the earth. (chapter 13)

frequency—The number of complete cycles that occur in one second of time; expressed as cycles per second or hertz (Hz); the number of times a radio ad airs during a given period of time. (chapters 2 and 7)

frequency-response curve—A plotted chart that shows how a microphone responds to various frequencies; called flat if the microphone treats all frequencies equally. (chapter 3)

gaffer's tape—Available in colors, gaffer's tape has the strength of duct tape without the sticky mess afterward. (chapter 3)

gain staging—Matching a microphone's maximum output voltage to the preamplifier's maximum input voltage to prevent overload. (chapter 3)

gateway ads—A brief ad that appears before the delivery of rich content on a website. For example, before you get to see a video clip you see a 10 to 12 second sponsor ad. (chapter 12)

gating—An audio processing technique that acts just as its name implies, as a gate. When audio is above a preset level, the gate allows it to pass. When the audio level drops below the preset level the gate closes, blocking the background noise. (chapter 5)

geosynchronous orbit—An orbit that places a satellite approximately 22,000 miles above the earth so that it rotates at the same speed as the earth, maintaining a fixed position in space. (chapter 13)

gooseneck—A piece of flexible metal conduit used to support a microphone. (chapter 3)

GPS—Global Positioning System. A satellite-based navigation system made up of a network of satellites that enables pinpoint location identification. (chapter 11)

groundwave propagation—The property of physics that allows amplitude-modulated radio carrier signals to follow the earth's surface through conduction. (chapter 13)

hard clipping—An audio condition in which the audio is driven beyond distortion to the point that the edges of the sine waves start breaking up. (chapter 4)

harmonic—A harmonic is a sound created by multiplying the fundamental frequency, or first harmonic. The second harmonic is two times the fundamental, the third harmonic is three times the fundamental, etc. (chapter 2)

headroom—The additional capacity of an audio device to handle level increases above average working levels to protect from transients, overload, and distortion. (chapter 4)

height above average terrain (HAAT)—Applied to an FM radio station, the term specifies in meters the height of the station's transmitting antenna above the average terrain that surrounds it. (chapter 13)

Hertz, Heinrich—Hertz (1857–1894) proved that electricity could be transmitted in electromagnetic waves, which travel at the speed of light. In 1933 his name was adopted as the international metric term used for radio and electrical frequencies (e.g., Hz, kHz, and mHz). (chapter 2)

high-pass filter—A filter that blocks audio frequencies below a specific frequency, allowing those frequencies above the specified frequency to pass; sometimes referred to as a low-cut filter, a term that is more descriptive of what the filter actually does. (chapters 4 and 6)

hollow—A sound that has a mid-frequency dip similar to when you cup your hands around your mouth and talk. (chapter 6)

hyper-cardioid—An elongated, heart-shaped cardioid pattern used primarily in shotgun microphones to allow the microphones to be moved farther from the sound source without increasing the apparent room reverberation. (chapter 3)

image transfer—When an audio track from a television commercial is used as a stand-alone radio commercial, causing the radio listener to visualize the images of the matching television commercial. (chapter 7)

imaging—Radio station imaging consists of the short live and recorded elements that are used between segments of a station's programming that tie everything together, creating a station's signature sound. For example, "Asheville's home for classic rock, Rock 107.9." (chapter 1)

infrasonic—Sounds or audio signals that fall below the average person's hearing range, typically 20 Hz and lower. (chapter 2)

integrated studio audio system—A digital audio system that links together all of the audio equipment in a radio station from the microphone to the transmitter using Audio over Internet Protocols (AoIP). Stand-alone control boards and audio wiring are replaced with digital audio over a local area network (LAN) linking all of the audio consoles and equipment together. The integrated system provides a seamless digital audio workflow to the transmitter. (chapter 1)

interactive—A generic term referring to anything involving a radio station website, such as advertising, website sponsorships, streaming audio, and video. (chapter 12)

internal marketing—Any marketing or promotion tool used on a radio station to retain the station's current listeners, such as contests, T-shirts, bumper stickers, etc. (chapter 1)

interruptible foldback (IFB)—A method for sending a monitor signal to the talent while also being able to talk to him or her through an earpiece or headphone; a sophisticated intercom system. (chapter 4)

Krispy Kreme donut—A Southern delicacy, wonderfully light donuts slathered in mouth-watering sugar frosting, often used to entice the crew to show up in the morning; a sugar-delivery device. (chapter 3)

lav (lavalier)—A microphone worn on the body. Lav is pronounced like lava from a volcano, just drop the last "*a*." Lavalier comes from the French word *lavaliere,* or pendant. (chapter 3)

limiting—An audio processing technique that automatically controls audio levels that approach a preset maximum audio level. When the preset level is reached, the audio is turned down so that it does not exceed the preset level. (chapter 5)

listener fatigue—A condition brought on by misadjusted audio processing, causing the listener to become consciously, or subconsciously, tired of listening to a musical selection or radio station. (chapter 6)

local area network (LAN)—A computer interconnection system for sharing data between or among computers within a company or media facility at one location. (chapter 5)

low-earth-orbit (LEO)—A term used to describe satellites orbiting earth at 300 to 500 miles high. (chapter 11)

low-pass filter—A filter that blocks audio frequencies above a specific frequency, allowing those frequencies below the specified frequency to pass; sometimes referred to as a high-cut filter, a term that is more descriptive of what the filter actually does. (chapters 4 and 6)

maximum sound-pressure level—The maximum dB-SPL that a microphone can reproduce before going into distortion and then failure. (chapter 3)

metrics—Data collected about the performance of a website, including such things as number of pages displayed or page views, session length, number of unique visitors, etc. Often referred to as website ratings. (chapter 12)

metro area—A metropolitan city's area as defined by the Office of Management and Budget; also generally accepted by the major rating services as a designated geographical area. (chapter 10)

mic—Pronounced "Mike"; industry abbreviation for microphone (mics is used for more than one mic). (chapter 3)

mic flag—A plastic nameplate, or cover, that goes on a microphone with the station's call letters or logo. (chapter 11)

micron—An outdated term still in common usage that has since been replaced with the term micrometer. A micron, or micrometer, is one millionth of a meter, or about 1/25,000th of an inch. (chapter 3)

middle/side microphone (M/S)—A stereo microphone incorporating a cardioid element facing forward and a bidirectional element facing out the sides of the microphone. The combination of the middle element and the separate halves of the bidirectional element create the left and right channels. (chapter 3)

modulation—The process of changing a silent carrier with either amplitude or frequency modulation. (chapter 13)

MP3—Moving Picture Experts Group Layer-3 audio compression compresses, or shrinks, CD-quality audio files by a factor of 12 or more, maintaining the original perceived CD-quality sound so that audio files can be transferred via the World Wide Web. (chapter 1)

muddy—Muffled-sounding audio, not clear, with too much reverb at the lower frequencies. (chapter 6)

multipath distortion—Multipath distortion occurs in a car radio as an FM signal and a reflected FM signal of the same frequency arrive at the radio at slightly different times. The radio has difficulty receiving the two signals and the result is an undesirable distortion of the audio's high frequencies. Multipath distortion can cause the radio signal to fade in and out rapidly. (chapter 13)

multiple outlet box (mult)—A mult is the same as a breakout box, which is designed to allow audio distribution at press conferences through multiple mic or line-level outputs. (chapter 11)

national rep—A national representative is a person or a company that represents a radio station to large national advertising agencies handling regional or national products. (chapter 1)

near-coincident pair—Stereo microphone placement in which two microphones are placed side by side facing the sound source. With the midpoint of the mic bodies serving as the rotation point, the right mic is aimed at the right side of the sound source, the left mic is aimed at the left side. (chapter 3)

near-field monitors—Smaller monitor speakers placed very near the mixing position, often at the front of the console. (chapter 4)

needle-drop basis—An old radio term used to describe paying for production music on a per cut used basis. (chapter 8)

NextRadio®—An app that takes advantage of the FM receiver chip that is in all current smart phones. Just like a regular radio, the chip receives the FM broadcast signals and the NextRadio® app integrates them into the smart phone. (chapter 13)

Nielsen Audio—The primary radio audience measurement service. Ratings are often referred to as "the book," since the ratings were originally published in book form each quarter. Stations now take electronic delivery of the data. (chapter 10)

noise—Electrical noise, which appears as hiss in an audio signal, measured in negative dB. The larger the negative figure, the lower the noise level. (chapter 3)

notch filter—An audio filter designed to eliminate a very narrow bandwidth of sound, creating a notch in the frequency. Generally used to eliminate unwanted sounds. (chapter 6)

off-air monitor—An actual air signal taken from a radio for monitoring purposes. (chapter 11)

off axis—Outside of the centerline of a microphone's directional pattern, usually measured in degrees. (chapter 3)

off-axis coloration—The change in the frequency response and sensitivity of a microphone as it is turned away from a sound source and aimed in a different direction. (chapter 3)

omnidirectional—A microphone pattern that allows the microphone to pick up sound from all directions equally. (chapter 3)

overload—The point at which a microphone stops functioning under extremely high sound-pressure levels. (chapter 3)

overtone—A sound or audio signal that occurs in between the harmonics or multiples of a fundamental frequency. (chapter 2)

page displays—The number of times a web page is successfully downloaded and displayed on a user's computer. (chapter 12)

pan (panoramic) control—A sonic positioning control that allows the input to a mixer channel to be assigned to the left channel or the right channel, or to remain in the center channel for a mono signal. (chapter 4)

passive crossover—A crossover network comprised of electronic components whose values are fixed, with the audio requiring no electrical power to pass through the network. (chapter 4)

pattern—The term used to describe the directional characteristics of a microphone. (chapter 3)

perceptual audio coding—A method of compressing, encoding, and decoding audio data files that takes advantage of the properties of the human perception of sound to achieve a size reduction of the data file with little or no perceptible loss of quality. (chapter 12)

phantom power—Power supplied to a condenser microphone from a source such as a control board, portable mixing board, or other outside source through the microphone cable. (chapter 3)

phone hybrid—An electronic device used to connect broadcast equipment with a telephone line so that audio can be sent and received on a phone line or cell phone. (chapter 11)

polar frequency response—The frequency-response curve of a microphone measured at various points in the microphone pattern. (chapter 3)

pool audio feed—A single-point audio feed for the press, usually accomplished with a mult or breakout box. (chapter 11)

pop filter—An acoustic foam filter within a microphone used to prevent popping plosive sounds from reaching the microphone element. (chapter 3)

popper stopper—A fine mesh screen placed a few inches off the face of a microphone to prevent popping plosive sounds from reaching the microphone element. (chapter 3)

pop-ups—Ads that open in front of a website as it opens and block the view of the website. A pop-under ad is the opposite. The ad loads behind the web page and displays when the page is changed or closed. Both are the most disliked type of advertising on the web. (chapter 12)

positioning statement—A unique selling proposition, or slogan, used by a station to separate itself from the competition. (chapter 10)

pot—Pot is the abbreviation for potentiometer, which is the technical term for a volume control. (chapter 4)

preamplifier—A voltage multiplier used to boost the weak signal of a microphone to a usable audio level. (chapter 3)

prosumer—Prosumer equipment is equipment that offers many professional features but is not built to rigid professional standards. It is generally priced in between consumer and professional-level equipment. (chapter 3)

proximity effect—The closer you move to a directional microphone, the more bass your voice appears to have. (chapter 3)

psychographic—Determining how a person or a group of people think about a particular product or service; what their image is of your product and how they see themselves relating to your product. (chapter 14)

qualitative research—Research centered on a person or group's lifestyle choices. For example, a radio station targets 18- to 34-year-old single males living in Jackson County who rent an apartment, make monthly car payments on a 2016 or newer vehicle, eat out three times a week, drink domestic beer, and go to the movies twice a month. (chapter 1)

quantitative research—Research centered on numerical or statistical data; most commonly used is audience ratings research to determine the number of listeners a radio station has. (chapter 1)

quantization—The process of assigning a binary number to each of the samples or voltage readings taken during the analog-to-digital audio conversion. (chapter 2)

release time—The fixed, or adjustable, length of time it takes for an audio processor, such as a compressor or limiter, to return a signal to its normal level after processing, measured in milliseconds. (chapter 6)

ribbon microphone—A member of the dynamic class of microphones; the ribbon uses a thin ribbon of metal as the sound pickup element. (chapter 3)

rubber ducky—A flexible radio antenna made from silicone-based rubber products that can be bent and flexed without damage to the antenna coil contained within. (chapter 3)

self-powered speakers—Speakers that have amplifiers built into them so that an audio feed from a mixer or control board can be fed directly into the speaker. (chapter 4)

sensitivity—An electrical measurement in either dBV (decibel volts) or millivolts that tells what the audio output of the microphone will be at a given sound-pressure input level. (chapter 3)

session length—The time a website viewer spends on a particular website or viewing a particular web page. (chapter 12)

shock mount—A device designed to hold a microphone and isolate it from vibrations that might enter the microphone body and cause undesirable sounds. (chapter 3)

shotgun microphone—A microphone designed with a long tube in front of the microphone element to focus the sound and to increase the microphone's direct-to-reverberant sound ratio. (chapter 3)

skywave propagation—A phenomenon that occurs as the long wavelengths of AM signals interact with the radically shifting layers of the ionosphere several miles above the earth's surface at night, causing AM signals to travel great distances. (chapter 13)

slate—The radio equivalent of the beginning of a scene in a television or motion picture shoot, when someone holds a slate, or clapboard, in front of a camera with the production name, scene number, take number, roll number, director, producer, and date to identify the material later. In radio, this is done audibly with a slate mic so that a production person can identify the audio cuts. (chapter 4)

Society of European Stage Authors and Composers (SESAC)—A membership-based performing rights organization that protects the copyrights of members' works by licensing and distributing royalties for their performance. (chapter 1)

software-defined radio (SDR)—A radio that converts the incoming radio signal into a digital data stream and then processes the data to extract the usable signals. (chapter 13)

Sony/Philips Digital InterFace (SPDIF)—A standard digital audio file transfer format using 75-ohm coaxial cable. (chapter 5)

sound—Acoustic energy traveling through a medium such as air or water. (chapter 3)

sound design—The overall sound quality of a radio station that sets it apart from others. (chapter 10)

SoundExchange—The nonprofit performance rights organization recognized by the US Copyright Royalty Board that collects statutory royalties from satellite radio, Internet radio, cable TV music channels, and similar platforms for digitally streaming sound recordings. (chapter 1)

spaced pair—Stereo microphone placement in which two microphones are spaced a distance apart following the three-to-one rule: three feet of separation for each foot of distance from the sound source. (chapter 3)

spectrum use fees—Fees charged by the FCC to broadcasters for the use of the public airwaves. (chapter 1)

spit screen—A fine mesh screen placed a few inches off the face of a microphone to prevent popping plosive sounds and spit from reaching the microphone element; same as a popper stopper. (chapter 3)

spot—Slang term for a radio or television commercial. Don't use this word. Although advertising salespeople call them spots, clients typically think in terms of ads or commercials. (chapter 7)

stop-set—A break in a station's music programming, composed of a jock talking, commercials, and promotional announcements. (chapters 7 and 9)

streaming—A term used to describe transmitting audio or video on the web. Usually used to describe a continuous, real-time audio feed, such as an on-air signal. (chapter 12)

subwoofer—A speaker designed to handle those sounds from 16 Hz to about 100 Hz. (chapter 2)

super-cardioid—A heart-shaped microphone pattern with maximum sensitivity at 0 degrees and maximum rejection at 180 degrees. Super-cardioid microphones are even less sensitive to the sides than a cardioid microphone. (chapter 3)

swag—Stuff we all get. Promotional items given away by a manufacturer, business, or radio station. (chapter 11)

sweet spot—The physical location in a concert hall or studio where there is the least phase cancellation of live sound and where the direct-to-ambient sound ratio is in proper proportion. (chapter 6)

T-1—The technical designation for a direct-leased Internet connection that is capable of transferring 1.544 Mbps. (chapter 13)

T-3—The technical designation for a direct-leased Internet connection that is capable of transferring 44.736 Mbps. (chapter 13)

tag—A 10- to 15-second space at the end of an agency ad over which the local station puts copy. (chapter 8)

target audience—A group of people that a radio station has selected to try to reach with the station's programming. Typically, this audience can be identified both quantitatively and qualitatively. (chapter 14)

terrestrial-based—Fancy term for earth-based. (chapter 13)

threshold—When used in relation to an audio processor, a threshold sets the audio level at which the device will become active and begin processing. (chapter 6)

Thunderbolt—A Thunderbolt 2 copper wire cable provides two digital channels on the same connector/cable at up to 20 gigabits per second and provides electrical power to external components. Thunderbolt 3 is able to deliver 40 gigabits per second over a fiber-optic cable. (chapter 5)

time spent listening—An audience ratings measurement indicating the number of quarter-hour time segments the average person spends listening to a radio station in a given time period, such as a week. (chapter 10)

tinny—A sound that has an almost telephone-like quality to it. A very narrow-band sound that sounds as if it were coming through a tin can or over a telephone. (chapter 6)

TOS link—The name given a system of fiber-optic cables and connectors developed by Toshiba and used for the transmission of SPDIF digital audio. (chapter 5)

transient response—How a microphone reacts to transients, or sounds, with a fast attack that trails off slowly. (chapter 3)

TRS connector—A one-quarter-inch diameter balanced connector with the tip as positive, the ring as negative, and the sleeve as shield or earth ground. (chapter 3)

T/S connector—A one-quarter-inch diameter connector with the tip as positive and the sleeve as negative and shield or earth ground. (chapter 3)

tune-out factor—Any psychoacoustic element of a radio station's signal that is annoying to the listener. (chapter 6)

UHF—Ultrahigh frequency; a band of frequencies assigned to television channels 14 through 83; 14 through 69 are also assigned to wireless microphones. (chapter 3)

ultrasonic—Sounds or audio signals occurring above the average person's hearing range, typically above 20 kHz. (chapter 2)

unbalanced microphone cable—A microphone cable or line that only has one positive conductor, with the shield acting as both the negative and the shield. (chapter 3)

unidirectional—A generic term for a directional microphone, meaning one direction. (chapter 3)

unity gain—A gain of one; the position on a fader at which the ideal operating level of 0 VU occurs. A device with unity gain does not raise or lower the volume, or gain, of a signal. (chapter 4)

Universal Serial Bus (USB)—The most widely used hardware interface for connecting equipment to computers. USB 2.0 can transmit up to 480 Mbps of data; USB 3.0 can transmit up to 10 Gbps of data. Virtually every portable device uses some form of USB connectivity for data transfer. (chapter 3)

vendor involvement—When the vendor of a product or service provides support to a local business in the form of product to give away or expert staff to help demonstrate the product. This kind of involvement with a local business can be quite varied from manufacturer to manufacturer. (chapter 11)

VHF—Very high frequency; a band of frequencies assigned to television channels 2 through 13 and to wireless microphones. (chapter 3)

wall-wart—A self-contained power supply for running portable equipment. It plugs directly into the wall and down-converts 110 volts AC to a DC voltage. (chapter 4)

wet signal—A processed audio signal that is returned to a control board after audio processing, such as reverb or delay. (chapter 4)

wide area network (**WAN**)—A computer interconnection system for sharing data among computers within a company or media facility at multiple geographic locations. (chapter 5)

windscreen—An acoustic foam cover for the head of a microphone used to block wind outdoors. (chapter 3)

XLR connector—A balanced microphone connector; sometimes referred to as a Cannon connector, after the man and the company that invented it. A connector in which there are pins or sockets for the positive, negative, and shield, or earth ground. (chapter 3)

X-Y stereo configuration—Stereo microphone placement in which the two microphones are placed in a horizontal position facing the sound source and aimed in either a coincident or near-coincident manner. (chapter 3)

Index